Development of *Trans*-free Lipid Systems and their Use in Food Products

Food Chemistry, Function and Analysis

Series editors:
Gary Williamson, *Monash University, Australia*
Alejandro G. Marangoni, *University of Guelph, Canada*
Graham A. Bonwick, *AgriFoodX Limited, UK*
Catherine S. Birch, *AgriFoodX Limited, UK*

Titles in the series:
1: Food Biosensors
2: Sensing Techniques for Food Safety and Quality Control
3: Edible Oil Structuring: Concepts, Methods and Applications
4: Food Irradiation Technologies: Concepts, Applications and Outcomes
5: Non-extractable Polyphenols and Carotenoids: Importance in Human Nutrition and Health
6: Cereal Grain-based Functional Foods: Carbohydrate and Phytochemical Components
7: Steviol Glycosides: Cultivation, Processing, Analysis and Applications in Food
8: Legumes: Nutritional Quality, Processing and Potential Health Benefits
9: Tomato Chemistry, Industrial Processing and Product Development
10: Food Contact Materials Analysis: Mass Spectrometry Techniques
11: Vitamin E: Chemistry and Nutritional Benefits
12: Anthocyanins from Natural Sources: Exploiting Targeted Delivery for Improved Health
13: Carotenoid Esters in Foods: Physical, Chemical and Biological Properties
14: Eggs as Functional Foods and Nutraceuticals for Human Health
15: Rapid Antibody-based Technologies in Food Analysis
16: DNA Techniques to Verify Food Authenticity: Applications in Food Fraud
17: Advanced Gas Chromatography in Food Analysis
18: Handbook of Food Structure Development
19: Mitigating Contamination from Food Processing
20: Biogenic Amines in Food: Analysis, Occurrence and Toxicity
21: Nutrition and Cancer Prevention: From Molecular Mechanisms to Dietary Recommendations
22: Health Claims and Food Labelling
23: Nutraceuticals and Human Health: The Food-to-supplement Paradigm
24: Nutritional Signalling Pathway Activities in Obesity and Diabetes
25: The Chemistry and Bioactive Components of Turmeric

26: Foodomics
27: Food Proteins and Peptides: Emerging Biofunctions, Food and Biomaterial Applications
28: Handbook of Antioxidant Methodology: Approaches to Activity Determination
29: Fats and Associated Compounds: Consumption and Human Health
30: Oral Processing and Consumer Perception: Biophysics, Food Microstructures and Health
31: Development of *Trans*-free Lipid Systems and their Use in Food Products

How to obtain future titles on publication:
A standing order plan is available for this series. A standing order will bring delivery of each new volume immediately on publication.

For further information please contact:
Book Sales Department, Royal Society of Chemistry, Thomas Graham House, Science Park, Milton Road, Cambridge, CB4 0WF, UK
Telephone: +44 (0)1223 420066, Fax: +44 (0)1223 420247,
Email: booksales@rsc.org
Visit our website at www.rsc.org/books

Development of Trans-free Lipid Systems and their Use in Food Products

Edited by

Jorge F. Toro-Vazquez
Universidad Autonoma de San Luis Potosi, Mexico
Email: toro@uaslp.mx

ROYAL SOCIETY OF **CHEMISTRY**

Food Chemistry, Function and Analysis No. 31

Print ISBN: 978-1-83916-220-6
PDF ISBN: 978-1-83916-653-2
EPUB ISBN: 978-1-83916-654-9
Print ISSN: 2398-0656
Electronic ISSN: 2398-0664

A catalogue record for this book is available from the British Library

The Royal Society of Chemistry is a charity, registered in England and Wales, Number 207890, and a company incorporated in England by Royal Charter (Registered No. RC000524), registered office: Burlington House, Piccadilly, London W1J 0BA, UK, Telephone: +44 (0) 20 7437 8656.

Visit our website at www.rsc.org/books

Printed in the United Kingdom by CPI Group (UK) Ltd, Croydon, CR0 4YY, UK

Preface

Since ancient times natural fats and oils have been used by humans for food and non-food purposes. The Egyptians, for example, used olive oil and fats as a lubricant for moving heavy building materials and to grease axles. Most ancient civilizations used fats and oils as fuel, typically in lamps and in the elaboration of candles that were the principal source of illumination. As understanding of the chemical nature of edible fats and oils increased, so did their commercial uses. Butter is even mentioned in the Bible, when Abraham and Sarah offer to three visiting angels a feast of meat, milk, and the "creamy yellow spread".[1] However, it was not until 1869 that the French chemist Hippolyte Mège-Mouriès invented and patented the process of elaborating margarine. The original process, developed in response to a prize offered by Napoleon III to the one who could find a substitute for butter, combined beef fat and skimmed milk (*i.e.* with the butter fraction removed). Later on, in 1871, Henry W. Bradley patented the process of creating margarine utilizing vegetable oil, mainly cottonseed oil, combined with animal fats. Subsequently, margarine production was limited by the availability of beef tallow, until in 1902 the German chemist Wilhelm Normann patented a process to harden oils through hydrogenation of the double bond of unsaturated fatty acids esterified to triacylglycerols (TAGs). This development greatly expanded the availability of margarine and widened the market opportunities for vegetable oils.[2,3] Hydrogenation of the double bond in fats and oils still constitutes a major industrial oleochemical process. This is mainly because the original research later led to the development of the partial hydrogenation process for vegetable oils that established the grounds for the shortening industry. In turn, the shortenings provided the functionality that supported the development of food products that exposed consumers to appealing

Food Chemistry, Function and Analysis No. 31
Development of *Trans*-free Lipid Systems and their Use in Food Products
Edited by Jorge F. Toro-Vazquez

Published by the Royal Society of Chemistry, www.rsc.org

new textures, mouthfeel, and flavors. Thus, partially hydrogenated vegetable oils (PHVO) became the major functional component of baked goods such as cakes, cookies, pie crust, and crackers; snacks like corn, chips, and microwaveable popcorn; refrigerated dough used for canned biscuits, cinnamon rolls, and frozen pizza crusts; and also non-dairy coffee creamers, confectionary spreads, stick margarines, and shortenings for frying foods.

Now we know that most of the functional properties associated with lipid-based foods are due to the strong function of the nano- and mesoscale organization of major constitutive bio-molecules, its phase changes, and rheological behavior. Since TAGs are the main component of fats, vegetable oils, and subsequently PHVO, the functional properties of lipid-based foods are directly associated with the phase changes, namely crystallization, melting, and polymorphism, of the TAGs. In particular, the physical properties associated with saturated TAGs and/or those containing *trans* fatty acids, the last one present in high concentrations in shortenings derived from PHVO, provide the solid fat content, melting, and textural properties requested by the consumers in several processed food products. Unfortunately, there is now a well-established link between the consumption of *trans* fatty acids and a higher risk of coronary heart disease[4] and the development of type II diabetes.[5,6] Within this context, and as a result of the government's recommendations issued to limit and even eliminate *trans* fatty acids in processed foods, palm oil and its fractions (*i.e.*, palm stearin) which are high in saturated TAGs (*i.e.*, palmitic and stearic acids), have become the edible oil most used by the food industry to replace *trans* fatty acids in food formulations. This is mainly because palm oil and its fractions provide functional properties and oxidative stability to lipid-based foods similar to those achieved by shortenings. Unfortunately, although the relationship of saturated fatty acid intake by humans with a higher risk for atherosclerotic cardiovascular disease remains controversial,[7] the consumer's response has been to move away from products high in saturated fatty acids; particularly those formulated with palm oil and its derivatives due to reports claiming that certain refined palm oil (*i.e.*, glycidyl fatty acid esters) components might be genotoxic and carcinogenic. These negative health issues associated with the consumption of *trans* and saturated fats, and the legislation issued by regulatory agencies, have put considerable pressure on the food industry to resolve this health issue associated with food consumption. Therefore, the development and use of *trans*-free oils and fats are of the utmost interest to research groups and industries worldwide. The challenge for the industry is to produce foods containing less "bad fats" while retaining most of the functional properties associated with saturated TAGs and/or TAGs with esterified *trans* fatty acids. Evidently, this situation indicates that food formulation must shift toward the use of healthy unsaturated oils, a goal not easily accomplished mainly because the physical and functional properties of unsaturated oils are significantly different from those provided by "bad fats". For instance, unsaturated edible oil remains essentially liquid under most of the temperature conditions used to produce food systems, and consequently, they do not positively contribute to food texture and mouthfeel. Additionally, the use of unsaturated edible oils in the formulation of food systems would require their protection against oxidation to avoid negative effects on the food quality and nutritional value. Thus, several research groups worldwide have investigated different mechanisms to provide structure and elastic properties to unsaturated oils. An important aspect to consider in the development of these new "structured unsaturated edible oils" is that, despite the health benefits associated with the consumption of unsaturated oils, consumers will not accept and consume a particular food if it does not meet their sensory expectations.

From the above is it understandable that the research associated with the development of strategies for structuring oils has grown extensively during the last 10 to 15 years. However, the implementation of "structured edible oils" that result in the commercialization of food systems, for instance through the use of oleogels, is still not a reality. Within this framework, there is a need for additional studies to evaluate and understand the digestion of "structured edible oils" in food systems, their oxidative stability, the interaction of the "structured edible oils" with food matrices under processing and storage conditions, and furthermore, trained sensory panels must evaluate the tentative development of off-flavors.[8] In this regard, this book encompasses the work of leading researchers discussing, from a scientific and technological perspective, the latest and most innovative approaches to structure edible oils without the use of *trans* fats. Additionally, the authors discuss the practical uses and technical limitations associated with the use of these "structured edible oils" in different food systems. We hope that the information discussed here will contribute to the successful development of a new generation of functional "structured edible oils" that once incorporated into food systems, will have higher rates of consumer acceptance.

Jorge F. Toro-Vazquez
Facultad de Ciencias Químicas
Laboratorio de Fisicoquímica de Alimentos
Universidad Autónoma de San Luis Potosí

References

1. A. Wetzel, *The History of Butter*, http://www.butterjournal.com/butter-history.
2. M. Vaisey-Genser, in *Encyclopedia of Food,* Sciences and Nutrition, ed. B. Caballero, Academic Press, 2nd edn, 2003, pp. 3704–3709.

3. R. R. Allen and A. H. Chen, World Conference on Soya Processing and Utilization, *J. Am. Oil Chem. Soc.*, 1981, 166–168.
4. A. Ascherio, M. B. Katan, P. L. Zock, M. J. Stampfer and W. C. Willett, *N. Engl. J. Med.*, 1999, **340**, 1994–1998.
5. J. Salmeron, F. B. Hu, J. E. Manson, M. J. Stampfer, G. A. Colditz and W. C. Willet, *Am. J. Clin. Nutr.*, 2001, **73**, 1019–1023.
6. U. Risérus, W. C. Willett and F. B. Hu, *Prog. Lipid Res.*, 2008, **48**, 44–51.
7. P. W. Siri-Tarino, Q. Sun, F. B. Hu and R. M. Krauss, *Curr Atheroscler Rep.*, 2010, **12**, 384–390.
8. C. Park and F. Maleky, *Front. Sustain. Food Syst.*, 2020, **4**, 1–8.

Contents

Chapter 1 Concentrated O/W Emulsions Stabilized by Proteins: A Route to Texturize Vegetal Oils Without Using *Trans* or Saturated Fats 1

W. Dridi, R. Kapel, S. Albe-Slabi and F. Leal-Calderon

1.1 Introduction 1
1.2 Experimental 3
1.2.1 Materials 3
1.2.2 Sunflower Protein Isolate 3
1.2.3 Methodologies 6
1.3 Results and Discussion 8
1.3.1 Impact of the Protein Content on the Average Droplet Size 8
1.3.2 Interfacial Coverage 9
1.3.3 Emulsions Concentrated by Centrifugation – Osmotic Resistance Measurements 11
1.3.4 Macroscopic Aspect of Concentrated Emulsions as a Function of the Droplet Fraction – Biliquid Foams 13
1.3.5 Emulsions Concentrated by Freeze-drying 15
1.4 Conclusions 17
Acknowledgements 18
References 18

Food Chemistry, Function and Analysis No. 31
Development of Trans-free Lipid Systems and their Use in Food Products
Edited by Jorge F. Toro-Vazquez

Published by the Royal Society of Chemistry, www.rsc.org

Chapter 2 Alternative Oil Structuring Techniques: Oil Powders, Double Emulsions and Oil Foams 21
Zijian Zhi, Koen Dewettinck and Filip van Bockstaele

2.1 Double Emulsions 21
2.1.1 Background 21
2.1.2 Preparation and Stabilization 22
2.1.3 Applications 26
2.2 Microencapsulated Oil Powder 27
2.2.1 Introduction 27
2.2.2 The Structure of Microencapsulated Oil 31
2.2.3 Emulsification and Drying Technology 32
2.2.4 Applications in the Food Industry 38
2.3 Oil Foams 38
2.3.1 Background 38
2.3.2 Preparation and Stabilization 39
2.3.3 Benefits and Application 45
2.4 Conclusions 46
Abbreviations 46
Acknowledgements 46
References 47

Chapter 3 Use of High-intensity Ultrasound to Structure Edible Fats 53
T. L. T. da Silva, A. Giacomozzi and S. Martini

3.1 Introduction 53
3.2 Effect of HIU on Fats 55
3.2.1 High-saturated Fats 55
3.2.2 Low-saturated Fats 62
3.3 Effect of HIU Application on the Physicochemical Properties of Oleogels 68
3.3.1 Binary and Ternary Oleogels 69
3.3.2 Mono-component Oleogels 72
3.4 Effect of HIU on Lipid Oxidation 78
3.5 Scale Up to a Pilot Plant 80
3.6 Application 84
3.6.1 Baked Products 84
3.6.2 Chocolate 84
3.6.3 Emulsions 86
3.7 Future Trends 86
Abbreviations 87
Acknowledgements 87
References 88

Chapter 4 Pulse and Oilseed Protein-based Oil Structuring for Baking Application 91
Yan Ran Tang, Manisha Sharma and Supratim Ghosh

4.1 Oil Structuring and Its Importance in Food 91
4.2 Potential Health Implication of Traditional Oil Structuring 92
4.2.1 Oleogel – A New Way of Oil Structuring 92
4.3 Different Types of Oleogelation Based on Processing – Direct and Indirect Approaches 93
4.4 Different Types of Oleogelation Based on Oleogelators – Small Molecules and Biopolymers 94
4.5 Indirect Oleogelation 96
4.5.1 Foam-templated Indirect Oleogelation 96
4.5.2 Emulsion-templated Indirect Oleogelation 96
4.5.3 Hydrogel-templated Indirect Oleogelation 97
4.6 Characterization of Biopolymer-based Oleogels for Food Application 97
4.6.1 Oleogel Stability in Terms of Oil Binding Capacity and Oil Loss 97
4.6.2 Rheology of Biopolymer-based Oleogels 98
4.6.3 Textural Analysis of Biopolymer-based Oleogels 99
4.6.4 Food Applications of Biopolymer-based Oleogels 100
4.6.5 Sensory Evaluation of Food Made From Biopolymer-based Oleogels 101
4.7 Utilization of Pulse and Oilseed Proteins for the Development of Emulsion-templated Oleogels 105
4.7.1 Characterization of Faba Bean and Canola Protein Isolate-based Oleogels 107
4.7.2 Cake Batter and Cake Properties Prepared with Faba Bean and Canola Protein Isolate-based Oleogels 109
4.7.3 Texture Analysis of Cakes Prepared with Different Fat Types 111
4.8 Summary 112
4.9 Future Prospects 114
References 114

Chapter 5 Physicochemical Properties and Polymorphic Behavior of Tropical Fats: Their Potential and Practical Use in *Trans*-free Food Systems **119**
M. R. Ramos-Ramos, V. A. Garcia-Londoño, V. Borroni and M. L. Herrera

5.1 Introduction 119
5.2 Chemical Composition 121
5.2.1 Fatty Acid Composition 121
5.2.2 Triacylglycerol Composition 123
5.3 Solid Fat Content 126
5.4 Crystallization Behavior 129
5.5 Microstructure 130
5.6 Rheological Properties 132
5.7 Polymorphic Behavior 133
5.8 Conclusions 136
Acknowledgements 136
References 136

Chapter 6 Edible Oleogels Produced with Fatty Alcohols: The Use of Policosanol as an Oleogelator **139**
F. R. Lupi, A. Shakeel, U. Farooq, N. Baldino and D. Gabriele

6.1 Introduction 139
6.1.1 LMW Oleogels: A General Overview 140
6.1.2 Fatty Alcohols in Foods and Their Use as Gelators 141
6.2 Rheological Characteristics and Modeling of Fatty Alcohol Oleogels 142
6.2.1 Policosanol as a Gelator: Rheological Properties 143
6.3 Microstructural Characteristics of Fatty Alcohol Oleogels 148
6.4 Application of Fatty Alcohols and Vegetable Wax Oleogels in Food Systems and as Delivery Agents for Bioactive Compounds 150
6.5 Conclusions 153
References 154

Chapter 7 On the Applicability of the Oryzanol–Sitosterol Structuring System **157**
M. Scharfe and E. Flöter

7.1 Introduction 157

7.2 β-sitosterol and γ-oryzanol 160
7.2.1 γ-oryzanol 160
7.2.2 β-sitosterol and Other Phytosterols 162
7.2.3 Legislation 163
7.2.4 Digestion 164
7.3 Physical Chemistry of the Binary Mixture 166
7.4 The Ternary Mixture 168
7.4.1 Oleogel Structure, Formation and Dissolution 168
7.4.2 Methods of Characterization 172
7.4.3 Influence of Solvent Composition 181
7.5 Food Applications 186
7.5.1 Confectionery Products 187
7.5.2 Meat Products 188
7.5.3 Margarine and Spreads for Direct Consumption 189
7.5.4 Dairy Products 190
7.6 Conclusion 192
References 193

Chapter 8 Structuring Vegetable Oils with Monoglycerides and Monoglyceride–Lecithin or Monoglyceride–Ethylcellulose Mixtures 201
Jorge F. Toro-Vazquez, Mayra Aguilar-Zárate, Araceli López-Martinez and Miriam Charó-Alonso

8.1 Introduction 201
8.1.1 Vegetable Oil Structuring Through Molecular Self-assembly 203
8.2 Monoglycerides as Gelator Molecules 205
8.2.1 Self-assembly of Monoglycerides in Vegetable Oils 206
8.2.2 Mesophases, Crystal Phases, and Polymorphic Transitions of Monoglycerides in Oleogels 210
8.3 Lecithin and Ethylcellulose as Gelators that Limit β Phase Formation 215
8.3.1 Molecular Self-assembly of Lecithin in Vegetable Oil 215
8.3.2 Ethylcellulose Effect on MG Crystallization in Vegetable Oils 222
8.4 Application of Oleogels in Food Systems and Conclusions 229

Acknowledgements 230
References 231

Chapter 9 Functionality of Natural Waxes in Hybrid Fat Crystal Networks 235
Filip Van Bockstaele, Marie Romanus, Ivana A. Penagos and Koen Dewettinck

9.1 Introduction 235
9.2 Waxes from Vegetal and Insect Sources 236
9.2.1 Waxes from Vegetable Sources 239
9.2.2 Waxes Obtained from Insects 240
9.2.3 Legislation 241
9.3 TAG–Wax Hybrid Fat Systems 243
9.3.1 Introduction 243
9.4 Conclusion 255
Acknowledgements 256
References 256

Chapter 10 Glycerolysis-structured Lipid Systems 261
R. A. Nicholson and A. G. Marangoni

10.1 Introduction 261
10.2 Effects of Glycerolysis Reaction Conditions 262
10.3 Effects of Oil Composition 266
10.4 Food Applications 275
10.5 Future Directions 278
References 279

Chapter 11 Modulating the Rheology of Oleogels to Mimic the Functionality of Commercial Fats 281
J. D. Pérez-Martínez and J. J. Ornelas-Paz

11.1 Introduction 281
11.2 Rheological Methods for Oleogel Characterization 283
11.3 Edible Oleogels Mimicking the Properties of Plastic Fats 286
11.3.1 Natural Wax Oleogels 286
11.3.2 Ethylcellulose Oleogels 298
11.3.3 Monoacylglycerol Oleogels 303
11.3.4 Oleogels of Fatty Acids and Fatty Alcohols 307
11.4 Conclusions 309
References 309

Chapter 12 Sensory Properties and Aromatics Profile of Edible Oleogels **315**
E. Yılmaz

12.1 Introduction 315
12.2 Physical Properties of Solid and Semi-solid Fats 316
12.3 Oil Structuring Through Oleogelation 318
12.4 Sensory Assessment of Edible Fats/Oils 319
12.4.1 Sensory Vocabulary Used for Fats/Oils and Oleogels 319
12.4.2 Sensory Techniques Used for Fats/Oils and Oleogels 322
12.5 Sensory Properties of Oleogels and Foods Containing Oleogels 324
12.5.1 Sensory Properties of Oleogel Products 324
12.5.2 Sensory Properties of Bakery Products Containing Oleogels 327
12.5.3 Sensory Properties of Comminuted Meat Products Containing Oleogels 329
12.5.4 Sensory Properties of Margarine/Spread Products Containing Oleogels 334
12.5.5 Sensory Properties of Chocolate/ Confectionery Products Containing Oleogels 335
12.5.6 Sensory Properties of Dairy Products Containing Oleogels 338
12.5.7 Sensory Properties of Other Food Products Containing Oleogels 339
12.6 Aromatics Profile of Oleogels and Food Products Containing Oleogels 340
12.7 Integration of Analytical and Sensory Data for Oleogel Optimization 343
12.8 Conclusions and Outlook 345
References 345

Subject Index **350**

CHAPTER 1

Concentrated O/W Emulsions Stabilized by Proteins: A Route to Texturize Vegetal Oils Without Using Trans *or Saturated Fats*

W. DRIDI,[a] R. KAPEL,[b] S. ALBE-SLABI[b] AND F. LEAL-CALDERON*[a]

[a] Université de Bordeaux, CNRS, Bordeaux INP, CBMN UMR 5248, 33600, Pessac, France; [b] Laboratoire Réactions et Génie des Procédés, Université de Lorraine, CNRS, LRGP, 54000 Nancy, France
*Email: fleal@enscbp.fr

1.1 Introduction

Public health and sustainable development objectives induce a pressing need for the substitution of saturated fats. Palm oil (PO) has been widely used owing to its advantageous features: texturizing ability linked to its solid-like behavior at room temperature, chemical stability, and high oil yield per hectare. The demand for PO has increased over recent decades, which has had profound consequences for the environment in terms of deforestation, land tenure, and community conflicts in countries where it is cultivated. This issue has led to the so-called "Roundtable on Sustainable

Food Chemistry, Function and Analysis No. 31
Development of *Trans*-free Lipid Systems and their Use in Food Products
Edited by Jorge F. Toro-Vazquez

Published by the Royal Society of Chemistry, www.rsc.org

Palm Oil" (RSPO) certification system, whose guidelines are sometimes considered insufficiently restrictive and are not always respected.

The interesting properties of PO are linked to its composition rich in saturated fatty acids (SFAs). Many countries are not PO producers but they produce vegetable oils from rapeseed or sunflower. One alternative to PO consists of submitting vegetable oils to hydrogenation. However, this process may induce the formation of *trans* double-bonds that have harmful effects on human health. In particular, they have a negative impact on the plasma lipoprotein profile as they increase the level of LDL (Low Density Lipoproteins, "bad cholesterol") and lower the level of HDL (High Density Lipoproteins, "good cholesterol"). Studies show that a daily intake of *trans*-fatty acid chains (TFACs) larger than 2% of the total energy intake (TEA) significantly increases the prevalence of cardiovascular diseases, which represent the major proportion of non-communicable diseases (40%), ahead of cancer (27%).[1] The World Health Organization (WHO) recommends limiting consumption of SFAs to 10% and that of TFACs to 2% of the TEA. As a matter of fact, WHO already foresees complete elimination of industrial TFACs.[2]

In this context, food grade systems based on liquid vegetal oils and exhibiting the rheological properties of soft solids are highly sought after. To achieve this goal, two different strategies can be envisaged. The first one consists of introducing compounds which are partially insoluble in oil. Very often, this involves crystallizing triacyglycerides rich in saturated long chains and aggregating the microcrystals to form fractal structures.[3] Other organogel-like systems consist of self-assemblies into long nanotubes that form a 3D entangled network.[4] The second strategy consists of forming oil-in-water (O/W) concentrated emulsions. For instance, Patel *et al.*[5] proposed an emulsion-templated approach for the preparation of oleogels by using a combination of a surface-active and a non-surface-active polysaccharide. The oleogels were prepared *via* a two-step process: (1) formulation of a concentrated O/W emulsion stabilized with a combination of cellulose derivatives and xanthan gum, and (2) selective evaporation of the continuous water phase to drive network formation. An analogous strategy will be considered in the present study, but using solely proteins to stabilize the emulsions. Our approach aims at conferring viscoelastic properties to liquid oils through the interfaces that oil droplets form with the aqueous phase. Texturizing will result from the emulsification of oil in water by vegetable proteins that adsorb at the oil/water interface, followed by droplet concentration through centrifugation or water evaporation.

At low volume fractions, emulsions comprise spherical droplets and behave like viscous fluids. Owing to droplet deformation, emulsions may be concentrated up to volume fractions much higher than the volume fraction corresponding to the close packing of a dispersion of hard spheres, ϕ^*. For randomly monodisperse spheres, $\phi^* \approx 64$ vol%. Above ϕ^*, the droplets can no longer pack without deforming. Emulsions then become predominantly solid at low deformations and exhibit considerable yield stress. As stated by Princen[6] and Mason *et al.*,[7] the elasticity (solid-like behavior) of concentrated emulsions is due to droplets that have been compressed by the application of an external osmotic pressure, Π. Actually, emulsions minimize their free energy by reducing droplet repulsion at the expense of deforming the droplet interfaces at each contact (films become flat), which creates additional surface area. The osmotic pressure Π is the energy per unit volume required to deform the droplets up to a volume fraction ϕ. Additional excess surface area created by a perturbative strain or stress determines the elastic shear modulus, G'. Although Π and G' represent different properties, they both depend on the degree of droplet deformation and therefore on ϕ.

In this study, we explored pathways to obtain very viscous fluids and highly elastic pastes by exploiting the intrinsic properties of concentrated emulsions. We used as emulsion stabilizers proteins extracted from by-products of the oil industry, namely press-cakes. These by-products were selected because they are abundant and have a large protein content. It has been established that in emulsions at droplet volume fractions above ϕ^*, both the elastic and viscous moduli (G' and G'') increase as the droplet radius decreases.[7] Conditions were thus adopted so as to minimize the droplet radius. Dilute emulsions of various average droplet diameters were first fabricated. We used high pressure homogenization coupled with the so-called limited coalescence process, at low protein content, to produce fine emulsions (average diameter <5 μm) with narrow size distributions.[8] The initially dilute emulsions were concentrated by centrifugation and evaporation. The osmotic compressibility, Π, was measured for two average droplet sizes and variable volume fractions above the random close packing, ϕ^*. The dimensionless osmotic compressibility $\Pi(\phi)/(\gamma/R)$, where R is the average droplet diameter and γ is the oil/water interfacial tension, was compared to that obtained in previous studies for equivalent surfactant-stabilized emulsions. Finally, we explored the possibility to fabricate redispersible dry-emulsion powders through a freeze-drying process.

1.2 Experimental

1.2.1 Materials

Rapeseed oil (density = 0.916 g cm^{-3} at 20 °C) was purchased from a local supermarket. All other reagents were from Aldrich. Hexadecane (density = 0.773 g cm^{-3} at 20 °C) was the oil phase in model systems. NaN_3 at a concentration of 0.01 mol L^{-1} was incorporated in all aqueous phases to prevent bacterial growth. Bradford reagent was used for protein quantification. Sodium dodecyl sulfate (SDS) helped to deflocculate emulsions during droplet size measurements. Different saccharides were probed as wall materials to protect emulsions from coalescence during the lyophilization step: Lactose ($C_{12}H_{22}O_{11}$), mannitol ($C_6H_{14}O_6$), and trehalose ($C_{12}H_{22}O_{11}$).

1.2.2 Sunflower Protein Isolate

All emulsions were fabricated with a sunflower protein (SP) isolate obtained from an industrial sunflower press-cake, made up of about 30% proteins, the two main families of which are globulins and albumins. Isolates contain 40 to 90% of globulins and 10 to 30% of albumins, depending on the variety of sunflower.[9] The remaining protein fractions are composed of prolamins, glutelins, and free polypeptide segments.

SP extraction was performed using a one liter reactor stirred at 400 rpm. The appropriate amount of cake and a 1 mol L^{-1} NaCl aqueous phase were added in order to obtain a 1 : 9 w/w (solid/liquid) mixture. A solution at 1 mol L^{-1} NaOH was used to adjust and maintain the pH at 7. The mixture was processed in the reactor at room temperature, for 30 min. Solid–liquid separation was carried out by centrifugation at 15 000 g (with g being the earth gravity constant), at room temperature for 30 min. The aqueous phase was filtered through pleated cellulose filters (Fisherbrand, France). During this step, sunflower albumins and globulins were extracted as well as other water-soluble compounds (free phenolics, oses, osides, amino acids, and free peptides). These compounds were removed by an ultrafiltration process. This process began with a diafiltration step on an Akta flux® 6 ultrafiltration pilot (GE Healthcare, USA) coupled with a 3 kDa cut-off, 4800 cm^2 hollow fiber. First, the liquid extract was concentrated by a 4.2 volume reduction factor, and the retentate was washed with 6 diafiltration volumes (DV) of 0.5 mol L^{-1} NaCl. Then, the pH was adjusted to 10 using a solution of 1 mol L^{-1} NaOH, and the retentate was washed additionally with 4 DV of ultrapure water. The final retentate was collected and freeze-dried. The sunflower isolate was analyzed by the Kjeldahl method[10] to determine its protein content. The obtained sample contained 91 wt% of proteins on a dry matter basis.

In order to determine the protein nature, the isolate was analyzed by high pressure liquid chromatography (HPLC, LC20 system from Shimadzu, Japan) and polyacrylamide gel electrophoresis in the presence of sodium dodecyl sulfate (SDS-PAGE). To perform chromatography, 10 μL of protein isolate solution at the concentration of 5 g L^{-1} were injected into a Biosep SEC s-2000 column (300×7.8 mm; 5 μm; Phenomenex, USA) and eluted using an acetonitrile/ultrapure water/trifluoroacetic acid (45 : 54.9 : 0.1 v/v) mixture flowing at 0.6 mL min^{-1}. Optical detection was performed at the wavelength $\lambda = 280$ nm.[11] Figure 1.1A shows the obtained chromatogram. The dotted lines represent the limits of the areas corresponding to globulins and albumins. Based on these areas and knowing the total protein content, it was possible to estimate the proportion of globulins and albumins in the isolate. Since the two peaks overlapped slightly, the area ratio only provides an estimate. The area corresponding to globulins is considered to be that under the curve at elution times between 8.0 min and 10.2 min, and the area corresponding to albumins is that between 10.2 min and 12.8 min. We deduce that the isolate is composed of 82% globulins and 18% albumins.

An SDS-PAGE technique under reducing conditions was used to confirm that the peaks correspond to sunflower globulins and albumins. For this purpose, a protein solution at a concentration of 5 g L^{-1} was diluted in 50 μL of Laemmli buffer containing 2% β-mercaptoethanol (v/v) and then heated at 95 °C over 5 min. Proteins were separated in a gel composed of 5% stacking and 17%

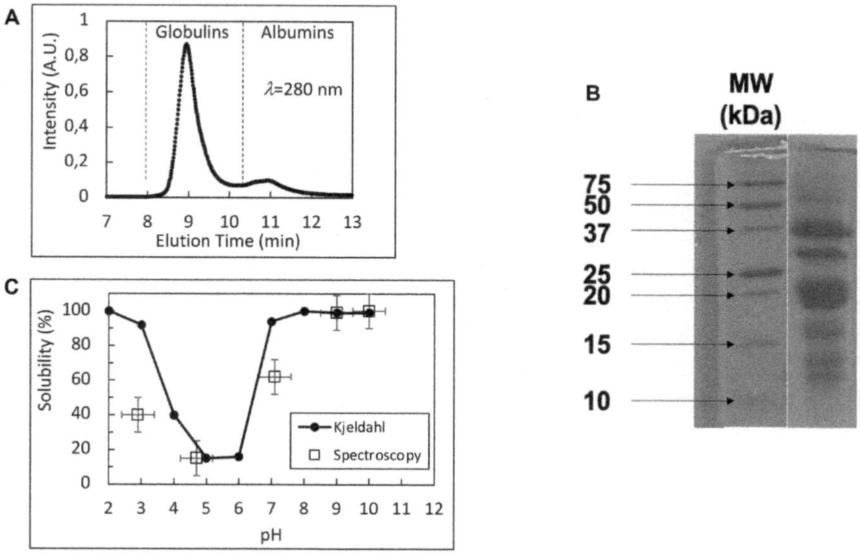

Figure 1.1 (A) HPLC chromatogram of the sunflower protein isolate measured at $\lambda = 280$ nm; (B) SDS-PAGE electrophoresis patterns of the protein isolate; (C) Protein solubility as a function of pH measured by the Kjeldahl method and UV-spectroscopy.

resolving gel by applying an electric current of 20 mA. A Precision Plus Protein Standard (10–250 kDa from Bio-Rad, USA) was used as a molecular weight marker. After migration, gels were stained with Coomassie Brilliant Blue and destained overnight in a 10% acetic acid solution (v/v). The obtained electrophoresis pattern is reported in Figure 1.1B. The left-hand side pattern was obtained with molecular weight markers to define the position scale. The right-hand side pattern is that of the whole isolate sample. The molar mass at 50 kDa is that of sunflower globulin monomer which is hexameric in its native state; the masses between 26 and 37 kDa correspond to acid polypeptides and those between 18 and 25 kDa to the basic polypeptides.[12–14] The molar masses from 14 to 18 kDa can be attributed to sunflower albumins.[15–17]

The solubility of the isolate as a function of pH was determined by two methods.

(i) The isolate was dispersed in reverse osmosis water at 5 $\mathrm{g\,L^{-1}}$. pH adjustment was made by adding small volumes of 1 $\mathrm{mol\,L^{-1}}$ solutions of HCl and NaOH. The mixture was stirred for 20 min and then centrifuged (11 000 g, 20 min, 20 °C). The concentration of proteins in the supernatant was measured by the Kjeldahl method.[10]

(ii) The solubility as a function of pH was also determined by UV-spectrometry. Measurements were carried out at the same ionic strength of 0.1 $\mathrm{mol\,l^{-1}}$, at pH 3 (acetate buffer), 5 (citrate buffer), 7 (phosphate buffer), and 10 (buffer carbonate).

The plots in Figure 1.1C show that the measured solubility is the same for the two protocols, except for the citrate buffer solution at pH 3, which appears to reduce solubility more than a hydrochloric acid solution at the same pH. As citric acid is a poly-anion, it is able to complex a fraction of the proteins. Full solubility of the sunflower isolate is obtained at pH > 7 and in highly acidic conditions (pH < 2). The solubility dip observed at 4 < pH < 6 is due to the isoelectric point of the proteins.[18]

The hydrodynamic radius of the proteins was characterized by dynamic light-scattering (Amtec goniometer (France) and Brookhaven BT9000 correlator (USA)) performed at a wavelength $\lambda = 532$ nm for different scattering angles between 40 and 120° with respect to the incident beam, at 20 °C.[19] The protein solutions were filtered through an ester cellulose filter (CME 0.22 mm, Karl Roth, Germany). The hydrodynamic radius deduced from the diffusion coefficient was $R_h = 10 \pm 1$ nm. This value also obtained by other authors[20] roughly corresponds to the dimension of globulin hexamers.[21]

1.2.3 Methodologies

1.2.3.1 Emulsion Preparation

Both hexadecane-in-water and rapeseed oil-in-water emulsions stabilized by sunflower proteins were prepared. Hexadecane was adopted as a model system in a series of experiments because of its high purity and chemical inertness, allowing a precise determination of the oil/water interfacial tension without artifacts linked to triglyceride hydrolysis. It was also used to generalize the concepts and to provide evidence that the main properties of concentrated emulsions are determined, not by the chemical nature of the oil phase, but by the interfacial properties.

O/W emulsions were prepared using a two-step procedure. First, 25 g of a coarse emulsion at 20 wt% of oil were fabricated using an Ultra-turrax® T-25 disperser (IKA, Germany) operating at 3000 rpm for 5 min. The aqueous phase contained SP from the isolate at a concentration from 0.25 to 2 wt%, and the pH varied between 8.5 and 9.2. This basic pH was due to the presence of NaOH from the extraction process. This pH assured full SP solubilization, as described in Section 1.2.2. The coarse emulsions were then submitted to a second homogenization step using a microfluidizer (Microfluidics M110S, Massachusetts, USA) at a pressure of 800 bars. Emulsions underwent 8 passes through the homogenizing chamber. To avoid sample warming, the emulsions passed through a metallic coil immersed in a cold-water bath (8 °C). Through this procedure, the emulsions obtained at the end of the emulsification process were close to room temperature.

1.2.3.2 Emulsion Characterization

Oil droplets were observed after emulsification and as a function of time using an optical microscope Olympus BX51 (Olympus, Germany) equipped with a digital camera (Leica, 2576×1932-pixel resolution, Germany). The droplet size distributions were measured by static light-scattering, using a Mastersizer 2000 Hydro SM from Malvern Instruments S.A (UK). The droplet size distributions were characterized in terms of their surface-averaged diameter, d_{32}, and polydispersity, P, defined as:

$$d_{32} = \frac{\sum N_i D_i^3}{\sum N_i D_i^2}, \quad P = \frac{1}{\bar{D}} \frac{\sum N_i D_i^3 (\bar{D} - D_i)}{\sum N_i D_i^3} \tag{1.1}$$

where $\bar{D}$ is the median diameter, *i.e.*, the value for which the cumulative undersized volume fraction is equal to 50%, and N_i is the total number of droplets with diameter D_i. Static light-scattering data were transformed into droplet size distribution using Mie theory.[22] The refractive index adopted for calculations was 1.43 for hexadecane, 1.47 for rapeseed oil, and 1.33 for water. Before analysis, emulsions underwent a 10-fold dilution in a 1 wt% SDS solution to disaggregate the droplets. The measuring cell was filled with pure water, and a small volume of the emulsion was introduced under stirring (1400 rpm).

1.2.3.3 Protein Quantification

After emulsification, a fraction of the proteins initially dissolved in the aqueous phase was irreversibly adsorbed at the oil/water interface. The residual protein concentration in the aqueous phase was measured using the Bradford assay.[23] The emulsions were centrifuged for 10 to 20 min at accelerations between 1000 *g* and 10 000 *g* (Eppendorf Centrifuge 5430, Germany). Under the effect of buoyancy, a cream of oil droplets was formed at the top of the centrifuge tube. The protein concentration in the subnatant, $C_{\mathrm{Pr}}^{\mathrm{f}}$, was measured, and the mass of proteins adsorbed at the interface, m_{a}, was determined by considering the following mass balance:

$$m_{\mathrm{a}} = (C_{\mathrm{Pr}}^{0} - C_{\mathrm{Pr}}^{\mathrm{f}}) V_{\mathrm{aq}} \tag{1.2}$$

where C_{Pr}^{0} is the protein concentration in the aqueous phase before emulsification, and V_{aq} is the volume of the aqueous phase.

1.2.3.4 Osmotic Pressure Measurements

Emulsions were concentrated by centrifugation using a Beckman Coulter (Avanti J-30I JS-24.15, USA) centrifuge equipped with JS-24.15 Swinging-Bucket Aluminum Rotor. Emulsions were processed for a time ranging from 90 min to 5 hours at accelerations from 1000 g to 100 000 g. Centrifugation was applied until constant droplet volume fraction was achieved in the cream. After creaming, if the thickness of the cream formed by the droplets is much lower than the length of the centrifuge lever arm, the spatial gradient in the acceleration can be neglected, and the osmotic pressure can be determined as:[24]

$$\Pi = L \Delta \rho \omega^2 \phi_i d \tag{1.3}$$

where L is the total liquid height in the centrifuge tube, ϕ_i is the initial oil volume fraction before centrifugation, $\Delta\rho$ is the density mismatch between oil and water, d is the length of the lever arm, and ω is the rotation speed of the centrifuge.

A small amount of the cream located at the top of the centrifugation tube was collected and left in a dry place at room temperature for approximately 12 h. During this time period, hexadecane evaporation was confirmed to be negligible due to its very low volatility. Samples weighing before and after water evaporation provided a direct measure of the oil volume fraction, ϕ.

1.2.3.5 *Interfacial Tension*

The oil/water interfacial tension γ was measured using the rising drop method (Tracker apparatus, Teclis Instruments, France). A droplet of hexadecane (10 μL) was formed at the tip of a steel needle in a transparent cell filled with an aqueous phase containing 0.2 g L^{-1} of SP, at pH 8.2. Images of the droplets were recorded at regular time intervals with a digital camera. The droplet profile was analyzed to calculate its area and surface tension by solving the Laplace–Young equation. The oil/water interfacial tension decreased over time until reaching an asymptotic value of 15 mN m^{-1} that will be used in further calculations.

1.2.3.6 *Emulsion Drying*

The emulsions were dried by lyophilization right after preparation. They were stored at least 12 h at −80 °C and then introduced into a freeze-dryer (FTS Systems, Dura-Dry, USA) at a pressure of 0.1 mbar for 24 h. The residual moisture in the powders was determined with a moisture analyzer (Ohaus MB120, USA). The obtained powders were stored at room temperature.

1.2.3.7 *Redispersion of Dried Emulsions*

The powders were gently redispersed by vigorous manual mixing for 1 min, until the samples became homogeneous, with the amount of distilled water required to restore the same composition as that of the parent emulsions. Their size distribution was measured after redispersion following the method described in Section 1.2.3.2.

1.3 Results and Discussion

1.3.1 Impact of the Protein Content on the Average Droplet Size

Hereafter, SP concentrations will be expressed in weight percent of the aqueous phase. Emulsions containing variable amounts of SP, from 0.25 to 2 wt%, were prepared. The data in Figure 1.2 were obtained for a hexadecane-in-water emulsion containing 1 wt% of SP. The emulsion was

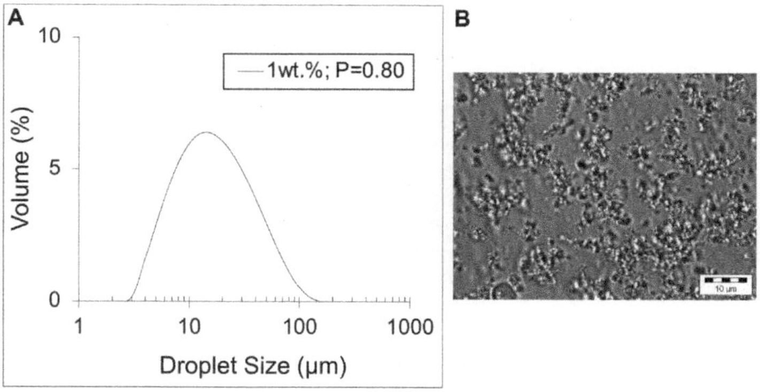

Figure 1.2 (A) Droplet size distribution of a hexadecane-in-water emulsion initially containing 1 wt% sunflower protein in the aqueous phase. Measurements were performed right after emulsification. (B) Micrograph of the same emulsion. Samples were analyzed without prior dilution in an SDS solution.

first analyzed by diluting it in pure water. The size distribution in Figure 1.2A is wide, as reflected by the relatively high value of the polydispersity index ($P = 0.8$). Actually, the size distribution does not reflect the size of individual droplets but that of droplet aggregates. Indeed, the micrograph of Figure 1.2B shows that all droplets are immobilized within large ramified clusters. The same aggregated state was obtained, irrespective of the oil nature (rapeseed oil or hexadecane) and of the protein content.

It is well known that the presence of SDS can disrupt protein aggregates and deflocculate emulsions. This is evidenced in the micrograph of Figure 1.3B, corresponding to an emulsion based on hexadecane, containing 0.25 wt% of SP. Prior to its observation under the microscope and analysis by light-scattering, the emulsions were diluted with a 1 wt% SDS solution. In this case, individual droplets submitted to Brownian motion are observed, and the size distributions reported in Figure 1.3A are quite narrow, with P values being lower than 0.40. Interestingly, the distribution shifts toward lower sizes as the protein concentration increases. The protein concentrations explored here are rather low, and it is likely that the shear applied by the microfluidizer is able to generate much more interfacial area than the proteins can cover. Since proteins are in general irreversibly adsorbed, conditions are thus met for the so-called limited coalescence process to occur: when the agitation is stopped, droplets recombine until the degree of interfacial coverage becomes sufficient to stabilize them against further coalescence. One of the main features of limited coalescence is its ability to generate narrow droplet size distributions.[8]

1.3.2 Interfacial Coverage

The total droplets interfacial area, A, can be expressed as:[8]

$$A = \frac{6V_\mathrm{d}}{d_{32}} = \frac{m_\mathrm{a}}{\Gamma} \tag{1.4}$$

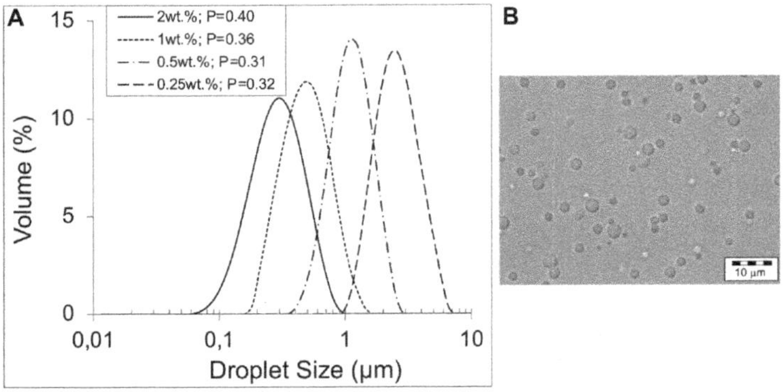

Figure 1.3 (A) Droplet size distribution of a hexadecane-in-water emulsion as a function of the initial sunflower protein concentration in the aqueous phase, expressed in weight percent of the aqueous phase. The initial sunflower protein concentrations varied from 0.25 wt% to 2 wt%. (B) Micrograph of the emulsion obtained with 0.25 wt% sunflower protein. Emulsions were diluted in a 1 wt% SDS solution before analysis.

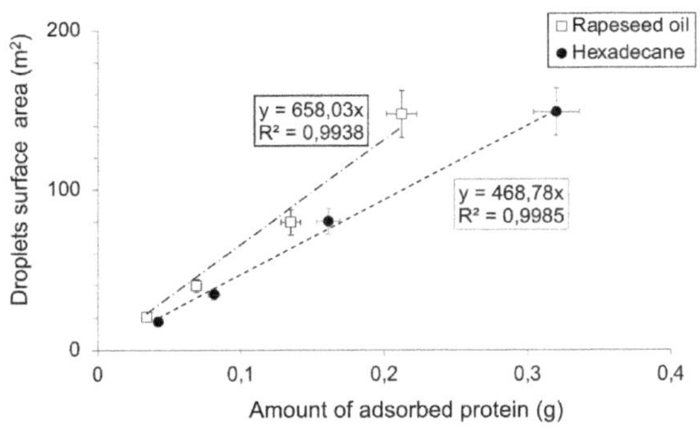

Figure 1.4 Droplet surface area as a function of the mass of sunflower protein adsorbed for emulsions with 20 wt% hexadecane or rapeseed oil. The total emulsion mass was 25 g.

where V_d is the volume of the dispersed phase and m_a is given by eqn (1.2). Assuming that the interfacial coverage adopts a unique value at the end of the limited coalescence process, Γ can be determined from the slope of the curve representing the evolution of the total surface area of the droplets, A, as a function of m_a. Figure 1.4 shows the corresponding plot. The linearity confirms the hypothesis about the existence of a unique interfacial coverage right after the emulsification process. From the slope, we deduce $\Gamma = 2.1 \pm 0.1$ mg m^{-2} and 1.5 ± 0.1 mg m^{-2} for emulsions based on hexadecane and on rapeseed oil, respectively. The emulsions were kinetically stable, with no apparent evolution of their size distributions over a time storage period of at least one month, at room temperature.

1.3.3 Emulsions Concentrated by Centrifugation – Osmotic Resistance Measurements

The emulsions based on hexadecane with average diameters $d_{32} = 0.24$ μm and 0.49 μm were centrifuged following the method described in Section 1.2.3.4 in order to obtain the Π *vs.* ϕ curves. Within the experimental conditions adopted, ϕ values from 65% to 95 vol% were obtained. The size distributions of the concentrated emulsions were identical to the initial ones (data not shown). In Figure 1.5A, we report the experimental data obtained for the two average droplet diameters. We would like to emphasize that the experimental points in our case refer to $\Pi(\phi)$ curves which are only valid for compression. Indeed, compressed emulsions did not redisperse spontaneously, even after being in contact with a large quantity of aqueous phase for a week. The compaction process probably involves some irreversibility. Since our emulsions did not expand when the applied pressure was set to zero, hereafter, the parameter Π defined by eqn (1.3) will be termed "osmotic resistance".

It is instructive to plot the data in $\Pi/(\gamma/R)$ with $R = d_{32}/2$, *versus* ϕ coordinates: this type of plot provides a possibility to compare the results with literature data excluding the effects of droplet size and interfacial tension. The first quantitative study on the properties of monodisperse emulsions was performed by Mason *et al.*[7] They investigated silicon O/W emulsions stabilized by SDS with different average diameters. The elastic shear modulus G' and the osmotic pressure Π had comparable values and were found to increase by nearly 4 decades as the droplet fraction, ϕ, increased from 50 to approximately 90 vol%. They both exhibited a universal

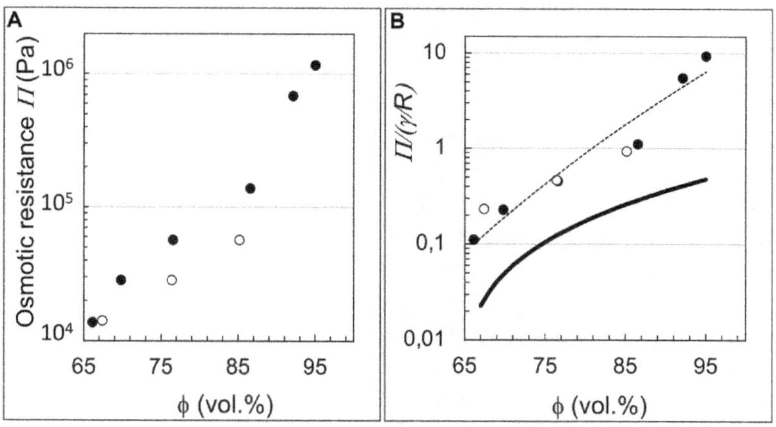

Figure 1.5 (A) Evolution of the osmotic resistance, Π, as a function of the droplet volume fraction ϕ for hexadecane-in-water emulsions. Open circles: $d_{32} = 0.49$ μm; full circles: $d_{32} = 0.24$ μm. (B) Evolution of the osmotic resistance, Π, normalized by γ/R. The dashed line just follows the trend of the experimental data. The solid line corresponds to eqn (1.5).

dependence on the oil volume fraction, when scaled with the Laplace pressure of the non-deformed droplets, γ/R:[25]

$$\Pi/\left(\frac{\gamma}{R}\right) \approx G'/\left(\frac{\gamma}{R}\right) \approx 1.7\phi^2(\phi - \phi^*) \tag{1.5}$$

The scaling with γ/R confirms that the elasticity and osmotic pressure of monodisperse emulsions depend only on the packing geometry of the compressed droplets and on Laplace pressure. In principle, it is necessary to consider the effective oil volume fraction in order to account for the repulsive interactions between the droplets. As highlighted by Princen *et al.*,[26] the effective oil volume fraction, ϕ_{eff}, integrates the thickness, h, of the thin films between the deformed droplets:

$$\phi_{\text{eff}} = \phi\left(1 + \frac{h}{d_{32}}\right)^3 \tag{1.6}$$

It is reasonable to assume that h is comparable to the protein size, *i.e.*, a few nanometers. For the emulsions that were probed, the droplet diameter ($d_{32} > 0.25$ μm) was very large compared to the protein size. Since $h/d_{32} < 5 \cdot 10^{-2}$, the corrective term in eqn (1.6) was neglected and the effective droplet volume fraction was assumed to be equal to the actual oil volume fraction, ϕ.

In Figure 1.5B, we report the evolution of the osmotic resistance normalized by γ/R, with $\gamma = 15$ mN m^{-1} (Section 1.2.3.5). The data fall onto a single curve within reasonable experimental uncertainty. Our results were compared to the normalized data obtained by Mason *et al.*[25] for emulsions stabilized by a low molecular-weight surfactant (SDS). The latter are represented as a solid line which corresponds to the best fit to the experimental points given by eqn (1.5). It clearly appears that the obtained values for protein-stabilized emulsions are much higher than the ones obtained in the presence of surfactant. This behavior indicates that, in our case, droplet deformability is not determined by surface tension. Dimitrova and Leal-Calderon[27] have measured $\Pi(\phi)$ and $G'(\phi)$ for hexadecane-in-water emulsions stabilized by proteins of animal origin. They found that emulsions stabilized by globular proteins like bovine serum albumin (BSA), β-lactoglobulin (BLG), or lysozyme possess remarkably high elasticity and osmotic resistance in comparison with equivalent emulsions stabilized by surfactants. They argued that this behavior may have two different origins: (i) adhesion or stickiness of the protein-stabilized films and (ii) viscoelasticity of the protein interfacial layers. Actually, the two factors cannot be considered separately, because both arise from specific interactions between the adsorbed proteins and are intrinsically interrelated.

It is widely accepted that proteins stabilize the emulsions, forming a viscoelastic film at the interface.[28] A protein layer can be considered as a two-dimensional body, which has its own rheological properties.[29,30]

The interfacial concentration is of the order of several mg m^{-2},[31] and this was confirmed by the interfacial coverages measured in Section 1.3.2. The adsorption of proteins is generally followed by gradual unfolding. In this process, proteins lose their secondary and higher structure in the adsorbed state. Unfolding uncovers different segments of adsorbed species, which facilitates lateral interactions of various types between the adsorbed molecules: ionic, hydrophobic, covalent (disulfide bridging), or hydrogen bonding. As a result, the adsorbed protein layers are viscoelastic, and almost tangentially immobile.[32,33] This is especially true for globular proteins like BLG, which can form disulfide bridges[34] and lead to a cohesive elastic network within the interface.[35]

Useful information concerning the properties of protein-stabilized emulsions can be obtained from model experiments on thin liquid films using a Scheludko-type cell.[36] Dimitrova and Leal-Calderon[27] performed this type of experiment with globular proteins in order to mimic the films present in concentrated emulsions. Films stabilized by BSA and BLG were found to aggregate forming rigid clusters that joined the interfaces and rigidified them. Dynamic light scattering from bulk solutions of BSA and BLG revealed no clustering in bulk solutions, while large aggregates at the surface were clearly observed during thin liquid film studies, most probably due to extensive interfacial protein unfolding.[37,38]

The same type of behavior has been found in emulsions stabilized by solid particles. Arditty *et al.*[39] measured the bulk properties of solid-stabilized concentrated emulsions and deduced some characteristic properties of the interfacial layers covering the droplets. The osmotic resistance, Π, of the emulsions was measured for different oil volume fractions above ϕ^*. The dimensionless osmotic pressure, $\Pi/(\gamma/R)$, was always substantially higher than the corresponding values obtained for surfactant-stabilized emulsions. The authors concluded that droplet deformation in solid-stabilized emulsions is not controlled by the capillary pressure, γ/R, of the non-deformed droplets but rather by σ_0/R, with σ_0 being a parameter characterizing the rigidity of the droplet surfaces induced by the very strong lateral interactions between the solid particles adsorbed at the interface. Their data were interpreted considering that the interfacial layers are elastic at small deformations and exhibit plasticity at intermediate deformations. σ_0 corresponds to the surface yield stress, *i.e.*, the transition between elastic and plastic regimes.

1.3.4 Macroscopic Aspect of Concentrated Emulsions as a Function of the Droplet Fraction – Biliquid Foams

Hexadecane-in-water emulsions were submitted to centrifugation until a dense cream was formed and the subnatant phase was fully transparent (Figure 1.6A). After centrifugation, the creams were collected and dried to estimate their droplet fraction. Figure 1.6B shows the macroscopic aspect of an emulsion whose droplet fraction is close to 75 vol%. The sample resembles a solid paste that sustains its own weight. The sample exhibits a smooth texture and is

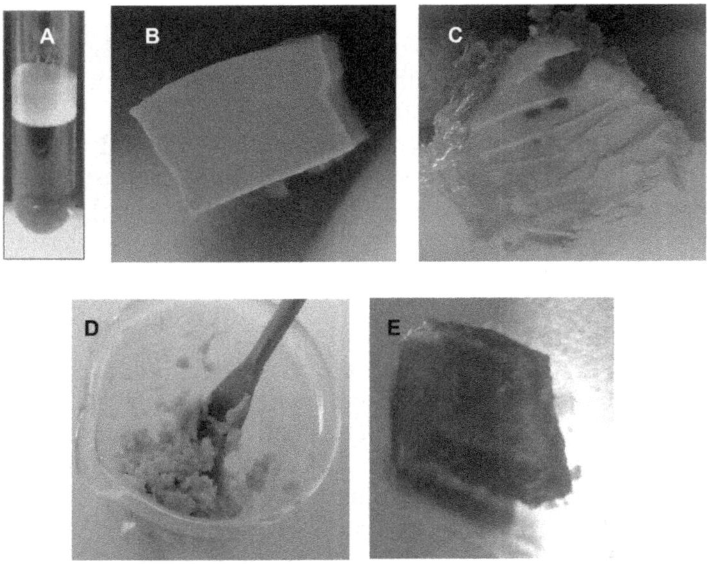

Figure 1.6 Hexadecane-in-water materials obtained by centrifugation (1 wt% proteins) and/or water evaporation. (A) Emulsion in the centrifuge tube; (B) and (C) Emulsions with 75% volume fraction; (D) Emulsions with 90% volume fraction; (E) Biliquid foam (~99 vol%).

spreadable upon application of shear, as illustrated in Figure 1.6C. At higher droplet fractions, of the order of 90 vol%, corresponding to larger acceleration rates in the centrifuge, the dense emulsions became brittle materials. They readily disintegrated into small pieces when they were manipulated. A photograph revealing this behavior is shown in Figure 1.6D. Such brittleness is the reason why rheological measurements were not performed, as it was not possible to load the sample cell properly at high droplet fractions. Osmotic resistance measurements were carried out preferentially, bearing in mind that Π and G' have been found to have comparable values in emulsions stabilized by surfactants[25] and solid particles.[39]

Some samples were allowed to evaporate for at least 24 hours after centrifugation. The evaporation of water led to the formation of a transparent solid-like material (Figure 1.6E) that contained about 99% of dry extract (oil + protein). Furthermore, within experimental error, the sample mass remained constant over several days. These observations provide evidence that almost all of the water content evaporated. In this case, droplets are expected to become polyhedral and these concentrated materials are referred to as "biliquid foams". They are the kinds of solids which, on gentle touch, exhibit a strong elastic response. However, they are fragile: at strong deformation, they release oil. The same observations were obtained with rapeseed-in-water emulsions (data not shown).

Biliquid foams based on low molecular-weight surfactants generally release oil when the stabilizing layers are dehydrated by free evaporation.[40]

On the contrary, our biliquid foams withstand evaporation without being destroyed. They exhibit remarkable resistance to coalescence. This property probably arises from the specific nature of the interfacial layers comprising unfolded proteins which are sufficiently thick and rigid to protect the globules against coalescence. It is worth noticing that this type of material has already been obtained in the presence of BSA and BLG.[27] Our results thus generalize the concept of using globular proteins that undergo interfacial unfolding to obtain emulsions with outstanding stability.

1.3.5 Emulsions Concentrated by Freeze-drying

Compared to conventional emulsions, dry emulsions have many technological assets: increased shelf life, lack of bacteriological growth owing to the very low water activities, facilitated transport, and improved oil stabilization against oxidation.[41–43] We implemented freeze drying as a means to obtain concentrated emulsions starting from dilute ones. Freeze drying induces cryo-concentration of the droplets, likely to compress them and to provoke coalescence.[44] To avoid oil droplets from merging into a macroscopic separated oil phase, either during the drying process or during the storage of the dry powder, a hydrophilic wall material is generally added to the aqueous phase. Typical examples of species used as wall materials are glucose, lactose, maltodextrin, starch, and cellulose. Once water is evaporated, this wall material, together with the emulsifier used to stabilize the emulsion, forms a solid phase that surrounds the droplets, thus avoiding their recombination (coalescence). The required amount of solid wall material generally varies from 30 to 80 wt% of the total final powder.[41]

Freeze-drying was implemented as described in Section 1.2.3.6 to dehydrate rapeseed O/W emulsions. The final dried systems were obtained in powder form. Parent emulsions were formulated in the presence of 3 saccharides: trehalose, mannitol, and lactose. Their aqueous phase initially contained 2 wt% of SP and 20 wt% of saccharide. The redispersible capacity of the powders in water was examined following the protocol described in Section 1.2.3.8. The behavior of emulsions containing wall material was compared to a control system devoid of saccharide.

After freeze-drying, the obtained powders contained about 53 wt% of rapeseed oil, 42 wt% of saccharide, 4 wt% of SP, and 1% of residual moisture. The powder deriving from the control system contained 92 wt% of rapeseed oil, 7 wt% of SP, and 1% of moisture. Irrespective of the saccharide nature, the moisture content was close to 1 wt%, within the prescribed regulations for dried powder in the food industry (<3–4 wt%).[41] The powders remained kinetically stable under quiescent conditions in sealed bags, with no apparent macroscopic oil leakage even after several months of storage at room temperature, reflecting efficient stabilization of the oil droplets by the solid component. However, the control system and the powder based on mannitol were clearly clumpier and greasier. Figure 1.7 shows the visual macroscopic aspect, as well as typical micrographs of the systems.

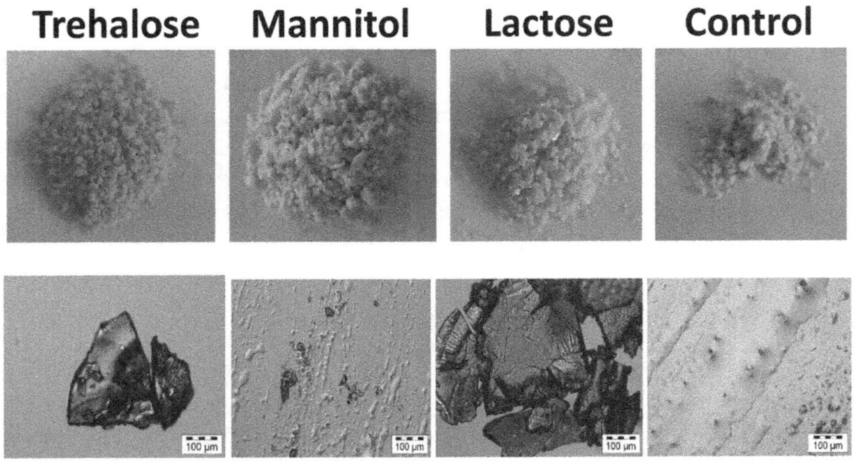

Figure 1.7 Morphologies of freeze-dried rapeseed oil-in-water emulsions containing different saccharides as wall material. The control system was devoid of saccharide. Top: Macroscopic aspect; Bottom: Micrographs.

Dry emulsions are made of large (>10 µm) irregularly shaped and aggregated solid grains. To carry on the observations with the microscope, powders were forced to spread on the glass slide with a spatula and it is apparent that free oil is present around the grains and wets the glass substrate in both the emulsion containing mannitol and in the control.

Powders were rehydrated the same day using pure water in order to restore the composition of the parent emulsions (method described in Section 1.2.3.8). To assess the ability of the drying process to preserve the state of dispersion, the size distributions of the oil droplets were compared before dehydration and after redispersion of the powders in an aqueous phase. Data are gathered in Figure 1.8 (A: size distributions; B: micrographs). Regardless of the presence or absence of wall material and of the saccharide nature, the average droplet size of parent emulsions is close to 0.25 µm and the droplet size distribution is quite narrow, with $P < 0.51$. All rehydrated emulsions have a size distribution broader than the one corresponding to their parent emulsion (the polydispersity index P is always larger than the initial one), reflecting partial droplet coalescence induced by the freeze-drying process. In the control system and in the emulsion based on mannitol, the peak of the initial droplet size has almost fully disappeared, at the expense of a mode between 10 and 100 µm. Thus, the emulsion droplets have undergone significant coalescence during freeze-drying. The size distributions are in accord with the greasy aspect of the powder and with the micrographs of Figure 1.7. Conversely, trehalose and lactose provide a good level of protection against coalescence. The intensity of the mode located between 0.1 and 1 µm remains significant, and the shoulder observed at larger sizes induced by coalescence represents less than 25% of the droplets volume (cumulated fraction). It is within the scope of future work to analyze

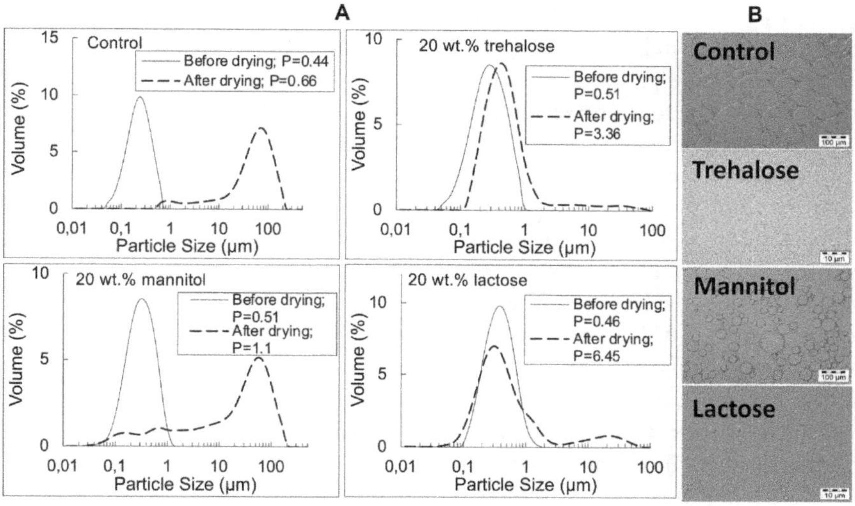

Figure 1.8 (A) Size distributions of parent and rehydrated emulsions for rapeseed oil-in-water emulsions containing different saccharides as wall materials. (—) Parent emulsion; (- - -) Rehydrated emulsion after freeze-drying. (B) Micrographs of rehydrated emulsions.

dehydrated emulsions by spray drying, a process that generally reduces the extent of droplet coalescence.[45]

To sum up, our results reveal that emulsions processed by freeze-drying are more prone to coalescence than those submitted to centrifugation and/or simple evaporation at room temperature that were described in Section 1.3.4. However, whatever the technique used to concentrate the emulsions, bulk or powdered solid materials can be obtained with no macroscopic oil leakage at rest.

1.4 Conclusions

Hydrophilic biopolymers like proteins or polysaccharides provide the chemical "scaffold" to produce water-based gels. Since they are insoluble in oil, finding a way to make them oil gelators would considerably expand their applications field. In this chapter, we have reported an approach to generate oleogels containing liquid oil and hydrophilic biopolymers. The approach is simple, versatile and is based on the preparation of a fine O/W emulsion, followed by removal of the water phase, resulting in the physical trapping of oil droplets in a protein network. A large variety of textures could be achieved by varying the droplet fraction: viscous fluids, viscoelastic pastes, and elastic "gums" (transparent biliquid foams), as ϕ increased from 65 to 99 vol%. Drying concentrated emulsions usually provokes coalescence of the oil droplets, resulting in the full separation of the two immiscible phases. This instability can be prevented if the droplet interface is stiffened. The concept of structuring liquid oils by using O/W emulsions as templates was first reported by Gao *et al.*[46] and by Romoscanu and Mezzenga.[47] In their cases, stabilization was achieved by either forming surfactant–protein complexes[46] or by crosslinking proteins at the oil–water interface with chemicals.[47] Patel *et al.*[5] found a route for emulsion stabilization without the need for any surfactant or a crosslinking step. Stabilization was obtained with only a combination of polysaccharides. Here, we went on step further by using solely globular proteins that deliver interfacial stiffness owing to unfolding. Following a previous work on concentrated emulsions stabilized by animal proteins,[24,27] vegetal proteins were used as emulsion stabilizers to build up solid properties while preserving the low viscosity and ease of manufacturing. This kind of system can find applications in food matrices that require texturized oils. They can be easily handled at room temperature at a volume fraction such that they remain easily spreadable (<75%), and then *in situ* evaporation transforms them into solids. Finally, we could formulate dried-emulsions in powder form after submitting initial dilute emulsions to a freeze-drying step in the presence of a wall material (saccharide). Advantageously, the obtained powders were fully redispersible in water.

Acknowledgements

This study was performed in partnership with the French Institute for the Energy Transition (Institut pour la Transition Energétique) PIVERT (www.institut-pivert.com) and supported by the French Government under the reference ANR-001-01. Alexandre Poirier, Laurence Ramos, Amélie Blanc, and Martin In are gratefully acknowledged for fruitful discussions.

References

1. R. Uauy, A. Aro, R. Clarke, M. R. L'Abbé, D. Mozaffarian, C. M. Skeaff, S. Stender and M. Tavella, *Eur. J. Clin. Nutr.*, 2009, **63**, S68.
2. https://www.who.int/news/item/14-05-2018-who-plan-to-eliminate-industrially-produced-trans-fatty-acids-from-global-food-supply (last accessed January 2021).
3. S. S. Narine and A. G. Marangoni, *Food Res. Int.*, 1999, **32**(4), 227.
4. A. Bot, R. den Adel and E. C. Roijers, *J. Am. Oil Chem. Soc.*, 2008, **85**, 1127.
5. A. R. Patel, N. Cludts, M. D. Bin Sintang, B. Lewille, A. Lesaffer and K. Dewettinck, *Chem. Phys. Chem.*, 2014, **15**(16), 3435.
6. H. M. Princen, *J. Colloid Interface Sci.*, 1983, **91**, 160.
7. T. G. Mason, A. H. Krall, H. Gang, J. Bibette and D. A. Weitz, in *Encyclopedia of Emulsion Technology*, ed. P. Becher, Marcel Dekker, New York, 1996, vol. 4, p. 299 and references therein.
8. W. Dridi, C. Harscoat-Schiavo, J. Monteil, C. Faure and F. Leal-Calderon, *Langmuir*, 2018, **34**, 9228.
9. S. Gonzalez-Perez and J. M. Vereijken, *J. Sci. Food Agric.*, 2007, **87**, 2173.
10. J. Kjeldahl, *Z. Anal. Chem.*, 1883, **22**(1), 366.

11. E. Layne, in *Method in Enzymology*, ed. P. S. Colowick and N. O. Kaplan, Academic Press, Inc., New York, 1957, vol. 3, p. 447.
12. R. A. Y. Haar, R. D. Allen, E. A. Cohen, C. L. Nessler and T. L. Thomas, *Gene*, 1988, **74**(2), 433.
13. M. Dalgalarrondo, J. Raymond and J.-L. Azana, *J. Exp. Bot.*, 1984, **35**(11), 1618.
14. A. A. Kortt and J. B. Caldwell, *Phytochemistry*, 1990, **29**(5), 1389.
15. A. A. Kortt and J. Caldwell, *Phytochemistry*, 1990, **29**(9), 2805.
16. J. Raymond, J. M. Robin and J. L. Azanza, *Plant Syst. Evol.*, 1995, **198**(3), 195.
17. I. N. Anisimova, R. J. Fido, A. S. Tatham and P. R. Shewry, *Euphytica*, 1995, **83**(1), 15.
18. S. Gonzalez-Perez and J. M. Vereijken, *J. Sci. Food Agric.*, 2007, **87**, 2173.
19. A. Poirier, *Propriétés fonctionnelles de protéines végétales, en volume et aux interfaces fluides*, PhD thesis, 2019, University of Montpellier.
20. D. Karefyllakis, S. Altunkaya, C. C. Berton-Carabin, A. J. van der Goot and C. V. Nikiforidis, *Food Hydrocolloids*, 2017, **73**, 326.
21. P. Plietz, G. Damaschun, J. J. Muller and K. D. Schwenke, *Eur. J. Biochem.*, 1983, **130**, 315.
22. G. Mie, *Ann. Phys.*, 1908, **330**(3), 377.
23. M. M. Bradford, *Anal. Biochem.*, 1976, **72**, 248.
24. T. D. Dimitrova and F. Leal-Calderon, *Adv. Colloid Interface Sci.*, 2004, **108–109**, 49.
25. T. M. Mason, J. Bibette and D. A. Weitz, *Phys. Rev. Lett.*, 1995, **75**, 2051.
26. H. M. Princen, M. P. Aronson and J. C. Moser, *J. Colloid Interface Sci.*, 1980, **75**, 246.
27. T. D. Dimitrova and F. Leal-Calderon, *Langmuir*, 2001, **17**, 3235.
28. E. Dickinson and A. Williams, *Colloids Surf.*, 1994, **88**, 317.
29. E. Lucassen-Reynders and J. Benjamins, in *Food Emulsions and Foams Interfaces, Interactions and Stability*, ed. E. Dickinson and J. M. Rodriguez Patino, Royal Society of Chemistry, London, 1999, p. 195.
30. D. E. Graham and M. C. Phillips, *J. Colloid Interface Sci.*, 1979, **70**, 415.
31. S. Tcholakova, N. D. Denkov, I. B. Ivanov and B. Campbell, *Langmuir*, 2002, **18**(23), 8960.
32. M. A. Bos and T. van Vliet, *Adv. Colloid Interface Sci.*, 2001, **91**(3), 437.
33. E. M. Freer, K. S. Yim, G. G. Fuller and C. Radke, *J. Phys. Chem. B*, 2004, **108**(12), 3835.
34. E. Dickinson and Y. Matsumura, *Int. J. Biol. Macromol.*, 1991, **13**(1), 26.
35. A. Williams and A. Prins, *Colloids Surf., A*, 1996, **114**, 267.
36. A. Scheludko, *Kolloid-Z*, 1957, **155**, 39.
37. T. D. Dimitrova, *Emulsions stabilisées par les protéines: forces collôïdales, rhéologie et crémage*, PhD thesis, University of Bordeaux, Bordeaux, France, 2000.
38. T. G. Gurkov, K. G. Marinova, A. Z. Zdravkov, C. Oleksiak and B. Campbell, *Prog. Colloid Polym. Sci.*, 1998, **110**, 263.

39. S. Arditty, F. Lequeux, V. Schmitt and F. Leal-Calderon, *Eur. Phys. J. B*, 2005, **44**, 381.
40. O. Sonneville-Aubrun, V. Bergeron, T. Gulik-Krzywicki, B. Jonsson, H. Wennerström, P. Linder and B. Cabane, *Langmuir*, 2000, **16**, 1569.
41. Y. S. Gu, E. A. Decker and D. J. McClements, *J. Agric. Food Chem.*, 2004, **52**(11), 3626.
42. U. Klinkesorn, P. Sophanodora, P. Chinachoti, D. J. McClements and E. A. Decker, *J. Agric. Food Chem.*, 2005, **53**(21), 8365.
43. L. A. Shaw, D. J. McClements and E. A. Decker, *J. Agric. Food Chem.*, 2007, **55**(8), 3112.
44. A. Marefati, M. Rayner, A. Timgren, P. Dejmek and M. Sjöö, *Colloids Surf., A*, 2013, **436**, 512.
45. C. Joseph, R. Savoire, C. Harscoat-Schiavo, D. Pintori, J. Monteil, C. Faure and F. Leal-Calderon, *LWT-Food Sci. Technol.*, 2019, **113**, 108311.
46. Z.-M. Gao, X.-Q. Yang, N.-N. Wu, L.-J. Wang, J.-M. Wang, J. Guo and S.-W. Yin, *J. Agric. Food Chem.*, 2014, **62**, 2672.
47. A. I. Romoscanu and R. Mezzenga, *Langmuir*, 2006, **22**, 7812.

CHAPTER 2

Alternative Oil Structuring Techniques: Oil Powders, Double Emulsions and Oil Foams

ZIJIAN ZHI,[a] KOEN DEWETTINCK[a,b] AND
FILIP VAN BOCKSTAELE*[a,b]

[a] Ghent University, Department of Food Technology, Safety and Health, Food Structure and Function (FS&F) Research Group, Coupure Links 653, Ghent, Belgium; [b] Vandemoortele Centre 'Lipid Science and Technology', Coupure Links 653, Ghent, Belgium
*Email: filip.vanbockstaele@ugent.be

2.1 Double Emulsions

2.1.1 Background

Double emulsions are complex emulsions containing small droplets which are also emulsions.[1] The two main types of double emulsions are: water-in-oil-in-water (W/O/W) emulsions and oil-in-water-in-oil (O/W/O) emulsions. Double emulsions have a high application potential in food, pharmaceutical, and cosmetic industries as their special structures allow reorganization of the lipid phase and control of the incorporation of biologically active compounds and chemical substances. Considering the alternative design of edible solid and semi-solid structured oil, in this part we only introduce the basic knowledge on O/W/O emulsions, which is an oil-continuous system containing water droplets with smaller oil droplets inside them (see Figure 2.1).

Food Chemistry, Function and Analysis No. 31
Development of *Trans*-free Lipid Systems and their Use in Food Products
Edited by Jorge F. Toro-Vazquez

Published by the Royal Society of Chemistry, www.rsc.org

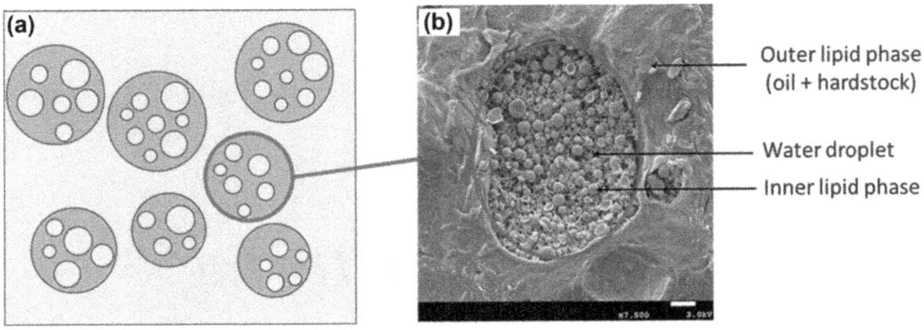

Figure 2.1 The microstructure of an oil-in-water-in-oil (O/W/O) double emulsion: (a) the model structure and (b) the cryo-SEM image of an O/W/O emulsion (FS&F, UGent).

The interest of food scientists and technologists in O/W/O systems has continued to grow steadily for two main reasons. First, it is possible to formulate healthier products with high energy by replacing traditional water-in-oil (W/O) emulsions with the equivalent O/W/O emulsions having a lower saturated fatty acid content but a similar in-mouth perceived texture. Second, O/W/O emulsions can be capable of encapsulating and protecting sensitive oil-soluble (or oil-dispersible) ingredients, such as fish oil, essential oils, or bioactive nutrients, followed by controlled release or delivery of these ingredients during consumption and/or digestion. The goal of food scientists is to achieve these formulation objectives using food-grade ingredients. Experience has shown that overcoming this challenge is not a straightforward matter. This is essentially because double emulsions, as has been described in the standard modern textbook on food emulsions,[2] are "more difficult and expensive to prepare than conventional emulsions" and are "highly susceptible to breakdown during storage or when exposed to environmental stresses." Indeed, the difference between the inner and outer phases may result in instability phenomena in double emulsions, such as creaming, flocculation, coalescence, and Ostwald ripening between droplets,[3,4] which requires an additional stability factor to control any swelling and rupture of the inner droplets induced by osmotic transport between different phases. Moreover, during the preparation of the outer emulsion, the inner emulsion is often partially disrupted due to mechanical forces. Therefore, the formulation and stability of double emulsions are key to overcoming and further expanding their applications.

2.1.2 Preparation and Stabilization

2.1.2.1 Preparation of O/W/O Double Emulsion

Generally, the standard procedure for making O/W/O emulsions consists of 2 separate stages (see Figure 2.2). A hydrophilic emulsifier is used in the 1st stage to prepare the primary O_1/W emulsion, and then a different lipophilic emulsifier is applied in the 2nd stage to fabricate the O_1/W/O_2 emulsion. It

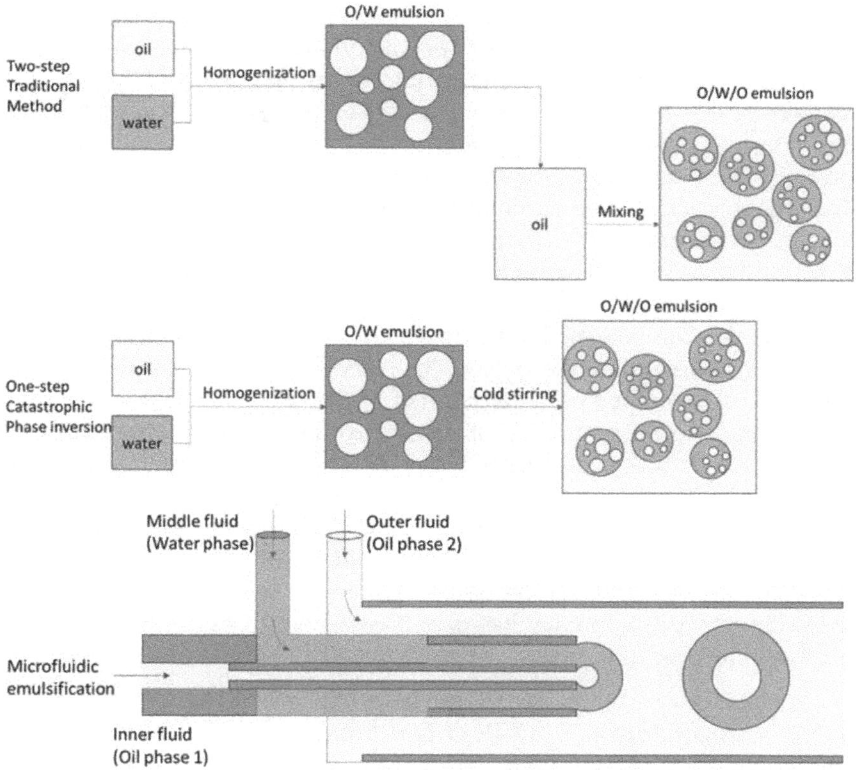

Figure 2.2 Popular methods for preparing O/W/O emulsions.
Adapted from ref. 3 with permission from Taylor & Francis, Copyright 2020.

has long been recognized that the methodological choice in the secondary emulsification step (dispersing O_1/W into O_2) is essential for the preparation of well-constructed formulations. If the applied hydrodynamic force is too intense during the 2nd stage, it may lead to unwanted rupture of the primary O_1/W emulsion, meaning that high-shear mixers and ultrasonic methods are generally unsuitable for the secondary stage of emulsification. While choosing the appropriate emulsification equipment and processing conditions for the 2nd stage, the researcher can address the challenging stability issues of double emulsion formulation in a series of different ways.[1] The sort of formulation strategies adopted are commonly the following: reduction of the mean O_1 droplet size, formation of an internal nano-emulsion, increasing the viscosity of the O_1 phase or the water phase, formation of an interfacial complex to enhance the stability of the inner emulsions, and incorporation of surfactants, polymers, and colloidal particles in the outer oil-soluble phase.[1]

The traditional two-step-emulsification method is simple but time-consuming and energy-intensive. Hence, new methods in constructing double emulsions have emerged in recent years, such as one-step homogenization[5,6] and microfluidic emulsification (see Figure 2.2). Besides, applying a one-step approach, such as a phase inversion[7] or microfluidics approach, can avoid the emulsion destabilization happening during the two-step process.[8] In particular, Groenweg *et al.*[9] described phase inversion as resulting from either a catastrophic event (because it suddenly appears in the form of a dramatic change in morphology), or coalescence (at an appropriately large volume fraction of the dispersed phase, coalescence dominates over the break-up event). However, there are still uncertainties as to the required dispersed volume fraction and the role of mixing in phase inversion (*i.e.*, time and intensity), as well as the true mechanism involved in double emulsion formation. Klahn *et al.*[10] stated that phase inversion was governed by the interplay between droplet coalescence and the escape of those internal droplets, which means that O/W emulsions would experience catastrophic phase inversion at appropriate shear rates. The phase inversion leads to the formation of nano and fine O/W/O double emulsions. Besides, there is also an advanced microfluidic emulsification method for preparing O/W/O emulsions.[11,12] In this method, the inner emulsion is injected into the outer emulsion using a capillary microfluidic device, making each emulsion droplet contain only one inner droplet. This approach was used to create monodisperse double emulsions with a high level of control of droplet size.[13] However, this approach has only been accomplished in pharmacy and chemical industries until now.

2.1.2.2 Stabilization of O/W/O Double Emulsions

Compared with the two bulk phases and single oil–water interface of traditional emulsions, double emulsions have three distinct bulk phases and two oil–water interfaces. The existence of these two kinds of interfaces means that two different emulsifiers are typically required in order to formulate double emulsions, one to stabilize the inner droplets (primary emulsion) and the other to stabilize the outer droplets (secondary emulsion). Under normal circumstances, due to the complex structure of double emulsions, their droplets are thermodynamically more unstable.[14] Several destabilization processes can occur both in the inner and the outer droplets of the emulsions.[15]

Due to the low stability, double emulsions often require high concentrations of synthetic surfactants, such as Tween and Span, to hinder instability and increase the shelf life.[1,16] However, this limits the applications of double emulsions in food and pharmaceutical industries.[1] Thus, biopolymers and their complexes are being investigated to replace the synthetic surfactants to stabilize the double emulsions and extend the shelf life. In addition, some new methods for preparing double emulsions have been found to improve their structural stability.[17] For example, the combination of sodium caseinate and lecithin could be used to replace the role of polyglycerol polyricinoleate (PGPR) in O/W/O double emulsions.[18–20] Similarly, a complex conjugation between whey protein and xanthan gum was used to stabilize O/W/O emulsions with flumethrin as the internal oil phase.[14] Also, the oil transport through the water layer can be restrained by using mixtures of whey protein–xanthan gum, whey protein–fenugreek gum,[14] or sodium caseinate–k-carrageenan,[20] which provides strong barriers based on repulsive interactions. These conjugates not only enhanced the stability of O/W/O emulsions but also improved their encapsulation efficiency.

In addition to biopolymers, small solid particles can be applied to stabilize the interfaces of double emulsions based on the Pickering stabilization mechanism. In these double emulsions, tiny particles surround the oil phase to form a rigid solid film at the W–O interface. Compared with conventional surfactant-stabilized double emulsions, Pickering emulsions tend to have better stability characteristics during long-term storage, but the wettability and contact angle of particles at the W–O interface need to be well controlled. In food applications, tiny fat crystals can be regarded as solid particles to stabilize Pickering emulsions.[21] In this case, it is very important to control the crystallization procedure (cooling and shear rate) affecting the crystal properties. This is because the size, shape, and morphology of fat crystals are the main factors stabilizing double emulsions.[22,23] For instance, β' fat crystals are favored due to their small size and their stability in storage.[24]

2.1.2.3 *Methods for Evaluating the Stability of O/W/O Double Emulsions*

The existing analysis approaches for evaluating the stability of O/W/O double emulsions are mainly based on microscopy, rheology, light scattering, and marker detection.[25] Optical methods can provide an impression of changes in droplet size or filling degree. Because of the limited amounts of droplets evaluated within an image, it provides evaluation on a qualitative level. In fact, microscopy can be applied to verify the existence of double emulsion structures, display the quantitative loss of the inner water phase (flow particle image analyzer),[26] visualize increasing inner droplet sizes (confocal laser scanning microscopy CLSM),[27] make general comparisons of different double emulsions (light microscopy),[28] and record the loss of single droplets over time (video-microscopy).[29] However, microscopy is also limited to length scales that can be approached optically.

The droplet size distributions (DSDs) of the inner and outer emulsions over time can also be used to evaluate the stability of double emulsions. The DSD of inner emulsions can be measured using the same methods available for single emulsions, *e.g.*, dynamic and static light scattering and image analysis. This measurement is usually performed before the second emulsification step, *i.e.*, in the O_1/W single emulsion. After that, CLSM can be applied as a technique to measure the DSD of internal water droplets over time. Similar to the measurements with marker substances, dyes are necessary, which may change the behavior of the emulsion.[30] Regarding the DSD of external emulsions, it is easier to measure the water droplets, but it must be performed carefully. The water droplet size will be influenced once the inner oil phase is lost in the second emulsification process due to demulsification or migration. The change of phase distribution will cause the viscosity ratio between the disperse O_1/W phase and the continuous O_2 phase to change. The viscosity ratio plays a certain influence in the breakage of water droplets.[20] Thus, the droplet size distribution of the primary O_1/W emulsion directly depends on both the stability of the double emulsion and the processing conditions.

The stability of an emulsion is also closely related to its rheological behavior which relies on its internal structure. Rheometers are mostly used for viscosity determination, which is closely related to the disperse phase fraction and its DSD.[31] Thus, by changing the emulsion formulation and structure, the viscosity of emulsions will be influenced, and this also applies to double emulsions.[32] In different models (*i.e.*, the Krieger–Dougherty equation), the viscosity of a liquid emulsion can be described as a function of the fraction of the disperse phase. For double emulsions, the encapsulation efficiency can be acquired by processing the viscosity data.[33]

Detecting the release of a marker is a common approach to measure the embedding stability of double emulsions. By surveying the encapsulation and targeted release of active ingredients, we can determine whether phase transfer occurs inside the double-layer emulsion, and thus obtain insight regarding its stability.[34] In this approach, marker substances are encapsulated that are easy to analyze. Previous publications have summarized a list of marker substances and their corresponding detection methods.[1,35] Detection methods range from simple conductivity[36] to more complex methods, *e.g.* nuclear magnetic resonance (NMR) spectroscopy.[25]

What's more, for a liquid double emulsion, the settling rate of water droplets can be used to evaluate the encapsulation efficiency. The settling velocity of water droplets is related to their density directly depending on the filling level of the drops.[37] Both the settling velocity and the phase fraction after settling can give information on the stability of the double emulsion. When inner oil is released by passing through the water phase, the total volume of the O/W droplets would reduce while the amount of outer oil phase increases. At the top of the analysis vessel, the increase in the volume fraction of the outer oil phase is the release amount of the inner oil phase. Under normal conditions, the settling phenomenon would be accelerated in the analytical centrifuge. Different disperse phase fractions result in a visible shift in the phase boundary, which makes it possible to measure the difference in phase distribution quickly and easily.

2.1.3 Applications

In the early food industry, the potential for O/W/O emulsion incorporation into low-calorie imitation spread and dressings was identified. Extensive patent literature indicates that the food industry was interested in the feasibility of double emulsion structures in commercial products. Gaonkar[38] described the possibility of double emulsion applications, such as adding W/O/W emulsions into a low-fat French dressing or using O/W/O emulsions containing an inner oil phase with buttery flavor to create a low-fat spread. However, it would appear that clear evidence of continued commercial success in the market is not matched by the bold claims for the potential benefits of O/W/O-based food structures. This may be due in large part to the difficulty of producing a perfect O/W/O-containing formula with long-term physical stability and storage stability, as well as the challenge of obtaining consumer's satisfaction in terms of product appearance, taste, and texture. Nevertheless, progress continues to be made in the development of ingredients and equipment. A recent report on sensory testing of low-fat foods asserted that replacing fat with internal water droplets would not significantly reduce the perceived intensity of fat-related attributes, which could encourage today's enthusiastic proponents of double emulsion technology.[39]

Employing double emulsions has many potential commercial benefits in the manufacture of low-fat foods and increases the delivery of nutrient and flavor delivery, which has been verified for several decades. As these emulsions possess excellent capacity in the controlled release of ingredients, drugs, colorants, minerals, and antioxidant particles, they are favored in the food and pharmaceutical industry. Some microencapsulation systems based on O/W/O emulsions in cosmetics, pharmaceutical, and food studies are presented (see Table 2.1). In fact, these multiple emulsions not only help control the release of drugs and bioactive compounds but also minimize the negative effect of high temperature in the preparation stages. Some recent studies have focused on formulating O/W/O double emulsions to deliver liposoluble nutrients or drugs and decrease the saturated fatty acids for spreading. Using double emulsions as fat replacers can help increase the water content in food products, without significantly affecting their sensorial properties. The special O/W/O structure could also offer a means to produce bakery fats with required plasticity but healthier ingredients.

2.2 Microencapsulated Oil Powder

2.2.1 Introduction

Marine, vegetable, and essential oils are widely favored by consumers as they provide several multidimensional functional properties, like antioxidant,[49] anti-inflammatory,[50] antifungal,[51] antihypertension,[52] anti-aging,[53] *etc.* Several applications in the food, agriculture, pesticide, cosmetic, and pharmaceutical industries using these oils have been widely studied due to their natural and safe status. However, they are chemically unstable and susceptible to deterioration by oxidation and loss of volatile compounds, especially when exposed to oxygen, light, moisture, and heat. The quality of food products containing these oils may decrease due to the oxidative degradation and the presence of unwanted tastes and off-flavors, which have a negative impact on the shelf-stability, sensory performance, and consumer acceptance of the developed products.[54] To create natural health products containing bio-functional oils, microencapsulated oil powder could be a viable technology to maintain their nutritional and bio-active properties and include them as a functional structure in food.

Table 2.1 Overview of reported O/W/O double emulsions and their applications.

Inner oil phase	Emulsifier 1	Middle water phase	Emulsifier 2	Outer oil phase	Production method	Application	Reference number
Oil soluble flavors	Hydrophilic emulsifier	Aqueous soluble/gellable polysaccharides mixed with metal cation	Lipophilic emulsifier	Edible triglyceride oils derived from seed oils	Two-step traditional method	Low-fat salad dressing or margarine-like spread	38
Retinol	Tween 20	Water	Span 80	Decyl alcohol containing hydroxypropyl cellulose and tetraethyl orthosilicate	Two-step traditional method	Cosmetics	40
Canola oil	Sodium caseinate	Water	Lecithin	Palm–cotton stearin (50 : 50)	One-step catastrophic phase inversion	Low solid fats	18
1-Octanol containing hydroxypropyl cellulose	Tween 20	Aqueous phase containing polyethylene glycol	Span 80	1-Octanol containing hydroxypropyl cellulose	One-step emulsification process	/	6
Retinol	Tween 20	NH_4OH solution and soluble polymers (hydroxypropyl cellulose, polyethylene glycol or polyvinylpyrrolidone)	Span 80	*N*-decyl alcohol including HPC and tetraethyl orthosilicate	Two-step traditional method	Cosmetics	41

Medium chain triglycerides and its mixture with a more hydrophilic oil–triacetin	Whey protein isolate/xanthan gum conjugate, whey protein isolate/xanthan gum conjugate	Water	/	Silicone or alkanic oils containing flumethrin®	Two-step traditional method (high-pressure homogenizer; Ultra Turrax homogenizer)	Drug delivery	14
Omega-3 oils	Sodium caseinate	Water containing NaCl and green tea extract	Polyglycerol Polyricinoleate	Palm oil and rapeseed oil, containing β-carotene and α-tocopherol	Two-step traditional method	Omega-3 enriched spreads	42
Fish oil	Succinic acid deamidated wheat gluten	Water	Span 80	Mineral oil	Two-step traditional method	Supplementation and functional foods	43
Mineral oil	Glycerol monoleate	De-ionized water	Glycerol monoleate	Mineral oil	One-step catastrophic phase inversion	Food, pharmaceutical, and cosmetic emulsions	5
Soybean oil and benzyl benzoate	Sodium alginate	Water containing $CaCO_3$ nanoparticles	Polyglycerol polyricinoleate	Soybean oil	Microfluidic	Carrier for drugs, chemicals, and nutrients	44
Pure oil	Tween 20	Mixing demineralized water containing dry glucose syrup and pectin	Polyglycerol polyricinoleate	Pure oil	Two-step traditional method (microfluidizer and stirring)	Supplementary, healthy food, and in pharmaceuticals	45
Fish oil	Sodium caseinate	Distilled water	Lactose monohydrate; polyglycerol polyricinoleate	Virgin olive oil	Two-step traditional method	Nutrient delivery	46

Table 2.1 (*Continued*)

Inner oil phase	Emulsifier 1	Middle water phase	Emulsifier 2	Outer oil phase	Production method	Application	Reference number
Paclitaxel	Gelatin gel	Water containing DMSO	Sorbitan sesquioleate	Linoleic acid	Two-step traditional method	Drug delivery	47
Sunflower oil	Gelatin	Water	Gelatin	Palm oil without and with different proportions of sunflower oil	Two-step traditional method	Pharmaceuticals, cosmetics, and foods	21
Sunflower oil	Sodium caseinate	Water containing k-carrageenan	Lecithin	Palm oil	One-step catastrophic phase inversion	Spreads and cosmetics	20
Fragrance	Tergitol 15-S-12	Aqueous phase composing of salt	Polyglycerol polyricinoleate 4150	Isopropyl myristate oil	Two-step traditional method	Cosmetics	48

Generally, microencapsulated oil powders can be described as oil products that use encapsulation technology to structure a suitable film surrounding the core oils, followed by a drying technology to build a functional solid barrier between the core and the wall material. The formulated solid powder structure can protect the core materials from chemical and physical impacts, maintain the biological, functional, and physicochemical properties of the inner oil materials, and even expand the applications of oil materials. Therefore, the advantages of the oil powder technique are (1) protecting core materials from external environmental factors such as oxygen, light, heat, and moisture; (2) masking of odor and taste of the core materials; (3) controlling the release of inner bio-active compounds; (4) separating incompatible components; (5) converting liquids to free-flowing solids.[55,56]

2.2.2 The Structure of Microencapsulated Oil

Microencapsulation is a technique that is often applied in the food industry. By covering the oil phase with a layer of coating material, tiny solid capsules can be obtained with certain physical or chemical properties. The structure of an encapsulated oil consists of a core and a wall. Generally, oil microcapsules are composed of a core material, which is also called the internal phase or fill, and a wall called the coating, carrier, shell, or membrane.[57] The wall material determines the stability of the microparticles, the process efficiency, and the degree of protection of the core. Wall materials used commonly for oil microcapsules are classified into synthetic polymers and biomaterials (usually carbohydrates and proteins or combined materials). Microcapsule models can be divided into three basic types: mono- or single-cored, poly- or multi-cored, and matrix types (see Figure 2.3). Single-core microcapsules include a single hollow chamber in the shell while the multi-core type includes many chambers with different sizes. Likewise, the matrix model of a microparticle represents active ingredients integrated within the matrix of the shell material.[55]

It has been found that the size and shape of the microcapsules rely on the wall materials and the preparation methods. The choice of wall materials is based on the properties of the core materials, and the characteristics of the

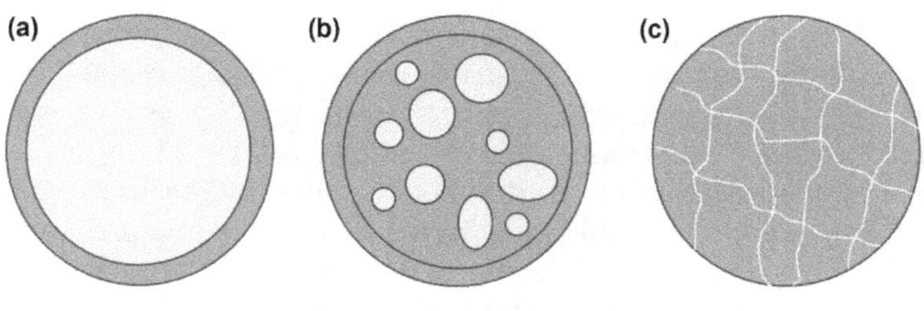

Figure 2.3 Three types of microcapsules: (a) single-core type, (b) multi-core type, and (c) matrix.

final products, such as food-grade requirements, processing costs, viscosity property, emulsification performance and stability, the ability of holding inner materials, the controlled release of the core material, and the protection from environmental conditions.[58] The commonly used wall materials of an oil microcapsule can be divided into three categories: carbohydrates, proteins, and lipids and wax.

2.2.3 Emulsification and Drying Technology

Many techniques have been used for producing oil powder, mostly combining emulsification and a drying process. The purpose of emulsification is to build the microcapsule structure while drying solidifies the liquid microcapsules, enhancing the stability of the oil-based products and expanding their application range.

2.2.3.1 Emulsification

Emulsification technology is a crucial step in preparing microencapsulated oil powder. It is generally used to encapsulate bio-active oil in an aqueous solution, after which the emulsified solution can be dried (spray- or freeze-drying) to form oil powder. Thus, it is the first key part in the micro-encapsulation process. For example, in spray- and freeze-drying microencapsulation, the core and shell materials can be formulated by emulsifying techniques before the final drying. Emulsion droplets can also be integrated into the matrix during, or act as templates for coacervation processing.

An emulsion is basically composed of at least 2 immiscible liquids, generally oil and water, one of which is dispersed in the form of small spherical droplets in the other one. In order to obtain a kinetically stable solution, the emulsion system usually contains added emulsifier or texture modifiers to improve the viscosity. The size of the emulsion droplets in food systems ranges from 0.1 to 100 μm.[59] The homogenization equipment (high shear mixer, high-pressure homogenizer, colloid mill, sonicator, or membrane homogenizer) can evenly mix and emulsify the oil, water, and emulsifier together to prepare the emulsion. The O/W emulsion refers to small droplets dispersed in an aqueous solution, while the oil droplets are enclosed by an amphipathic interfacial layer composed of emulsifiers. The advantages of these structured oils are that they are relatively easy to produce and at low cost, and their disadvantages are also obvious, for example, when exposed to heating, chilling, freezing, drying, pH extremes, and high mineral concentrations, the system would exhibit poor physical stability, as well as limited controlled release.[60] Additional functionalities of the emulsions can be achieved by modification of the emulsifiers, such as the use of Maillard reaction products.[61] Oil could be dispersed in aqueous mixtures composed of carbohydrates (oligosaccharide, polysaccharides, dried glucose syrup, and glucose) and proteins (whey protein isolate, soy protein isolate, casein, or gelatin) and then spray-dried to formulate oil powders.

2.2.3.2 Spray-drying

Spray-drying is an important microencapsulation technology most widely used on an industrial scale, which has many attractive advantages in preparing microcapsules, like simple, feasible and continuous operation, low cost, and high controllability over the final products, compared to other microencapsulation techniques.[59] This technique has been successfully used for several decades to encapsulate various oils in the food, cosmetics, and pharmaceutical industry.[62] The formulation and processing conditions of different oil powders using the spray-drying technique have been summarized (see Table 2.2). It can be noticed that many different inner and outer materials have been investigated at very different spray-drying process conditions resulting in diverse encapsulation efficiencies.

The spray-drying process involves atomization of emulsions into a drying chamber at a relatively high temperature, which results in a very fast evaporation of water. Afterwards, crusts are generated at a fast speed, and the oil is captured quasi-instantaneously.[62] Spray-drying is the most widely applied approach to remove water in the food industry, improving the microbiological stability of products. Furthermore, it is conducive to obtain a product with specific functional properties, preventing risks of chemical and/or biological degradation and lowering the total storage and transport costs.[63] The production of microencapsulated oil powder by spray-drying includes 4 steps: (i) formulation of the mixture, (ii) homogenization of the emulsion, (ii) atomization of the emulsion, and (iv) water evaporation and the collection of oil powder.[57] Generally, in the first step the shell materials are dissolved in distilled water while mixing. The solutions are kept in a refrigerator at 4 °C overnight to ensure that the polymer molecules are fully saturated and to avoid any changes because of temperature. During the spray-drying process, the diameter of the initial emulsion droplets ranges from 0.1 to 100 μm. Besides, the generated emulsion must have a certain physical and storage stability before the spray-drying process. Its viscosity should be low enough to prevent air inclusion in the sample, and oil droplets should be rather small. Otherwise, because of high viscosity, it is possible that elongated and large droplets are generated, which adversely affect the drying rate and the atomization process. In addition, spray-drying needs well-optimized processing conditions, as well as a suitable composition of the solution that contains the active components.[83]

In order to acquire a high encapsulation efficiency, optimized spray-drying conditions must be applied. The main parameters of spray-drying, such as the feed temperature, air inlet temperature, and air outlet temperature, should be optimized.[63] Feed temperature has a key impact on the viscosity and fluidity of the emulsion, thereby negatively affecting its capacity to spray homogeneously. When the feed temperature is increased, the viscosity and droplet size will be decreased.[105] However, too high temperatures may lead to some heat-sensitive ingredients to volatilize and deteriorate. Furthermore, the feed rate to the atomizer should be well controlled to ensure that each sprayed droplet can reach the desired drying level before entering into the

Table 2.2 Studies on different oil microencapsulation by spray-drying.

Inner material	Outer material	Inlet temperature	Outlet temperature	Feed flow rate	Feed atomization pressure	Encapsulation efficiency	Reference number
Vegetable oil	Maltodextrin DE12/acacia gum	200 or 220 °C	100–130 °C	24–68 mL per min	3 bar	/	64
Roasted coffee oil	Gum Arabic Instantgum BA®	158 °C	/	0.8 $L h^{-1}$	/	82.57%	65
Linseed oil	Methyl cellulose-Gum Arabic/ Maltodextrin 10 DE	175 ± 5 °C	75 ± 5 °C	15.0 mL per min	2.8 bar	91.4%	66
α-tocopherol and olive oil	Maltodextrin DE12, agave inulin and acacia gum	180 °C	90 °C	57.6 g per min	5 bar	/	63
Gac oil	Whey protein concentrate and gum Arabic	155 °C	80 °C	16 mL per min	2 bar	87.22%	67
Sunflower oil	Hydroxypropyl methylcellulose and maltodextrin	163 °C	80 °C	/	6 bar	79.43%	68
Nigella sativa seeds oil and thymoquinone	Modified starch and maltodextrin mixture	140 °C	95–98 °C	/	6 bar	90.10%	69
Brucea javanica oil	Arabic gum/Gelatin	151 °C	/	1.32 mL per min	/	82.9%	70
Rambutan seed oil	Whey protein isolate	180 °C	60 °C	/	3 bar	73.6%	71
Ginger Essential Oil	Whey Protein isolate-Inulin/ Maltodextrin Blends	170 °C	/	13.33 mL per min	/	61.64%	72
Chia oil	Whey protein concentration/ pectin + Maltodextrin 10 DE	170 °C	/	5.5 mL per min	6 bar	99.95%	73
Squalene oil	Maltodextrin and polymerized sodium caseinate	140 °C	80–90 °C	25 mL per min	/	/	74
DHA Algal oil	Whey protein isolate and maltodextrin	180 °C	/	34 mL per min	/	88–96%	75
Fig seed oil	Gum arabic and maltodextrin	150 °C	/	/	5 bar	74.78%	76

Hazelnut oil	Hydroxypropyl methylcellulose/ maltodextrin	140–220 °C	/	75 L h^{-1}	5.5 bar	75–80%	77
Walnut oil	Skim milk powder	180 °C	/	30 ± 0.7 g per min	3 bar	91.01%	78
Drumstick (Moringa oleifera) oil	Maltodextrin, gum Arabic, and whey protein concentrate	180 °C	80 °C	10 g per min	/	66.23–91.05%	79
Nigella sativa oil	Sodium caseinate and maltodextrin DE10	160 °C	/	16.67 mL per min	4.5 bar	95.43%	80
Extra virgin olive oil	Maltodextrin and whey protein isolate	200 °C	80 °C	8.8 g per min	/	/	81
Pumpkin (Cucurbita spp.) seed oil	Whey protein/maltodextrin	160 °C	60 °C	13.3 g per min	3 bar	94.5%	82
Walnut, peanut, and pecan oils	Gum Arabic (3.1%), maltodextrin (4.7%), and gelatin (2.2%).	180 °C	80 °C	17 mL per min	/	35–45%	83
Grape seed oil	Maltodextrin 10DE/gum Arabic	180 ± 3 °C	105 ± 8 °C	5.83 g per min	1.8 bar	63.47%	84
Refined kenaf (Hibiscus cannabinus L.) seed oil	β-cyclodextrin/gum arabic/ sodium caseinate	160 °C	/	8 ± 2 g per min	/	95.3%	85
Corn oil	Brea gum	150 ± 5 °C	60 ± 5 °C	/	1 bar	91.72%	86
Essential avocado oil	OSA-maltodextrin	160 °C	90 °C	5 mL per min	/	61.7%	87
Peony seed oil	Whey protein isolate, corn syrup, and soy lecithin	180 ± 5 °C	90 ± 5 °C	5 mL per min	1.5 bar	93.63%	88
Soybean oil	Soybean protein isolate	150–170 °C	/	3 mL per min	/	84–90%	89
Winter squash (Cucurbita moschata Duchesne) seed oil	Maltodextrin 10DE/gum Arabic	172 °C	90–100 °C	10 mL per min	4 bar	77.4%	90
Fish oil	Soybean protein isolate	180 ± 2 °C	96 ± 8 °C	3.1 mL per min	6–8 bar	88–75%	91
Buriti oil	Chickpea protein-pectin matrix	160 °C	/	2 mL per min	/	91.6%	92

Table 2.2 (*Continued*)

Inner material	Outer material	Inlet temperature	Outlet temperature	Feed flow rate	Feed atomization pressure	Encapsulation efficiency	Reference number
Silver Carp Oil	Hi-cap®100/Inulin	170 °C	80–85 °C	9.17–10.83 mL per min	/	92.78%	93
Pomegranate Seed Oil	Arabic gum and Xanthan gum	170 ± 5 °C	85 ± 5 °C	20 mL per min	/	93.3–96.6%	94
Tuna oil	Whey protein isolate–agar gum and gellan gum complex coacervates	160 °C	94 °C	/	/	95.8%	95
Anchovy oil	Gelatin–sodium hexametaphosphate complex	160 °C	94 °C	/	/	99.83%	96
Almond oil	Taro starch spherical aggregates	145 °C	/	10.6 g per min	/	56.0%	97
Oregano essential oil	Gelatin and chia mucilage	160 °C	/	5 g per min	/	95.6%	98
Matcha-tuna oil	Mactha–maltodextrin mixtures	185 °C	80 °C	12 g per min	1.5 bar	84.9–89.3%	99
Macadamia oil	Macadamia protein isolate/chitosan hydrochloride	160 °C	85 °C	5 mL per min	/	87.0%	100
Sour cherry oil	Gum Arabic–Maltodextrin	195 °C	/	8 mL per min	/	90.10%	101
Orange essential oil	Gelatin and lignin	150 °C	/	2.5 mL per min	/	97.0%	102
Sesame seed oil	Tamarind seed mucilage	135 ± 5 °C	80 ± 5 °C	40 mL per min	4 bar	91.05%	103
Mentha spicata essential oil	Inulin and gum Arabic	110 °C	/	5 mL per min	/	91.0%	104

drying chamber. What's more, it is important to appropriately adjust the air inlet temperature and flow rate.

2.2.3.3 Freeze-drying

Freeze-drying, also named lyophilization or cryodesiccation, is a simple physical process, which is commonly applied to dehydrate almost all heat-sensitive materials and aromatic oils. This technique has been successfully used for the microencapsulation of oils such as fish, flaxseed, nut, and krill oil (see Table 2.3). Before freeze-drying, samples need to be fully frozen at −20 °C. Then the ice can be sublimed from the solid state under vacuum (sublimation step), after which unfrozen (bound) water is removed by evaporation under reduced pressure (desorption step).[106] There are some advantages of freeze drying in terms of oil powders, as listed below:

a) To avoid deterioration of heat-sensitive compounds (unsaturated fatty acids, tocopherols *etc.*);
b) To control the moisture of the final products;
c) To improve the reconstitution or application of freeze-dried oil powders.[59]

Despite these advantages, the high operation costs, long processing time, and open porous structure of the end products limit the application of this

Table 2.3 Recent studies on oil powder by freeze-drying.

Inner material	Outer material	Condenser temperature	Pressure	Encapsulation efficiency	Reference number
Fish oil	Soyafibe and maltodextrin	−50 °C	0.12 mbar	0.4158	109
	Soyafibe and OSA-starch			0.4768	
	Soyafibe, maltodextrin, and hydroxypropyl betacyclodextrin			0.2861	
Shea nut oil	Modified quinoa starch granules	−50 °C	0.01 mbar	/	110
Flaxseed oil	Whey protein isolate–Maltodextrin–sodium alginate	−20 °C	40 Pa	0.9544	111
Chia seed oil	Sodium caseinate	/	/	0.863	112
Krill oil	Maltodextrin–whey protein concentration–gum Arabic	−50 °C	0.04 mbar	51.9–58.2%	108
Sunflower oil	Whey protein isolate and soy protein isolate	−55 °C	0.04 mbar	/	113

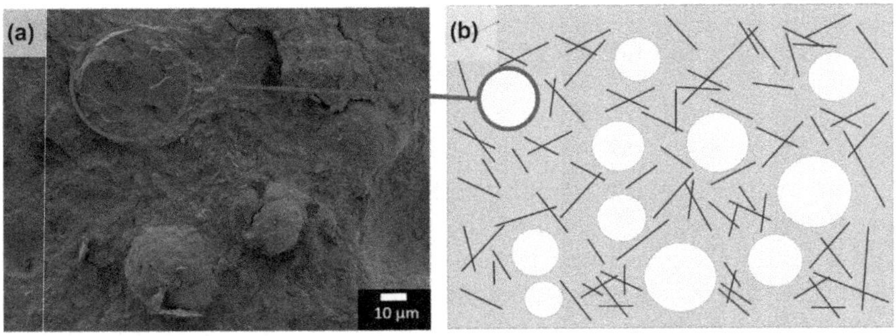

Figure 2.4 Cryo-SEM of protein-stabilized oil powder dissolved in a PO/POSt continuous fat matrix, picture from FS&F, UGent (a) and its corresponding O/O model (b).

technology in the food industry.[107] In addition, encapsulation efficiency by freeze drying is not as high as by spray drying.[108]

2.2.4 Applications in the Food Industry

Currently, there is a trend toward a healthier way of living, including a growing awareness of what a consumer should eat and which kinds of ingredients in food are beneficial to maintaining good health. Oil powder is considered as a unique processing approach for the delivery of functional substances and prevention of inner materials from light and oxidation degradation, off-flavor formation, and color change. Microencapsulation could be used to overcome all these challenges by providing viable texture blending, appealing aroma release, and taste, odor, and color masking.[55] The technology enables food companies to incorporate minerals, vitamins, flavors, and essential oils. In addition, it has been shown that oil powder technology can be applied to isolate liquid oil from the crystalline fat matrix by partly transferring the liquid oil to an internal fat phase, resulting in a more concentrated outer phase and less solid fat to be structured. Dispersing the oil powder into a fat phase would lead to an innovatively structured food with O/O dispersion, like low-fatty acid shortening, and increased nutritional value (see Figure 2.4).[114] Therefore, oil powder could be recognized as an effective technique in the food, cosmetic, and pharmaceutical industries owing to its low cost, ease of production, and high stability.

2.3 Oil Foams

2.3.1 Background

Foams are complicated colloidal systems, in which gas is dispersed in a continuous phase with surfactants.[115] Liquid foams can be divided into aqueous and non-aqueous foams depending on the continuous phase which can be water or oil respectively. Although aqueous foams are more

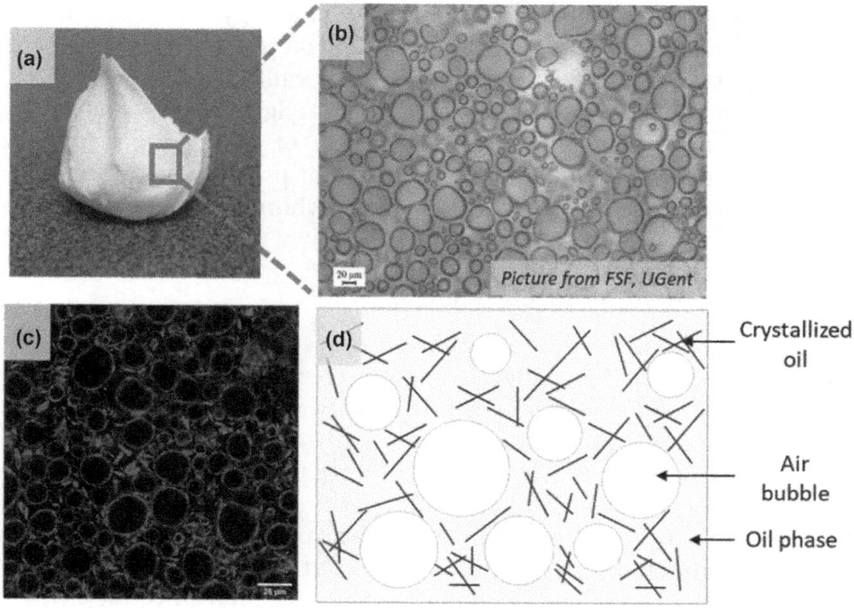

Figure 2.5 The structure of oil foams. (a) Macroscale (mm–cm); (b) microscale (μm), picture from FS&F, UGent; (c) CLSM of 10% monoglyceride oil foam (nile blue as the fluorescent dye), picture from FS&F, UGent; (d) structure model.

abundant, there is an increasing interest in oil foams. Generally, edible oil foams are stabilized by solid particles (see Figure 2.5), *i.e.*, Pickering stabilization, which provides new opportunities for food scientists to formulate uniquely structured and healthier oil-containing products with a decreased fat content and less saturated fatty acids.[116]

There is quite limited research on oil foams compared to aqueous foams, especially for edible applications. Only recently have edible-gas-in-oil systems stabilized by particles been explored, because of their special structure, in the food industry,[117] such as in aerated chocolate bars and cakes. Also, oil foams using Pickering stabilization are being successfully applied in edible model systems,[118–121] but most food researchers pay more attention to the characterization of oil foams formulated with different components (*e.g.* gelatin, xanthan gum and canola oil,[122] myristic acid and high oleic sunflower oil,[123] tricaprylin, tristearin, and coconut oil,[120] and monoglyceride and rapeseed oil[118]).

2.3.2 Preparation and Stabilization

2.3.2.1 Preparation of Oil Foams

Generally, the production method of oil foams is relatively simple as it comprises whipping a semi-crystalline lipid phase. In brief, edible oil foams are produced by heating the oil phase containing a high-melting component (*e.g.* fatty acids, acylglycerols, *etc.*) at low concentrations (up to 10%), after which a specific static or dynamic time-temperature crystallization protocol is applied which results in a partially crystalline oil phase. Depending on the type of high-melting component, this phase is called an oleogel (*e.g.* fatty acids or monoacylglycerols) or shortening (triacylglycerols), which can be (semi-)liquid depending on the concentration of the high-melting component and crystal morphology. Then, the final production step is air incorporation which is typically achieved with a whipping device.[116]

2.3.2.2 Stabilization of Oil Foams

In many food applications, air bubbles are stabilized by surface-active components, such as low-molecular weight surfactants, amphiphilic polymers, proteins, and particles. Historical overview reveals that there have been two main kinds of stabilization mechanisms including Pickering particles and bulk gelation.[124] An example of a complex food foam is whipped cream in which particles (semi-crystalline fat droplets) and surfactants stabilize air bubbles. These droplets interact because of mechanical shearing and air inclusion and form a rigid layer of partially coalesced fat droplets surrounding the air bubble.[125] Similar to Pickering emulsions, Pickering oil foams refer to foams stabilized by solid particles that adsorb onto the interface between the air and oil phase. The particles can adsorb at the air bubble surface either alone or with surfactants to decrease the gas – oil interfacial energy and prevent bubble coalescence.[126] Generally, oil foams stabilized by particles and surfactant molecules exhibit different adsorption kinetics. Under diffusion-controlled conditions, the adsorption of surfactant molecules shows a fast and dynamic equilibrium between the interface and the bulk phase, but the particle adsorption exhibits slower kinetics due to the larger size compared to surfactant molecules. But once attached, Pickering oil foam would have increased stability owing to their higher detachment energy (eqn (2.1)) and maximum capillary pressure (eqn (2.2)).[127] Obviously, as particles are more strongly adsorbed, more Gibbs energy will be required to disrupt the particle layers.

$$\Delta G_{\text{remove}} = \pi R^2 r_{\text{A/L}}(1 \pm \cos\theta)^2 \tag{2.1}$$

$$P_{\text{c}}^{\max} = P\frac{2r_{\text{A/L}}}{R}\cos\theta \tag{2.2}$$

Eqn (2.1) is given on the premise of ignoring buoyancy and gravity impacts, where θ refers to contact angle, R is the radius of spheric particles, and $r_{\text{A/L}}$ means surface tension. The positive sign refers to the detachment energy for removing a particle into the air while the removal of a particle into the liquid phase is given by the negative sign. Thereby, contact angle, particle size and shape, and aspect ratio should be considered to evaluate the Pickering stabilization. Also, the molecular structure and the polymorph of the fat crystals should be studied since they impact the properties of the crystals.

In foam systems, there are lots of air bubbles interacting and affecting foam stability. Researchers thereby need to consider the maximum capillary pressure. A higher maximum capillary pressure means that a thin liquid film between bubbles can resist a higher pressing force, indicating that foams can have better stability. Based on eqn (2.2), the maximum capillary is directly influenced by the contact angle (θ) and inversely proportional to the particle size (R), which is different from the Pickering stabilization theory based on particle detachment energy.

2.3.2.3 *The Influence of Crystal Properties on the Stability of Oil Foam*

The properties of crystallized particles play a vital role in regulating the stability of the oil foams, such as the molecular structure, crystal size and shape, polymorph, location, arrangement, network formation, and processing method.[128] However, because there are only a few research studies on the effect of crystal characteristics on oil foam stability, a relevant study on Pickering emulsions will be useful to better understand oil foams.

1) Molecular structure

   Wettability of particles and their influence on oil foam stabilization was discussed by Binks *et al.*[129] Generally, particles with contact angles ranging from 40 to 90°, measured in oil, can stabilize the air–oil interface. Binks *et al.*[123] formulated crystal-stabilized oil foams by whipping mixtures of high oleic sunflower oil (HOSO) containing myristic acid crystals, demonstrating that the contact angle through the oil phase was around 40°. Similar results were reported for a layer of pure monoglycerides between HOSO and air.[130] Contact angle measurements of commercial mixtures of mono-, di-, and triglycerides with a drop of sunflower oil in air were between 45 and 47°, which was consistent with Binks's result.[123]

   Besides, it was reported that headgroup size and hydrophobic chain length within the particles play a significant role in the oil foam by comparing different monoglycerol fatty acid esters.[131] The foam stability decreased with the reduced hydrocarbon chain length of fatty acid esters that directly affects the shape and phase behavior of the crystals. Monocaprin (C10)-oil foams contain large bulky flat solid particles while the monolaurin (C-12)-oil system includes fine, needle-like crystals which could adsorb more efficiently at the interface generating a rigid film to improve the stability of the foam.

2) Size and shape of crystals

   Generally, smaller size particles can form a more stable W/O interface than large particles owing to their better surface coverage.[23] A similar result has been reported for the oil foam system. It has been found that particle size directly affects the foamability of the suspension. Micro-rod-shaped crystals, with a length of 20 μm and a diameter of 5 μm, emerged to form stable oil foams, but for bigger particles, they accumulated to generate huge clusters and sedimented in the oil phase, restraining foam generation. Gunes[130] demonstrated that foamability could be improved efficiently if crystals broke up into smaller ones. In addition, particle shape affects the packing efficiency and the wetting behavior at the interface.[132,133] Medivala (2009) reported that particles with a high aspect ratio (ratio of major dimension and minor dimension) were beneficial to stabilize emulsions, while spherical particles were not able to stabilize them.[134] Also, Binks (2006) performed a research study on rod- and disk-like particles and demonstrated that higher aspect ratios were related to higher desorption energies since the potential contact area of particles with the interface was increased.[116]

The size and shape of particles are mainly regulated by the cooling rate, shear rate, and crystallization method. Firstly, a higher cooling rate can be attributed to a faster nucleation rate, resulting in the generation of many small crystals. Mishima *et al.* used several tempering procedures to crystallize melting triglycerides in salad oil to acquire crystals with different size ranges.[135] They found that a higher cooling rate led to the formation of smaller crystals (2–6 μm). Different crystal faces, like flat, well-defined faces, rounded-off, roughened faces, or spherulitic faces, can be formed by controlled growth rates. Secondly, shear rate also has an important influence on crystal size and shape. Sonwai *et al.* reported that the crystal size rose at low shear rates ($3\ s^{-1}$) and reduced at high shear rates ($>100\ s^{-1}$), which was consistent with the crystallization property of milk fat.[136] Thirdly, different crystallization methods can be applied to change the crystal size and shape, as has been shown by a related study on emulsions. W/O emulsions with post-crystallized wax crystals would be more stable because of smaller crystals formed by more nuclei. A thin layer of small plate-like crystals would be generated at the interface when direct crystallization is applied in the preparation of a butter emulsion.[137]

3) Polymorphism and phase behavior

Polymorphism refers to the existence of different crystalline types with the same chemical structure, which differ in physical properties, such as melting point and density, but can yield the same liquid phase after melting. Unstable polymorphs are formed because they nucleate more rapidly due to their structure that is close to that of the liquid phase.[138] Metastable crystals can transfer to the stable state by solid-state (*i.e.* rearrangement of the structural unities) or melt-mediated phase transition (*i.e.* melting and recrystallizing).[137]

Research on oil foams stabilized by crystals gives information on the importance of phase behavior and crystal polymorphism. Lok Kumar Shrestha (2012) showed that diglycerol monolaurate foams possessed lower stability compared with diglycerol monomyristate foams because of their different crystal dispersion.[131] Under the circumstances, α-crystals of diglycerol monomyristate or β-crystals after polymorphic transition were attributed to stronger foam stabilization compared to the stabilization by lamellar liquid crystals of diglycerol monolaurate. Mishima *et al.* (2016) reported that salad oil was whipped with overrun values over 200% by using high-melting triglycerides in a β-polymorph.[135] They emphasized the significance of the tempering procedure to induce the formation of tiny β-crystals. Stable salad oil foams were hardly produced if they contained large β-crystals or β′-crystals. Therefore, the desired crystal polymorph can be acquired by using special processing conditions, and then the foamability and stability can be regulated.

The chemical composition and processing conditions, like time, temperature, and shear rate, affect polymorphism formation and polymorphic transition of the crystallites.[139] Diglycerides can exist as two β-polymorphs, while 1-monoglycerides can also crystallize in a sub-α, indicating the complexity of polymorphism.[140] Besides, process parameters play an important role in the appearance of certain polymorphs.[141] Time regulates polymorphic transformations since crystals would finally reach the thermodynamically most stable state. Temperature can control the nucleation rate, meaning that the lower temperature would induce more small nuclei owing to the higher driving force.[142] Many tiny crystals of the most stable polymorph will be produced. Also, shear rate absolutely affects the crystallization behavior, which could enhance primary crystallization influencing the generation of certain polymorphic types.[143] In addition, crystallites can be formed from the melt which speeds up solid-state phase transformations by using a shear flow.[144] Regulating the polymorphic transitions can be important for the preparation of edible oil foams as transition to a more stable form could result in a change in crystal shape and packing type at the interface.

Furthermore, tempering can be applied to control the formation of the desired polymorphic forms, which is used in the chocolate industry in particular.[121,145] Heymans *et al.* (2018) found that different tempering protocols can lead to certain polymorphs of monoglyceride (MG)-oleogels, like α-, sub-α, and β-polymorphs, and the degree of crystallization, which further influences the crystallization behavior of MG-oleogels and further foamability.[121] Based on the differential scanning calorimetry (DSC) curve of the MG–oil mixture, five tempering protocols were applied to bring about different MG–oleogel types and overruns after whipping (see Figure 2.6). It was reported that PAC MG–oleogels had the highest foamability, because in comparison with AC MG–oleogels, the reduced viscosity of PAC samples showed more flow and turbulence when whipping, thereby leading to better air incorporation. Almost no air bubbles were included for samples with sub-α crystals, suggesting that the crystalline packing of sub-α crystals does not allow efficient air incorporation.[121] During the initial crystallization, MG were present in an inverse lamellar phase which could be

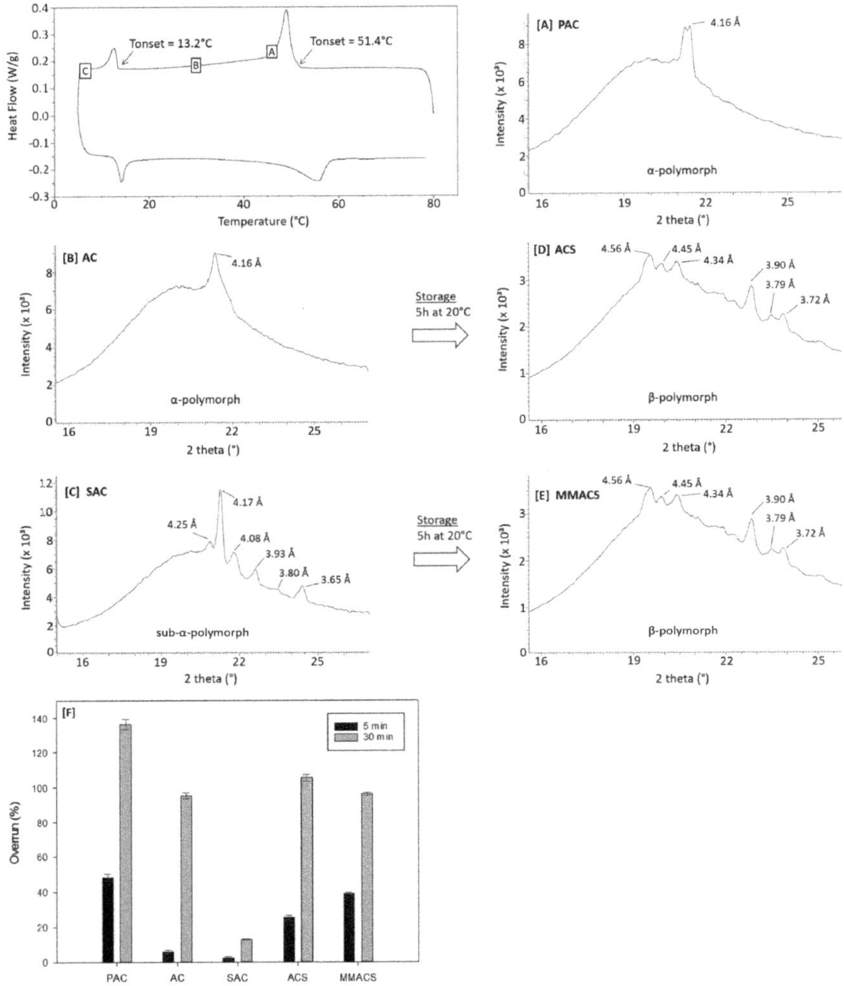

Figure 2.6 Upper-left figure: DSC profile of an MG–oil mixture upon cooling from 80 °C to 5 °C, holding isothermally for 1 min, followed by heating to 80 °C at a rate of 5 °C per min. (A–E) Wide-angle X-ray diffractogram (XRD) of the MG–oil mixture under the same experimental conditions. The letters presented on the thermal profile correspond to the letters on the wide-angle XRD. (PAC = Partially α crystallized; AC = α crystallized; ACS = α crystallized and stored; SAC = Sub-α crystallized; MMACS = Melt-mediated α crystallized and stored.) (F) Overrun (%) after 5 minutes (black color) and 30 minutes (gray color) of whipping the MG–oleogels.
Reproduced from ref. 121 with permission from the Royal Society of Chemistry.

more efficient in capturing air compared to sub-α (SAC MG-oleogels) or β crystals (ACS and MMACS MG-oleogels) due to their different microstructures.[146]

Furthermore, partially crystallized mixtures (PCMs) could incorporate more air but formed weaker structures with lower stability (*i.e.* lower G'_{LVR}). This is because crystals would preferentially be positioned at the interface contributing to the firm stabilization, and thus less crystals are present in the bulk phase decreasing the bulk stabilization due to lower gelation of the continuous oil phase. Besides, they captured larger bubbles and were thus less stable and firm. By adjusting the tempering protocol, fully crystallized mixtures (FCMs) could generate firmer (*i.e.* higher G'_{LVR}) and more stable networks containing less air and smaller bubbles in the continuous phase retaining higher elastic moduli. Therefore, Heymans proposed that the rheological properties of PCMs were characterized by interfacial contribution while FCMs by bulk gelation contribution.[121]

4) Crystal network generation

Various research studies have clearly demonstrated that oil foams are primarily network stabilized. In a study on edible oil foams, Mishima *et al.* (2016) found that tiny fat crystals constructed an oleogel through crystal–crystal links which was subsequently whipped to obtain a stable foam.[135] They also used synchrotron radiation microbeam X-ray diffraction (SR-μ-XRD) measurements to prove that the lamellar planes of β-fat crystals near the air–oil surface were arranged almost parallel to the air–oil surface plane. Similar results showed that the stability of oil foams of the ascribed myristic acid crystals absorbed at the interface and the network formation creating a gel film in the oil phase.[118,123,135] Therefore, network generation can enhance the stabilization of oil foam in application production, and cluster formation by coagulation should be avoided because of its destabilization mechanism.[147]

2.3.3 Benefits and Application

Up to now, few research studies have been performed on oil foams for edible applications although many fat-continuous products can serve as a host for air incorporation. Special textures can be developed, allowing healthier food products with less calorific content.[148]

Oil foams can be regarded as an innovative approach to modify the sensorial properties of food products. Involving air bubbles could definitely decrease the density of products, and change the texture and rheological properties, generating a special mouthfeel and appearance.[116] Similar to whipped cream and mousses, edible oil like sunflower oil or olive oil can be whipped to create a particular structure. Besides, foamed oil structures can improve flavor delivery by filling aromatized air into the system. Also, oil foams can be used to develop healthier food products. They can lower both the concentration of saturated and *trans* fatty acids in food products and the calorie content by incorporating air into the fat phase. For example, aerated chocolate contains less fat content, decreasing calorie consumption, without sacrificing the sensorial properties.[149] In addition, oil foams can be considered as a fat replacer, thereby reducing the saturated fat and *trans* fatty acid ester contents which are related to an increased risk of cardiovascular disease.[148,150] It's worth noting that oil foams can restrain microbial spoilage as they do not contain water.[151] As a consequence, oil foam systems need less or no microbial preservatives compared to aqueous foams. However, they may be more prone to oxidation than their saturated fat counterparts owing to the higher content of unsaturated fatty acid,[116] requiring anti-oxidants to be added to the formulation.[151]

2.4 Conclusions

Nowadays, there is a clear trend toward the replacement of *trans* and saturated fats with unsaturated oil. To structure this oil in food products, alternative oil structuring techniques, like O/W/O emulsions, oil powders, and oil foams, have been investigated. This chapter comprehensively summarized the preparation methods, stabilization mechanisms, and possible applications of these three oil structuring systems. However, although some articles on O/W/O emulsions have been published, their stabilization mechanism is still not very clear and use in edible applications is still limited. Oil capsules have been shown to be applicable in the bulk fat phase but reports are scarce and also research on oil foams is still emerging. Besides, there is a big gap between scientific research, usually performed at the lab scale and industrial production. Further research in oil structuring should address the issues of processing and product stability and scaling up toward edible applications.

Abbreviations

CLSM	Confocal laser scanning microscopy
DSD	Droplet size distribution
DSC	Differential scanning calorimetry
FCMs	Fully crystallized mixtures
HOSO	High oleic sunflower oil
MG	Monoglyceride
NMR	Nuclear magnetic resonance
O/W/O	Oil-in-water-in-oil
PCMs	Partially crystallized mixtures
PGPR	Polyglycerol polyricinoleate
SR-μ-XRD	Synchrotron radiation microbeam X-ray diffraction
TAG	Triacylglycerol
W/O/W	Water-in-oil-in-water
W/O	Water-in-oil

Acknowledgements

Vandemoortele Lipids NV is acknowledged for financially supporting the UGent Vandemoortele Centre on 'Lipid Science and Technology'. The Hercules foundation is recognized for its financial support in the acquisition of the scanning electron microscope JEOL JSM-7100F equipped with the cryo-transfer system Quorum PP3000T (grant number AUGE-09-029).

References

1. G. Muschiolik and E. Dickinson, *Compr. Rev. Food Sci. Food Saf.*, 2017, **16**, 532.
2. D. McClements, *Ann. Rev. Food Sci. Technol.*, 2010, **1**, 241.
3. C. Huynh Mai, T. Thanh Diep, T. Le and V. Nguyen, *J. Dispersion Sci. Technol.*, 2019, **41**, 479.
4. E. Dickinson, *Food Biophys.*, 2010, **6**, 1.
5. M. Pradhan and D. Rousseau, *J. Colloid Interface Sci.*, 2012, **386**, 398.
6. C. Oh, J. Park, S. Shin and S. Oh, *J. Dispersion Sci. Technol.*, 2004, **25**, 53.
7. S. Sajjadi, F. Jahanzad, M. Yianneskis and B. Brooks, *Ind. Eng. Chem. Res.*, 2003, **42**, 3571.
8. A. Utada, E. Lorenceau, D. Link, P. Kaplan, H. Stone and D. Weitz, *Science*, 2005, **308**, 537.
9. F. Groeneweg, W. Agterof, P. Jaeger, J. Janssen, J. Wieringa and J. Klahn, *Chem. Eng. Res. Des.*, 1998, **76**, 55.
10. J. Klahn, J. Janssen, G. Vaessen, R. Swart and W. Agterof, *Colloids Surf., A*, 2002, **210**, 167.
11. W. Gao and Y. Chen, *Int. J. Heat Mass Transfer*, 2019, **135**, 158.
12. P. Lu, L. Wu and X. Liu, *Micromachines*, 2017, **8**, 268.
13. Z. Fang, X. Cao, Y. Yu and M. Li, *Colloids Surf., A*, 2019, **570**, 282.
14. A. Benichou, A. Aserin and N. Garti, *Colloids Surf., A*, 2007, **297**, 211.
15. S. Abbott, *Surfactant Science Principles and Practice*, DEStech Publications, USA, 2016.
16. J. Weiss, I. Scherze and G. Muschiolik, *Food Hydrocolloids*, 2005, **19**, 605.
17. A. Benichou, A. Aserin and N. Garti, *Adv. Colloid Interface Sci.*, 2004, **108–109**, 29.
18. F. Jahaniaval, Y. Kakuda and V. Abraham, *J. Am. Oil Chem. Soc.*, 2003, **80**, 25.
19. F. Jahanzad, G. Crombie, R. Innes and S. Sajjadi, *Chem. Eng. Res. Des.*, 2009, **87**, 492.
20. T. Diep, T. Dao, H. Vu, B. Phan, D. Dao, T. Bui, V. Truong and V. Nguyen, *J. Dispersion Sci. Technol.*, 2018, **39**, 1752.
21. A. R. Patel, *MRS Adv.*, 2017, **2**, 1095.
22. T. Tran, S. Ghosh and D. Rousseau, *Cryst. Growth Des.*, 2014, **14**, 6383.
23. S. Ghosh and D. Rousseau, *Curr. Opin. Colloid Interface Sci.*, 2011, **16**, 421.
24. I. Aini and M. Miskandar, *Eur. J. Lipid Sci. Technol.*, 2007, **109**, 422.
25. N. Leister and H. Karbstein, *Colloids Interfaces*, 2020, **4**, 1.
26. K. Frank, E. Walz, V. Graf, R. Greiner, K. Kohler and H. Schuchmann, *J. Food Sci.*, 2012, 77, N50.

27. A. Pawlik, P. Cox and I. Norton, *J. Colloid Interface Sci.*, 2010, **352**, 59.
28. M. Kanouni, H. Rosano and N. Naouli, *Adv. Colloid Interface Sci.*, 2002, **99**, 229.
29. A. Schuch, L. Leal and H. Schuchmann, *Colloids Surf., A*, 2014, **461**, 336.
30. S. Schuster, R. Bernewitz, G. Guthausen, J. Zapp, A. Greiner, K. Köhler and H. Schuchmann, *Chem. Eng. Sci.*, 2012, **81**, 84.
31. A. Malkin, I. Masalova, P. Slatter and K. Wilson, *Rheol. Acta*, 2004, **43**, 584.
32. R. Lutz, A. Aserin, L. Wicker and N. Garti, *Colloids Surf., B*, 2009, **72**, 121.
33. U. Schmidt, R. Bernewitz, G. Guthausen and H. Schuchmann, *Colloids Surf., A*, 2015, **475**, 55.
34. D. McClements, *Adv. Colloid Interface Sci.*, 2015, **219**, 27.
35. H. Lamba, K. Sathish and L. Sabikhi, *Food Bioprocess Technol.*, 2015, **8**, 709.
36. A. Schuch, A. Tonay, K. Köhler and H. Schuchmann, *Can. J. Chem. Eng.*, 2014, **92**, 203.
37. K. Pays, J. Giermanska-Kahn, B. Pouligny, J. Bibette and F. Leal-Calderon, *Langmuir*, 2001, **17**, 7758.
38. A. Gaonkar and V. Hills, *U. S. Pat.*, US005332595A, 1994.
39. A. Oppermann, B. Piqueras-Fiszman, C. Graaf, E. Scholten and M. Stieger, *Food Res. Int.*, 2016, **85**, 215.
40. M. Lee, S. Oh, S. Moon and S. Bae, *J. Colloid Interface Sci.*, 2001, **240**, 83.
41. Y. Hwang, C. Oh and S. Oh, *J. Controlled Release*, 2005, **106**, 339.
42. S. Dwyer, D. O'Beirne, D. Eidhin and B. O'Kennedy, *J. Food Sci.*, 2012, 77, N58.
43. L. Liao, Y. Luo, M. Zhao and Q. Wang, *Colloids Surf., B*, 2012, **92**, 305.
44. L. Liu, F. Wu, X. Ju, R. Xie, W. Wang, C. Niu and L. Chu, *J. Colloid Interface Sci.*, 2013, **404**, 85.
45. R. Bernewitz, U. S. Schmidt, H. P. Schuchmann and G. Guthausen, *Colloids Surf., A*, 2014, **458**, 10.
46. E. Jiménez-Martín, A. Gharsallaoui, T. Pérez-Palacios, J. Ruiz Carrascal and T. Antequera Rojas, *Food Bioprod. Process.*, 2015, **96**, 52.
47. E. J. Lee and K. H. Lim, *Bioprocess Biosyst. Eng.*, 2017, **40**, 1701.
48. M. Stasse, T. Ribaut, V. Schmitt and V. Héroguez, *Polym. Chem.*, 2019, **10**, 4154.
49. H. Ali, A. Al-Khalifa, A. Aouf, H. Boukhebti and A. Farouk, *Sci. Rep.*, 2020, **10**, 2812.
50. R. Avola, G. Granata, C. Geraci, E. Napoli, A. C. E. Graziano and V. Cardile, *Food Chem. Toxicol.*, 2020, **144**, 111586.
51. P. Pongsumpun, S. Iwamoto and U. Siripatrawan, *Ultrason. Sonochem.*, 2020, **60**, 104604.
52. R. Waghmare, J. Moses and C. Anandharamakrishnan, *Crit. Rev. Food Sci. Nutr.*, 2021, DOI: 10.1080/10408398.2021.1873730.
53. A. Elshamy, N. Ammar, H. Hassan, S. Al-Rowaily, T. Ragab, A. Gendy and A. Abd-ElGawad, *Ind. Crops Prod.*, 2020, **148**, 112272.

54. Y. Miyagawa and S. Adachi, *Biosci., Biotechnol., Biochem.*, 2017, **81**, 625.
55. R. Dubey and B. Rao, *Def. Sci. J.*, 2009, **59**, 82.
56. P. Zhang, F. Weng, E. Koranteng, D. Guo, P. Cao and Q. Wu, *J. Appl. Polym. Sci.*, 2020, **138**, e49963.
57. A. Bakry, S. Abbas, B. Ali, H. Majeed, M. Abouelwafa, A. Mousa and L. Liang, *Compr. Rev. Food Sci. Food Saf.*, 2016, **15**, 143.
58. Z. Aksoylu and P. Ergönül, *Celal Bayar Üniv. fen bilim. derg.*, 2017, **2**, 293.
59. Z. Fang and B. Bhandari, *Trends Food Sci. Technol.*, 2010, **21**, 510.
60. D. McClements, E. Decker, Y. Park and J. Weiss, *Crit. Rev. Food Sci. Nutr.*, 2009, **49**, 577.
61. J. Wu, S. Chen, A. Doost, Q. A'yun and P. Meeren, *Food Hydrocolloids*, 2021, **112**, 106342.
62. C. Encina, C. Vergara, B. Giménez, F. Oyarzún-Ampuero and P. Robert, *Trends Food Sci. Technol.*, 2016, **56**, 46.
63. C. Turchiuli, M. Munguia, M. Sanchez, H. Ferre and E. Dumoulin, *Powder Technol.*, 2014, **255**, 103.
64. M. Fuchs, C. Turchiuli, M. Bohin, M. Cuvelier, C. Ordonnaud, M. Peyrat-Maillard and E. Dumoulin, *J. Food Eng.*, 2006, **75**, 27.
65. E. Frascareli, V. Silva, R. Tonon and M. Hubinger, *Food Bioprod. Process.*, 2012, **90**, 413.
66. G. Gallardo, L. Guida, V. Martinez, M. López, D. Bernhardt, R. Blasco, R. Pedroza-Islas and L. Hermida, *Food Res. Int.*, 2013, **52**, 473.
67. T. Kha, M. Nguyen, P. Roach and C. Stathopoulos, *Powder Technol.*, 2014, **264**, 298.
68. P. Roccia, M. Martínez, J. Llabot and P. Ribotta, *Powder Technol.*, 2014, **254**, 307.
69. A. Abedi, M. Rismanchi, M. Shahdoostkhany, A. Mohammadi and H. Hosseini, *Int. J. Food Sci. Technol.*, 2016, **51**, 2280.
70. L. Hu, J. Zhang, Q. Hu, N. Gao, S. Wang, Y. Sun and X. Yang, *J. Drug Delivery Sci. Technol.*, 2016, **36**, 46.
71. N. Ton, T. Tranand and V. Le, *Int. Food Res. J.*, 2016, **23**, 123.
72. R. Fernandes, E. Silva, S. Borges, C. Oliveira, M. Yoshida, Y. Silva, E. Carmo, V. Azevedo and D. Botrel, *Food Bioprocess Technol.*, 2016, **10**, 115.
73. C. Noello, A. Carvalho, V. Silva and M. Hubinger, *Food Res. Int.*, 2016, **89**, 549.
74. A. Ghani, K. Matsumura, A. Yamauchi, H. Shiga, S. Adachi, H. Izumi and H. Yoshii, *Drying Technol.*, 2016, **34**, 1726.
75. W. Shao, M. Yan, Q. Xie and X. Pan, *International Conference on Biological Engineering and Pharmacy*, 2016.
76. N. Icyer, O. Toker, S. Karasu, F. Tornuk, T. Kahyaoglu and M. Arici, *J. Food Meas. Charact.*, 2016, **11**, 50.
77. F. Kalkan, S. Vanga, R. Murugesan, V. Orsat and V. Raghavan, *Drying Technol.*, 2016, **35**, 527.

78. S. Shamaei, S. Seiiedlou, M. Aghbashlo, E. Tsotsas and A. Kharaghani, *Innovative Food Sci. Emerging Technol.*, 2017, **39**, 101.
79. M. Premi and H. Sharma, *Int. J. Biol. Macromol.*, 2017, **105**, 1232.
80. N. Mohammed, C. Tan, Y. Manap, A. Alhelli and A. Hussin, *Powder Technol.*, 2017, **315**, 1.
81. A. Bastıoğlu, M. Koç and F. Ertekin, *J. Food Meas. Charact.*, 2017, **11**, 1295.
82. H. Le, T. Tran, N. Ton, T. Tran, T. Huynh, N. Nguyen, S. Quang and M. Le, *Int. Food Res. J.*, 2017, **24**, 1227.
83. J. Luna-Guevara, C. Ochoa-Velasco, P. Hernández-Carranza and J. Guerrero-Beltrán, *Food Struct.*, 2017, **12**, 26.
84. B. BÖGer, S. Georgetti and L. Kurozawa, *Food Sci. Technol.*, 2018, **38**, 263.
85. S. Chew, C. Tan and K. Nyam, *J. Food Eng.*, 2018, **237**, 78.
86. V. Castel, A. C. Rubiolo and C. R. Carrara, *Food Res. Int.*, 2018, **103**, 76.
87. M. Sotelo-Bautista, L. A. Bello-Perez, R. A. Gonzalez-Soto, J. Yañez-Fernandez and J. Alvarez-Ramirez, *J. Dispersion Sci. Technol.*, 2019, **41**, 235.
88. S. Wang, Y. Shi and L. Han, *J. Food Eng.*, 2018, **231**, 1.
89. J. Benetti, J. Silva and V. Nicoletti, *Food Res. Int.*, 2019, **122**, 383.
90. J. Pino, O. Sosa-Moguel, E. Sauri-Duch and L. Cuevas-Glory, *J. Food Process. Preserv.*, 2019, **43**, e14136.
91. L. Giorgio, P. Salgado and A. Mauri, *Food Hydrocolloids*, 2019, **87**, 891.
92. P. Moser, S. Ferreira and V. Nicoletti, *Food Bioprod. Process.*, 2019, **117**, 183.
93. T. Nawas, M. Azam, A. Ramadhan, Y. Xu and W. Xia, *J. Aquatic Food Product. Technol.*, 2019, **28**, 49.
94. N. Yekdane and S. Goli, *Food Bioprocess Technol.*, 2019, **12**, 1614.
95. B. Wang, B. Adhikari and C. Barrow, *Powder Technol.*, 2019, **358**, 79.
96. B. Wang, B. Adhikari, M. Mathesh, W. Yang and C. Barrow, *Powder Technol.*, 2019, **358**, 68.
97. J. Hoyos-Leyva, L. Bello-Perez, J. Agama-Acevedo, J. Alvarez-Ramirez and L. Jaramillo-Echeverry, *LWT - Food Sci. Technol.*, 2019, **101**, 526.
98. R. Hernández-Nava, A. López-Malo, E. Palou, N. Ramírez-Corona and M. T. Jiménez-Munguía, *Food Hydrocolloids*, 2020, **109**, 106077.
99. M. Shi, D. Ying, M. Hlaing, J. Ye, L. Sanguansri and M. Augustin, *Food Res. Int.*, 2020, **132**, 109050.
100. Y. Zhong, L. Yang, D. McClements, X. Wang, J. Ye and C. Liu, *Food Chem.*, 2020, **342**, 128380.
101. B. Başyiğit, H. Sağlam, Ş. Kandemir, A. Karaaslan and M. Karaaslan, *Powder Technol.*, 2020, **364**, 654.
102. M. C. S. Aguiar, M. F. das Gracas Fernandes da Silva, J. B. Fernandes and M. R. Forim, *Sci. Rep.*, 2020, **10**, 11799.
103. E. Alpizar-Reyes, V. Varela-Guerrero, J. Cruz-Olivares, H. Carrillo-Navas, J. Alvarez-Ramirez and C. Perez-Alonso, *Int. J. Biol. Macromol.*, 2020, **145**, 207.

104. M. Mehran, S. Masoum and M. Memarzadeh, *Ind. Crops Prod.*, 2020, **154**, 112694.
105. G. Marques, R. Fernandes, A. Lago, S. Borges, S. Bertolucci, A. Lima and D. Botrel, *Drying Technol.*, 2021, **1**, DOI: 10.1080/07373937.2021.1871917.
106. E. Jakubowska and J. Lulek, *J. Drug Delivery Sci. Technol.*, 2021, **62**, 102357.
107. X. Duan, M. Zhang, A. Mujumdar and R. Wang, *Drying Technol.*, 2010, **28**, 444.
108. T. El-Messery, U. Altuntas, G. Altin and B. Özçelik, *Food Hydrocolloids*, 2020, **106**, 105890.
109. S. Anwar and B. Kunz, *J. Food Eng.*, 2011, **105**, 367.
110. A. Marefati, M. Rayner, A. Timgren, P. Dejmek and M. Sjöö, *Colloids Surf., A*, 2013, **436**, 512.
111. S. Fioramonti, A. Rubiolo and L. Santiago, *Powder Technol.*, 2017, **319**, 238.
112. E. Rodriguez, L. Julio, C. Henning, B. Diehl, M. Tomas and V. Ixtaina, *J. Sci. Food Agric.*, 2019, **99**, 1682.
113. B. Erdem and S. Kaya, *Food Chem.*, 2021, **339**, 127976.
114. I. Tavernier, B. Heyman, P. Meeren, T. Ruyssen and K. Dewettinck, *J. Food Eng.*, 2019, **244**, 136.
115. A. Salonen, *Curr. Opin. Colloid Interface Sci.*, 2020, **50**, 101381.
116. R. Heymans, I. Tavernier, K. Dewettinck and P. Meeren, *Trends Food Sci. Technol.*, 2017, **69**, 13.
117. S. Friberg, *Curr. Opin. Colloid Interface Sci.*, 2010, **15**, 359.
118. M. Brun, M. Delample, E. Harte, S. Lecomte and F. Leal-Calderon, *Food Res. Int.*, 2015, **67**, 366.
119. A. Fameau, S. Lam, A. Arnould, C. Gaillard, O. Velev and A. Saint-Jalmes, *Langmuir*, 2015, **31**, 13501.
120. B. Binks and I. Marinopoulos, *Food Res. Int.*, 2017, **95**, 28.
121. R. Heymans, I. Tavernier, S. Danthine, T. Rimaux, P. Meeren and K. Dewettinck, *Food Funct.*, 2018, **14**, DOI: 10.1039/C8FO00536B.
122. M. Abdollahi, S. Goli and N. Soltanizadeh, *Eur. J. Lipid Sci. Technol.*, 2019, **122**, 1900196.
123. B. Binks, E. Garvey and J. Vieira, *Chem. Sci.*, 2016, 7, 2621.
124. I. Lesov, S. Tcholakova, M. Kovadjieva, T. Saison, M. Lamblet and N. Denkov, *J. Colloid Interface Sci.*, 2017, **504**, 48.
125. E. Fredrick, P. Walstra and K. Dewettinck, *Adv. Colloid Interface Sci.*, 2010, **153**, 30.
126. Y. Huang, J. Yang, L. Chen and L. Zhang, *ACS Sustainable Chem. Eng.*, 2018, **6**, 10552.
127. T. Hunter, R. Pugh, G. Franks and G. Jameson, *Adv. Colloid Interface Sci.*, 2008, **137**, 57.
128. T. Horozov, *Curr. Opin. Colloid Interface Sci.*, 2008, **13**, 134.
129. B. Binks, A. Rocher and M. Kirkland, *Soft Matter*, 2011, 7, 1800.
130. D. Z. Gunes, M. Murith, J. Godefroid, C. Pelloux, H. Deyber, O. Schafer and O. Breton, *Langmuir*, 2017, **33**, 1563.

131. L. Shrestha and K. Aramaki, in *Foam Engineering: Fundamentals and Applications*, ed. P. Stevenson, John Wiley & Sons, Chichester, 1st edn, 2012, ch. 9, pp. 201–203.
132. D. Rousseau, *Curr. Opin. Colloid Interface Sci.*, 2013, **18**, 283.
133. J. Guevara, A. Mejia, M. Shuai, Y. Chang, M. Mannan and Z. Cheng, *Soft Matter*, 2013, **9**, 1327.
134. B. Madivala, S. Vandebril, J. Fransaer and J. Vermant, *Soft Matter*, 2009, **5**, 1717.
135. S. Mishima, A. Suzuki, K. Sato and S. Ueno, *J. Am. Oil Chem. Soc.*, 2016, **93**, 1453.
136. T. Truong, C. Lopez, B. Bhandari and S. Prakash, Springer Nature, Switzerland, **1**, 2020.
137. K. Sato and S. Ueno, *Curr. Opin. Colloid Interface Sci.*, 2011, **16**, 384.
138. P. Cruz, F. Rocha and A. Ferreira, *Org. Process Res. Dev.*, 2019, **23**, 2592.
139. P. Ramel, R. Campos and A. Marangoni, *Cryst. Growth Des.*, 2017, **18**, 1002.
140. J. Vereecken, W. Meeussen, I. Foubert, A. Lesaffer, J. Wouters and K. Dewettinck, *Food Res. Int.*, 2009, **42**, 1415.
141. C. Fang, W. Tang, S. Wu, J. Wang, Z. Gao and J. Gong, *Ultrason. Sonochem.*, 2020, **68**, 105227.
142. J. Lee, R. Silva, V. Gibon and S. Martini, *J. Food Sci.*, 2018, **83**, 902.
143. V. Nguyen, T. Rimaux, V. Truong, K. Dewettinck and F. Bockstaele, *Food Res. Int.*, 2020, **137**, 109412.
144. K. Sugioka, *Handbook of Laser Micro- and Nano-Engineering*, Springer Nature, Switzerland, 2020.
145. H. Pirouzian, N. Konar, I. Palabiyik, S. Oba and O. Toker, *Food Chem.*, 2020, **321**, 126718.
146. G. Richardson, M. Langton, P. Faldt and A. Hermansson, *Cereal Chem.*, 2002, **79**, 546.
147. A. Pombo, *Int. Dairy J.*, 2020, **117**, 104948.
148. L. Du, Q. Jiang, S. Li, Q. Zhou, Y. Tan and Z. Meng, *Food Hydrocolloids*, 2021, **112**, 106245.
149. J. Haedelt, S. Beckett and K. Niranjan, *J. Food Sci.*, 2007, **72**, E138.
150. S. Chaireh, P. Ngasatool and K. Kaewtatip, *Int. J. Biol. Macromol.*, 2020, **165**, 1382.
151. M. Saltmarsh, *J. Sci. Food Agric.*, 2015, **95**, 649.

CHAPTER 3

Use of High-intensity Ultrasound to Structure Edible Fats

T. L. T. DA SILVA,[b] A. GIACOMOZZI[c] AND S. MARTINI*[a]

[a] Department of Nutrition, Dietetics, and Food Sciences, Utah State University, 8700 Old Main Hill, Logan, UT 84322-8700, USA; [b] Science des Aliments et Formulation, Gembloux Agro-Bio Tech, ULiège, Gembloux 5030, Belgium; [c] Departamento de Ingeniería Química, Universidad Nacional del Sur (UNS). Bahía Blanca, Argentina. Planta Piloto de Ingeniería Química – PLAPIQUI (UNS-CONICET), Bahía Blanca 8000, Argentina
*Email: silvana.martini@usu.edu

3.1 Introduction

Acoustic techniques have been used in lipids and lipid-based products to change their crystallization behavior and to monitor structural changes in the material formed. These acoustic techniques can be classified according to the properties of the acoustic waves used, such as frequency (cycles per second), wavelength (distance between cycles), and amplitude (height of the wave, power level).[1]

High-frequency ultrasound techniques (100 KHz–1 MHz) are non-invasive analytical tools used to monitor physical or chemical changes in materials.[2–4] More invasive techniques, such as power ultrasound or high-intensity ultrasound (HIU), use low-frequency acoustic waves (20–100 kHz) to induce physicochemical changes in the material, making it ideal for inducing crystallization (sonocrystallization).[5,6] High-intensity ultrasound techniques have received a lot of attention not only because of the changes in physicochemical properties that they promote in the sample but also because they are

Food Chemistry, Function and Analysis No. 31
Development of *Trans*-free Lipid Systems and their Use in Food Products
Edited by Jorge F. Toro-Vazquez

Published by the Royal Society of Chemistry, www.rsc.org

economically viable, relatively easy to incorporate into industrial operations, reproducible, non-thermal, and environmentally clean.[1,7,8]

Crystallization is an important process in the lipid industry and can be affected by many internal and external factors. Internal factors include fat composition and the presence of impurities, while external factors include crystallization temperature, supercooling, agitation rate, use of ultrasound waves, and others.[9] The effect of HIU on inducing crystallization is due to a phenomenon called cavitation which involves the formation of a vapor cavity or bubble due to localized changes in acoustic pressure. Bubbles formed during cavitation can either be stable (stable cavitation) or can collapse, generating high shear forces and high localized temperatures (inertial cavitation).[10] Bubbles generated during cavitation events may remain in the system even after sonication is stopped and may play a role in the induction of crystallization.[11] Acoustic waves can be applied using different frequencies, acoustic power levels, pulse *vs.* continuous modes, and pulse duration. All these variables can affect the crystallization behavior of lipids and the physical properties of the materials in different ways.[12]

The use of HIU in fats was first reported to tailor polymorphic transformations and to promote the formation of stable polymorphic forms.[13] Over the years, studies have explored the effect of sonocrystallization on the physical properties of fats.[5,14] The use of HIU as an additional technology to tailor the crystallization and physical properties of fats can be useful for two groups of fats: low-saturated fats including oleogels and high-saturated fats.

In most of the studies, sonication of fats was explored by using acoustic waves operating at 20 kHz frequency and by evaluating the effect of processing conditions such as crystallization and sonication conditions on the physical properties of the material.[5,6,13–20] Recently, the World Health Organization (WHO)[21] has suggested that the consumption of saturated and *trans* fats should be minimized or eliminated due to their association with chronic diseases such as obesity, high blood cholesterol, cancer, and coronary heart disease.[22,23] Moreover, in 2015 the U.S. Food and Drug Administration eliminated the "Generally Recognized as Safe" (GRAS) status of partially hydrogenated fats.[24] In this context, low-saturated fats and oleogels have become an alternative for the consumer and food industries as sources of "healthy fats" due to their low levels of saturated fatty acids (SFA) and absence of *trans*-fat. Nevertheless, this replacement is not trivial since the elimination of *trans*-fats and the reduction of SFA are related to the loss of desired physical properties such as mouthfeel, hardness, and adhesiveness. Following this worldwide requirement, the use of HIU to improve the physical properties of low-saturated fats, structured oils, and emulsions has become an interesting technology that can be used to improve the quality of healthier lipid systems.

In this chapter we describe research performed on the use of HIU to improve the physical properties of fats with low and high levels of SFA and of oleogels formulated with various types and concentrations of gelators. In most of the studies described in this chapter, HIU was operated at a frequency of 20 kHz.

3.2 Effect of HIU on Fats

Fats are complex systems composed of a combination of triacylglycerols (TAGs) that interact with each other to form a semi-solid material when cooled. Fat crystallization is a very complex process that involves four main stages: (1) generation of a driving force, (2) nucleation, (3) crystal growth, and (4) recrystallization.[25] These steps can be affected by the processing conditions (cooling rate, initial and final temperatures, agitation rate, sonication) and lipid composition (fatty acids and triacylglycerol composition, presence of additives).[26] Considering the importance of chemical composition on lipid crystallization, this chapter is organized to describe the effect of sonication on lipid crystallization according to the material's chemical composition. The chapter first describes the effect of HIU on fats with high levels of saturated fatty acids, followed by the effect on fats with low levels of saturated fatty acids, and a description of the effect of HIU on oleogels, and finishes with the effects of HIU on lipid oxidation and other food applications.

3.2.1 High-saturated Fats

The effect of sonication on fats with at least 50% of SFA is discussed in this section. Among these fats, cocoa butter (CB), palm oil (PO), palm kernel (PKO), and anhydrous milk fat (AMF) are the ones mainly used in the food industry and therefore, sonocrystallization becomes a very interesting technology to improve the crystallization of these samples. The main parameters studied in this topic are summarized in Table 3.1.

3.2.1.1 Cocoa Butter and Pure Triacylglycerols

The first study that evaluated the effect of HIU on fat crystallization was performed in high-saturated fats.[13] The goal of this study was to evaluate the effect of ultrasound power on tripalmitin (PPP, 100 W) and cocoa butter (CB, 300 W) crystallization. Sonication in these cases increased the crystallization rate and the formation of more stable polymorphic forms, even though the very high-power level used resulted in a significant increase (~30 °C) in temperature (higher than the sample's melting point). The reduction in the induction time and increase in the number of crystals suggested that ultrasound affected crystallization during the nucleation process. Authors explained these results with the following theories for HIU mechanisms: (1) violent collapse of cavitation bubbles may form active sites for primary nucleation, (2) enhanced agitation may affect profound mobility of crystallizing molecules, (3) cooling caused by evaporation of liquid from the surface of the bubbles during the growth of a cavitating bubble may increase supercooling, and (4) local pressure may increase the melting point in the vicinity of a collapsing cavity resulting in increased supercooling.[13] When polymorphism was closely evaluated in the sonicated samples using *in situ* time-resolved SAXS/WAXS for PPP and LaLaLa (trilaurin), the same group observed that HIU significantly reduced the induction time required for

Table 3.1 Studies of sonication effect on high-saturated fats.

Lipid material	What was evaluated	HIU conditions	Crystallization conditions	Main findings	References
AMF	Supercooling Pulse duration Power level	20 kHz 5 or 10 s 5–50 W	22, 24, 26, 28, 30 °C	↓ Induction time Smaller crystals	18
	Supercooling When HIU should be applied	20 kHz 10 s 50 W	22, 24, 26, 28 °C	↓Induction time ↑Crystallization rate Smaller crystals Higher hardness	5
	Non-isothermal supercooling	24 kHz 200 W 7 mm tip 5 s	14, 22, 26, 30 °C 3 or 0.5 °C/min	↓Induction time ↑Crystallization rate Smaller crystals Higher SFC Higher hardness	20
	Degree of saturation Continuous system	20 kHz 40 W 0, 2.5, 5, 7,10 min	28 °C 0, 10, 20, 30% blend with rapeseed oil	↑Crystallization rate Smaller crystals More efficiency with high content of oil in the blend	17
Palm oil	Acoustic threshold	30–45 dB	23 °C	Smaller crystals Higher viscosity	30
	Supercooling Power level Duration	20 kHz 6 mm tip 47.5, 95, 270, 475 W; 20, 60, 120, 240 s (5 s pulses)	20, 25, 30, 36 °C	↓Induction time ↑Crystallization rate Avoid phase separation Reduced crystal size	6
	Continuous flow Power level Pulses duration	20 kHz 3.8 cm^2 tip 75, 110, 180 W 5 min (0, 5, 10 and 15 s pulse)	35 °C	↓Induction time ↑Crystallization rate Smaller crystals size More crystals Higher G'	15
All-purpose shortening	HIU in the presence of fewer crystals or not	20 kHz 10 s 50 W	30 °C	↓Induction time ↑Crystallization rate Smaller crystals; Higher hardness	5

	Supercooling and tip size Cooling rate HIU during nucleation or crystal growth	20 kHz 12.7 mm or 3.2 mm	20 or 30 °C 0.5 or 1 °C/min	Smaller crystals Higher G' ↑ Efficiency in nucleation	27
	HIU effects over long storage (24 weeks)	20 kHz 3.2 mm 45 W	5 or 25 °C	HIU speed up more stable crystalline form	28
Palm kernel oil	HIU in the presence of fewer crystals or not	20 kHz 3.2 mm 10 s 50 W	30 °C	↓Induction time ↑Crystallization rate Smaller crystals Higher hardness	5
	HIU effect on oil migration	20 kHz 3.2 mm tip 97 W 10 s	30 °C	Less oil migration Smaller crystals and different morphologies Hardness material No changes on TAGs composition;	35
Cocoa butter	Nucleation on stable polymorphic forms	20 kHz 300 W 3 or 15 s	32.3 °C	↓Induction time More nuclei formed Stable polymorph nucleated	13
TAGs	PPP Nucleation on stable polymorphic forms	20 kHz 100 W 15 s	Range 45–57 °C	↓Induction time More nuclei formed Stable polymorph nucleated	13
	PPP and LaLaLa: Controlling polymorphism Nucleation rate	20 kHz 100 W 2 s	50 °C (PPP) 20 or 30 °C (LaLaLa)	↓Induction time Stable polymorph nucleated	16
Coconut oil	Supercooling and power level	40 kHz 12, 36, 60 W	13, 15, 17, 19 °C	↓Induction time ↑Crystallization rate Smaller crystals More crystals and nuclei	33
	Sonication with and without emulsifiers	40 kHz 50 W 10 s 12.7 mm tip	21 °C	↓Induction time ↑Crystallization rate Smaller crystals No changes in G' or SFC due to sonication	34

nucleation. However, sonication was not the only factor to induce more stable polymorphic forms. The crystallization temperature (and degree of supercooling) also determined the obtained polymorphic form.[16] HIU was also used in a mango fat where sonication induced nucleation and accelerated the crystallization rate, and also accelerated the polymorphic transformation from β′ to β-form. This finding shows that HIU could be used for tempering chocolate using cocoa butter alternatives.[27]

Reduced induction times, increased crystallization rates, formation of smaller crystals, and induced nucleation were all common results among studies using cocoa butter, pure TAGs, and all other studies on lipid materials with high levels of SFA such as milk fat,[5,17,18,20] palm oil,[6,15] all-purpose shortenings,[5,28,29] and palm kernel oil.[5] In the food industry, accelerating crystallization is significant because it could shorten processing times and consequently reduce processing costs.[20]

3.2.1.2 *Anhydrous Milk Fat (AMF)*

The relationship between reducing the induction time of crystallization through sonication and supercooling was first explored in AMF.[18] For example, after sonocrystallizing AMF under different crystallization temperatures ($T_c = 22–30$ °C), power levels (5–50 W), and times (5 or 10 s), reductions in the induction time of crystallization and crystal size were observed only when lower supercoolings were used ($T_c \geq 24$ °C). The lack of HIU's effect on nucleation at 22 °C was attributed to (1) nucleation being more affected by supercooling due to the lower temperature rather than by HIU and/or (2) formation of non-inertial bubbles due to higher viscosity at higher supercooling (lower temperatures). Cavitation generated by acoustic waves can be *inertial*, involving a variety of bubble sizes that tend to grow and collapse; or *non-inertial* where bubbles experience small amplitude oscillations with little or no growth over time.[30] Bubble collapse is associated with the invasive nature of this technique and has been proposed as the reason to alter crystallization. In short, the lack of sonication effect at higher supercooling can be attributed to the stability of non-inertial cavitation bubbles generated during sonication and to the lack of inertial cavities.[18]

Similar studies were performed on AMF exploring the moment at which HIU should be applied *versus* supercooling during isothermal[5] and non-isothermal conditions,[20] and later on in a system with continuous flow.[17] Frydenberg and others[20] observed an effect of sonication at 22 °C, but not at lower temperature (14 °C). The authors attributed these results to sonication having been being applied after nucleation occurred and to the high viscosity of the system, which made the generation of cavitation at 14 °C difficult. However, when samples were crystallized under non-isothermal crystallization conditions, such as the ones used by Frydenberg *et al.*[20] it seems that HIU can affect crystallization even at 22 °C, which is different from what was previously observed.[18] Suzuki and others[5] also observed a better effect of sonication on AMF crystallized at lower supercooling when

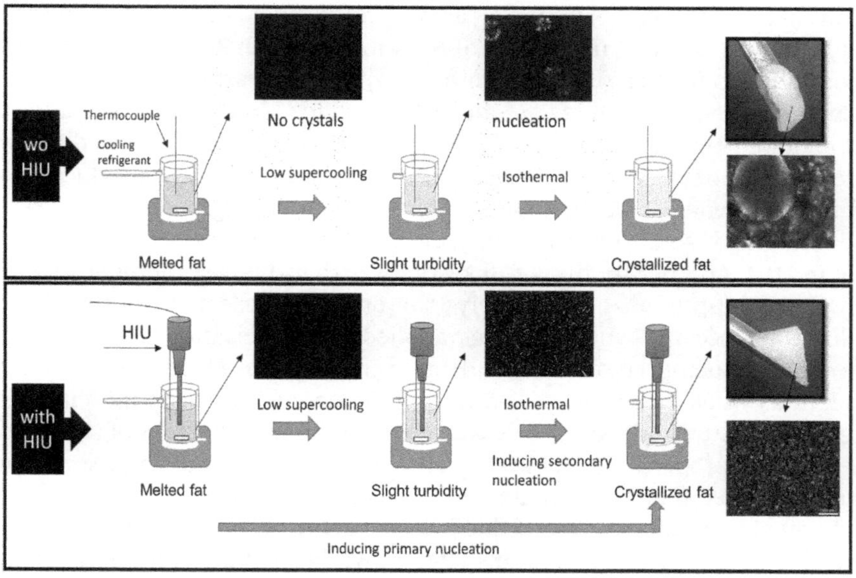

Figure 3.1 Schematic representation of fat crystallization in a batch system with and without high-intensity ultrasound (HIU).

the sample was crystallized at 24 °C or higher temperatures under isothermal conditions. Moreover, these authors observed that HIU induced more nucleation if applied in the presence of fewer crystals rather than when no nucleation had begun, confirming what Martini *et al.*[18] found only for 30 °C. With these findings the authors suggested another mechanism to the ones Higaki *et al.*[13] suggested a few years earlier: sonication could also induce secondary nucleation.[5] In this context, it can be assumed that if HIU is applied in the absence of crystals, bubbles and shear forces associated with cavitation events might induce primary nucleation. However, if sonication is applied in the presence of crystals, high shear forces generated by sonication are responsible for disrupting the growing crystals and for inducing primary and secondary nucleation.[12] Both effects and a set-up of sonication in batch are outlined in Figure 3.1.

In addition to observing changes in AMF at the nanoscale (nucleation) due to sonication, changes were also observed at the macroscopic level through an increase in the amount of solid fat formed and its hardness.[5,20]

3.2.1.3 All-purpose Shortening

The effect of HIU on AMF was compared to other fats such as all-purpose shortening at low supercooling (30 °C), and a similar effect was observed at the nano-, micro-, and macroscopic level.[5] Additional effects of sonication due to tip size (12.7 or 3.2 mm), acoustic wave amplitude (108 or 216 μm), temperature (20 or 30 °C), pulse duration (10, 5, or 2.5 s) and the step of crystallization in which HIU was applied were also evaluated in an all-purpose shortening. HIU was more efficient at generating smaller clusters when it was applied using the 12.7 mm tip at 20 °C compared with 30 °C, but pulse duration did not affect crystal size significantly,[28] in contrast to the supercooling results reported by Suzuki *et al.*[5] and Martini *et al.*[18] in the AMF. The authors also showed that although HIU induced crystallization and reduced the induction time of crystallization, the maximum solid fat content produced and its melting behavior were not affected, suggesting that HIU did not increase the amount of solid material; however, a more elastic material was formed due to HIU. Another finding of this work was that the best moment to apply HIU is in the presence of fewer crystals (ongoing nucleation, onset of crystallization), because when HIU was applied after nucleation took place the effect on reducing crystal size and on increasing the elasticity was not as evident as when HIU was applied during the ongoing nucleation.[28] Interestingly, a later study by Lee *et al.*[29] evaluated the effects of 24 weeks of storage on the physical properties generated by HIU through an induction of secondary nucleation. Lee *et al.*[29] reported that HIU accelerated the crystallization of the more stable crystalline form of this fat. Non-sonicated samples continued to crystallize during storage and only after 12 weeks of storage did they achieve elastic modulus (G') values similar to the ones obtained in sonicated samples after 1 h of crystallization at 25 °C.

3.2.1.4 Palm Oil

The crystallization of palm oil in combination with HIU was also evaluated. When palm oil was sonicated using an acoustic power lower than the one needed to reach the cavitation threshold (30–45 dB or 1–31 W), a reduction in crystal size and a change in the fat's viscosity were found.[31] Sonocrystallization of palm oil was further investigated in batch[6] and continuous flow systems.[15] A marked increase in the crystallization rate, reduction in the induction time of crystallization, and reduction in crystal size due to HIU application also happened using only 47.5/75 W.[6,15] Both studies evaluated low and high power levels at different durations (Table 3.1) and verified different magnitudes of temperature increase in the media, dependent on the combination of both parameters (sonication duration and power). Interestingly, when palm oil was crystallized in a flow cell a significant increase in the heat generated by sonication was also found.[15] This suggests that although higher ultrasonic power for long periods could have a greater impact on crystallization, the thermal effect generated by HIU can be prejudicial if HIU is applied for too long or at power levels that are too high, and a balance between benefit and undesirable heat in the crystallization process must be taken into account.[6,15] The increase in temperature observed during sonication was attributed to cavitation and to bubbles generated during sonication, which affect lipid crystallization. The number of bubbles generated during sonication and their life cycles are affected by acoustic power levels, pulse duration, and the temperature of the media. More bubbles are generated at higher power levels which generate higher temperatures; however, after sonication is stopped, a faster dissolution of the bubbles is observed at higher temperatures and the negative effect on crystallization of these high-power levels can be attributed either to melting nuclei or the lower efficiency of cavitation due to dissolution of the bubbles.[11,32,33]

An alternative found by Ye and Martini[15] for using higher power levels was to use pulses instead of continuous application. Chen *et al.*[6] and Ye and Martini[15] agree with the mechanism described by Higaki *et al.*[13] and with the additional mechanism of secondary nucleation suggested by Suzuki *et al.*[5] They both concluded that faster and higher nucleation promotes the formation of smaller crystals that interact closely with each other. These smaller crystals increased the elasticity of the sample and reduced phase separation due to a more organized crystalline network and orderly arrangements of TAGs after cavitation.[6,15]

These studies also reported other benefits of HIU on the physical properties of palm oil such as no increase in solid fat content (SFC) after storage due to sonication,[6,15] no phase separation,[6] higher elasticity (G'), and sharper melting profiles.[15]

3.2.1.5 Lauric-based Fats

In lauric-based fats, different materials were sonocrystallized such as palm kernel oil,[5] coconut oil,[34,35] and a palm kernel-based shortening.[36]

An effect of HIU on inducing secondary nucleation when HIU was applied after primary nucleation (in the presence of a few crystals) and crystallized under low supercooling was also found for lauric-based fats as previously discussed for AMF and all-purpose shortening.[5] Regarding the power level, HIU showed an effect when applied at power levels as low as 12 W, and an increased crystallization rate, reduced induction time, and reduced crystal size were found when the HIU power level increased,[34,35] even though no changes in physical properties such as melting behavior and elasticity were found after sample sonication.[35]

The effect of HIU on the oil migration of palm kernel-based shortening was also evaluated. Da Silva *et al.*[36] was able to reduce oil loss in a palm kernel-based sample processed with sonication in batch from 48% (for the non-sonicated sample) to only 3% (for the sonicated sample). The effect of HIU was attributed to higher shear forces induced by cavitation and bubble dynamics that broke the crystals into more and smaller clusters forming a more organized crystalline network. Sonication in this case also increased the sample's hardness and changed the melting peak without changing the polymorphic form and the SFC after storage for 48 h. It is important to point out that in general neither SFC nor the polymorphic form was affected by HIU after storage, and only a few studies found a change in the melting profile. The chemical composition of the crystalline material remaining after oil migration was also evaluated by Silva *et al.*[36] and confirmed that the same crystalline material was formed with and without HIU in terms of chemical composition of TAGs.[36] This similarity in the chemical composition of the crystalline material was also found in the AMF crystallized with and without HIU.[20] These results suggest that the effect of HIU

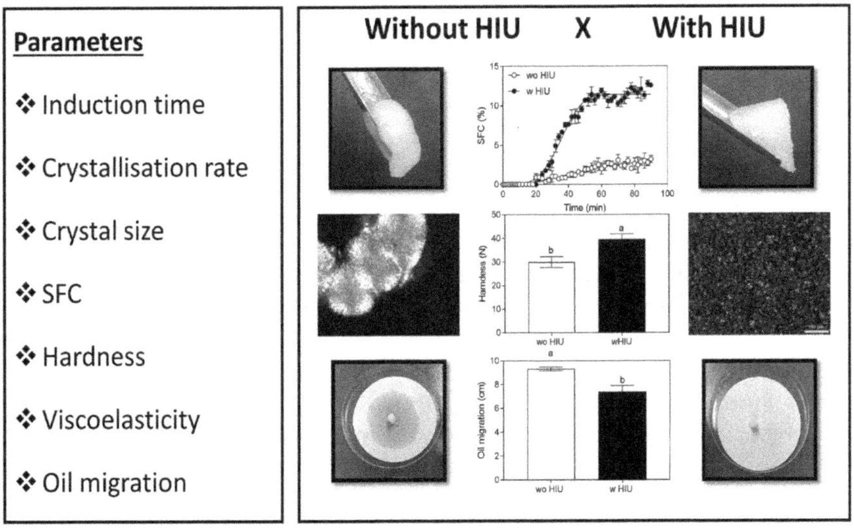

Figure 3.2 Crystallization parameters and physical properties affected by high-intensity ultrasound (HIU) in fats.

on high-saturated fats is related to physical interactions among the TAGs rather than to changes in the chemical composition of the crystalline portion.[20,36] Figure 3.2 summarizes the effects on crystallization due to sonication in high-saturated fats as previously discussed.

3.2.2 Low-saturated Fats

In response to new regulations by the Food and Drug Administration (FDA) in 2015, recommendations by the WHO,[37] HHS, and USDA,[38] and changes in consumer preferences regarding fat consumption, finding healthier fat alternatives to reduce saturated- and/or *trans*-fat content has drawn the attention of food scientists and food producers. Healthier fat sources can be obtained if *trans*-fats are replaced by SFA. However, a reduction in SFA is also sought to improve the nutritional properties of fats even further. A decrease in SFA is associated with an increase in unsaturated fatty acids affecting the final structure of products, their technological characteristics, and the consumer expectations (functional and sensorial properties). This section will discuss the use of HIU not only to accelerate the crystallization process but also to improve the physical properties of fats with low SFA content such as hardness, elasticity, and oil binding capacity (Table 3.2).

3.2.2.1 Soybean-based Fat

The first study performed on this type of fat was published in 2011.[14] Interesterified soybean fat (~33% saturated) was crystallized at different temperatures (26–32 °C) and power levels (44–101 W). Even with low amounts of crystalline material, this sample also showed induced nucleation when

Table 3.2 Studies of sonication effect on low-saturated fats.

Lipid material	What was evaluated	HIU conditions	Crystallization conditions	Main findings	References
IE soybean base	Supercooling Power level When applying HIU Physical properties	20 kHz 3.2 mm tip 10 s 110, 72, 61, 54, 44 W	26, 28, 30, 32 °C 5 °C per min	Faster nucleation Smaller crystals Higher hardness Higher G'	14
	HIU parameters: Frequency Speed agitation Amplitude Pulses	20–40 kHz 24 or 108 μm 5, 10, 15 s 12.7 mm tip Or High speed agitation: Small probe Big probe 6000–24 000 rpm 10 s	44 °C	↓Induction time ↑Crystallization rate Smaller and more crystals Higher G' (20 kHz)	39
	Agitation after sonication	20 kHz 3.2 mm tip 10 s 216 μm	Non-agitation *versus* continuous (100 rpm) after HIU 29, 34 and 35 °C	↓Induction time ↑Crystallization rate Smaller and more crystals Higher G' (no agitation after HIU)	44
	Removing StStSt	20 kHz 3.2 mm tip 10 s 216 μm	27 °C	Smaller and more crystals Higher G' ↑ Melting enthalpy	51
	Different amount of St St on sn-2	20 kHz 3.2 mm tip 5 s 216 μm	ΔT under melting point 3, 6, and 9 °C	↑ Crystallization rate Smaller and more crystals Higher G'	50

Table 3.2 *(Continued)*

Lipid material	What was evaluated	HIU conditions	Crystallization conditions	Main findings	References
	Degree of saturation; Supercooling	20 kHz 24 or 108 μm 10 s 12.7 mm tip	44, 32, and 25 °C 0, 50, 80% of HOSO	↓Induction time ↑Crystallization rate Smaller and more crystals Higher G'	46
	HIU effect on oil migration and supercooling	20 kHz 3.2 mm tip 10 s 216 μm	26, 28, 30, 32 °C	Lower oil migration Crystallization of low MP TAGs	19
	HIU effects over long storage (24 weeks)	20 kHz 3.2 mm 45 W	24 and 34 °C	Higher amount of crystalline material formed and kept over storage	38
Soft stearin	Supercooling Sonication effect Physical properties	20 kHz 1.98 mm tip 10 s 0.5 W	17, 18, 19, 20, 21 °C	↓ Induction time ↑Crystallization rate Changed crystal morphology Smaller G'	45
IE palm base	Different amount of P P on sn-2	20 kHz 3.2 mm tip 5 s 216 μm	ΔT under melting point 3, 6 and 9 °C	↑Crystallization rate Smaller and more crystals Higher G' ($P = 30\%$)	47
	Palm olein	40 kHz 50 W 10 s 12.7 mm tip	19 °C	↓ induction time ↑ Crystallization rate; smaller crystals No changes in G' or SFC due to sonication	34

crystallized with HIU. In this case, the supercooling used showed different effects on nucleation: at lower supercooling primary nucleation was induced, while at high supercooling both primary and secondary nucleation were induced by sonication. This effect was observed even when a very low power 44 W was used, but the greatest induction was observed using the highest power level of 110 W. The use of 110 W was further studied when HIU was applied before and after nucleation starts, to evaluate which nucleation process is more effectively induced by sonication. These authors reported that sonication showed more influence on the crystallization behavior when applied in the presence of few crystals, suggesting that sonication has a greater effect on physical properties when secondary nucleation is induced. Overall, under this condition where samples were sonicated in the presence of few crystals, the material obtained was harder and more elastic.[14] A recent studied performed by Lee *et al.*[39] observed that the improved physical properties obtained in this type of fat due to sonication are maintained even after a long storage time (6 months).

A similar interesterified base but with a higher amount of saturated fatty acid (45%) was evaluated using a higher frequency HIU (40 kHz) and high-speed agitation to compare with the results obtained using a 20 kHz sonicator,[40] which is the most commonly used frequency employed in fat sonocrystallization studies (Tables 3.1 and 3.2). The authors observed that all three processes were able to induce fat crystallization and more elastic materials were formed. However, the magnitude of the effects was dependent on processing conditions. In terms of crystallization rate, 40 kHz HIU increased the crystallization rate and induced nucleation but a higher power level was necessary to achieve the same results as those obtained at 20 kHz. Interestingly, the results for samples treated with high-speed agitation were like those obtained with 20 kHz HIU. Differences between the conditions tested were attributed to the number of bubbles and cavitation events generated. Sonication at 20 kHz visibly produced more cavitation and very small bubbles associated with shear and microstreaming events. Sonication at 40 kHz produced fewer and larger bubbles that did not dissipate easily in the media. This condition also showed lower shear and microstreaming and a lower increase in temperature. Regarding high-speed agitation, no cavitation was generated under this condition. Instead, a much stronger shear was experienced by the sample. Similar effects were observed for the crystal sizes and morphology. That is, 20 kHz sonication and high-speed agitation formed more and smaller crystals than 40 kHz sonication. Differences between 20 kHz and high-speed agitation were observed in the melting behavior and elasticity. Although both induced crystallization and formed more and smaller crystals, the crystalline materials formed were different. A sharper melting peak and higher elasticity were observed for samples sonicated with 20 kHz due to the smaller crystal sizes observed in these samples.[40] Although crystallization induced by shear has been widely studied by the scientific community[41–44] and showed a similar benefit in terms of crystallization, Silva *et al.*[40] showed that crystallization and the physical properties can be improved further by combining shear and cavitation through the generation of a more connected and strong crystalline network.

With these results, the effect of HIU in combination with agitation became an interesting combination to evaluate. Agitation in combination with HIU did not improve the physical properties of an interesterified soybean oil with 33% SFA.[45] Kadamne and Martini[45] showed that continuous agitation without HIU was not able to change the phase behavior of the sample when crystallized under low supercooling. The sample remained in the metastable zone for a long time without any stable nucleation. However, if the sample was crystallized under the same conditions and HIU was applied, stable nucleation was generated and an increase in crystallization rate was observed. Nevertheless, cavitation's effect on forming a stronger crystalline network was more pronounced if no agitation was applied after sonication. Perhaps continued agitation after sonication favored dissolution of the bubbles, annulling the cavitation effect. Furthermore, continuous agitation might also break the interactions between crystals formed due to sonication.

In addition to hardness and elasticity, oil binding capacity is another critical parameter that must be considered in fats with low levels of SFA. The use of sonication on a soybean-based fat showed a significant reduction in oil migration in all supercooling conditions tested (Table 3.2). This result was attributed to secondary nucleation promoted by sonication. In this case, sonication induced the crystallization of low melting point TAGs, since those TAGs were found in the crystalline portion remaining after oil migration of the sonicated samples and not on the non-sonicated ones.[19] This mechanism was not previously observed in fats with high SFA content[20,36] due to the natural crystallization of the high melting point TAGs present in those samples, but became an important event in fats with low levels of saturation. In Figure 3.3, a schematic of cavitation events in both types of fats, with high and low levels of

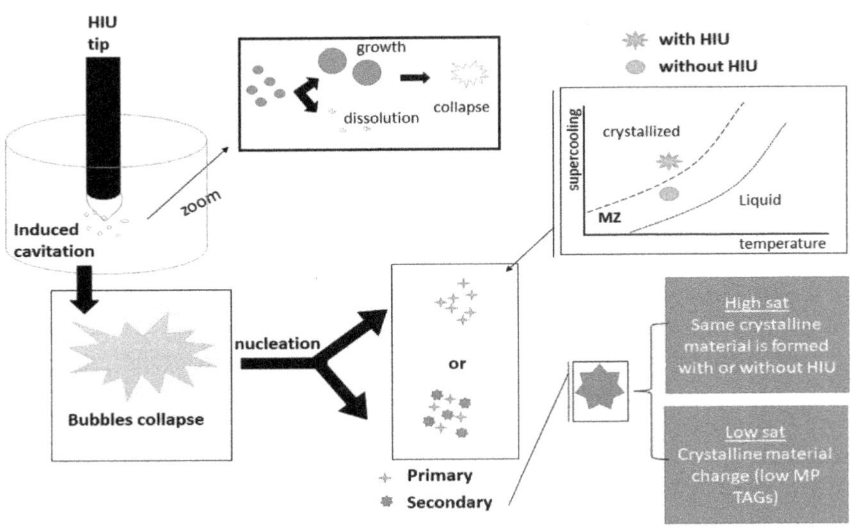

Figure 3.3 Illustration of the mechanism of cavitation in sonocrystallization for a low and high-saturated fat.

SFA, illustrates a comparison of the mechanisms described above. Both types of fats were induced to crystallize and leave the metastable zone, and they also presented a more compact crystalline network formed by small crystals and had low oil migration. Although HIU induced secondary nucleation in both fats, no crystallization of low-melting point TAGs was induced by HIU in fats with high levels of SFA due to their fast crystallization. This was different from the behavior of fats with low levels of SFA, where the non-crystalline phase was present in high amounts and the induction of secondary crystallization caused by HIU was followed by the crystallization of low melting point TAGs.

3.2.2.2 Soft Stearin

Rincón-Cardona *et al.*[46] evaluated the effect of HIU on a sunflower stearin sonocrystallized under very low power (0.5 W). Even with such low power, HIU induced faster crystallization, and changes in crystal morphology were observed. The crystallization rate obtained in sonicated samples crystallized at 21 °C was as high as the one obtained at 17 °C without HIU, indicating that HIU can be used to achieve the same crystallization rate as that obtained at much higher supercooling. However, no changes in elasticity were observed due to sonication which was attributed to the changes in crystal morphology and polymorphism.

The effectiveness of the low power level evaluated by Rincón-Cardona *et al.*[46] was very positive, because high power levels and pulse durations can create excessive heat with a consequent retardation of crystallization as was also observed in a soybean-based fat.[40] However, if the degree of saturation is too low the use of a higher power level is necessary to induce crystallization. This means that the degree of saturation also affects sonication efficiency.[47]

3.2.2.3 Palm-based Fat

Few studies explored the effect of HIU on a low-saturated, palm-based fat.[35,48] Kadamne *et al.*[48] evaluated the effect of different amounts of palmitic acid (20 and 30%) in an interesterified palm-based fat with and without sonication under different supercooling conditions (Table 3.2). The authors observed an increase in crystallization rate and in the amount of crystalline material, followed by a reduction in crystal size for both palmitic acid contents due to sonication. However, the rheological parameters, G' and viscosity, were significantly changed only when at least 30% of palmitic acid was present in the sample. This suggests that crystallization kinetics can be tailored by sonication even in palm fats with low amounts of saturated fatty acids; however, to cause significant changes in the macroscopic physical properties the palm-based fat has to contain at least 30% palmitic acid.[48] A study performed in a palm olein[35] showed a slight increase in the amount of smaller crystals after sonication; however, no significant changes in the crystallization rate, melting behavior, or elasticity were found.

The authors attributed this lack of effect of HIU to the high supercooling used (19 °C), but it could also be an effect of the slow crystallization of palm-based fats.[49] The very slow crystallization rate could not form enough nuclei needed for HIU to have an effect, as was discovered later for a similar palm-based sample reported by da Silva *et al.*[50]

3.2.2.4 Effect of Composition on HIU Efficiency

As briefly described above, the effects of HIU were dependent on the amount of saturation and on the type of fatty acid and TAG present. Kadamne *et al.*[48,51,52] showed that HIU efficiency was affected by fat composition and the type of TAGs. Kadamne and others explored the effect of HIU on interesterified fats with the same amount of saturated fats but with different TAG compositions. Palm-based[48] and stearic-based samples were evaluated.[51] These authors also evaluated the effect of sonication in the same stearic-based sample after removing the main tri-saturated TAG, tristearin (StStSt).[52] They showed that HIU affects the crystallization behavior and rheological properties of fats with low content of saturation by generating small crystals and by inducting the crystallization of certain TAG fractions in all samples. When StStSt was present, the crystallization process was driven by this triacylglycerol due to its high melting point, and sonication of these samples promoted crystallization of low melting point TAGs by incorporating them into the StStSt crystalline network as suggested by the authors. The interesterified samples with palmitic acid naturally did not present the StStSt in high amounts and showed a slower crystallization compared to the stearic sample. In this case HIU was not as effective, mainly when the amount of palmitic was lower (20%). The effect of having a low TAG content with very high melting points such as StStSt was even more pronounced when the interesterified sample without StStSt was evaluated because although sonication improved the properties of the crystalline network formed, this effect was not as strong as when the StStSt was in the system. These results suggest that HIU is more effective in stearic-based samples and that a low TAG content paired with high melting point in the sample makes sonication a more efficient method of forming a stronger crystalline network.

3.3 Effect of HIU Application on the Physicochemical Properties of Oleogels

Section 3.2 described the use of HIU as a versatile technology to change the crystallization process in lipids since several parameters can be accurately controlled such as sonication time, tip size, and power level. The importance of the chemical composition of the sample used in terms of type and content of TAGs was also described in the previous section. Other lipid-based materials can benefit from the use of HIU, including oleogels.

Table 3.3 Studies that focus on the use of high-intensity ultrasound (HIU, 20 kHz) in oleogels.

Gelator type[a]	Oil source	Gelator concentration (wt%)	Mode of application	Duration (s)	Reference
PW	Olive	10	Continuous	120	55
CW/MG/HF	HOSO	6	Continuous	60, 180	54
MG	HOSO	3, 4.5, 6	3 Pulses	10 (ON) 5 (OFF)	56, 65
MG/HF	Rapeseed	6	Pulses or continuous	10, 30 or 10 (ON/OFF)	64

[a]PW: propolis wax, CW: candelilla wax, MG: monoglycerides, HF: fully hydrogenated crambe oil; HOSO: high oleic sunflower oil.

Structuring edible oils by oleogelation is one of the most interesting strategies to produce low-saturated and zero *trans* fats.[53,54] Researchers continue to search for oleogels with tailored physicochemical properties using gelator concentrations as low as possible to decrease the amount of SFA in edible lipids and to obtain appropriate sensory properties.

In the past few years, research performed on the application of HIU in lipids triggered interest in using sonication during the preparation of oleogels.[55–57] This section presents an overview of the main findings reported in the use of HIU as an additional processing tool to change the physicochemical properties of oleogels. A summary of the studies published on this topic is provided in Table 3.3.

Considering that the crystallization process in oleogels is driven by the type and concentration of the gelator,[58] the application of ultrasound in these materials would result in different physicochemical properties depending on the gelator.

Oleogels obtained by structuring edible oils should have functionality similar to fats and a nutritional profile similar to liquid oils. Achieving proper textural properties is one of the major concerns related to the replacement of solid fats. The direct replacement of solid fats with structured oils is challenging because the textural properties of food products are mainly due to saturated fats.[59] Moreover, knowledge of the melting behavior of oleogels is essential to addressing their physical properties and is commonly measured by differential scanning calorimetry (DSC). Other characteristics used to understand the physicochemical behavior of oleogels are their rheological properties, SFC, and microscopic structure. All these properties will be discussed in the sections below.

3.3.1 Binary and Ternary Oleogels

The effect of HIU on the physical properties of oleogels formed by structuring high-oleic sunflower oil (HOSO) with 6 wt% of gelators [2 wt% of monoglycerides (MGs), 2 wt% of candelilla wax (CW), and 2 wt% of a fully hydrogenated crambe oil (HF)] was evaluated by da Silva *et al.*[55] These authors investigated ultrasound performance using two different sized tips (diameter 3.2 or 12.7 mm) when HIU was applied for 1 or 3 min.

Crystal size and shape are the key factors influencing fat functionality.[60] To understand how ultrasound affects crystallization behavior and therefore, the macroscopic properties of oleogels, their microstructure has been studied by polarized light microscopy (PLM). In accordance with the findings previously reported for several lipids systems,[12] da Silva *et al.*[55] found that crystal sizes and the percentage of crystallized area (C_a) were significantly reduced by the use of ultrasound in oleogels from HOSO structured by CW, MGs, and HF. However, crystal morphology was not affected by sonication. The greatest reduction in crystal diameter (D_m) was found for 3 min of HIU application using the 3.2 mm tip (amplitude: 216 μm) or 12.7 mm tip (amplitude: 60 μm) (Figure 3.4).

A reduction in crystal size and changes in the morphology of oil/wax systems as a consequence of using ultrasound were first reported by Jana and Martini.[61] These authors applied HIU for 10 s in the presence of crystals, at a vibration amplitude of 180 μm using a 2 mm probe. They found that crystal morphology changed, evidencing the presence of spherulites in non-sonicated samples, which were not observed in the sonicated ones. It was hypothesized that HIU application breaks down the spherulites, thus

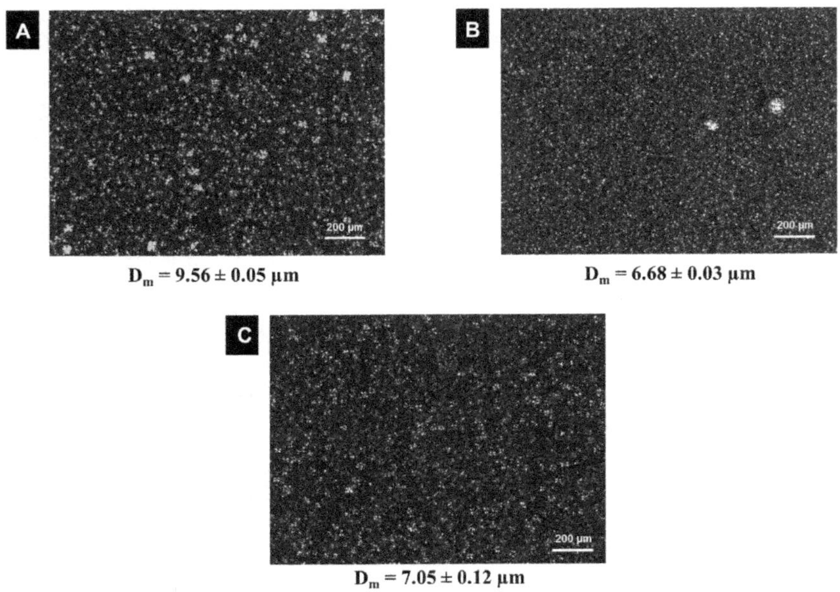

Figure 3.4 Polarized light microscopy images of oleogels from high oleic sunflower oil structured with candelilla wax, monoglycerides, and fully hydrogenated crambe oil, obtained: (A) without high-intensity ultrasound (wo HIU); (B and C) with HIU application for 3 min at 60 and 216 μm, respectively. Effect of HIU application on the mean diameter of crystals (D_m).
Adapted from ref. 55 with permission from John Wiley & Sons, Copyright © 2019 AOCS.

forming small needle-like crystals, which could be responsible for better physical stability and strengthening of the oleogel's structure.[62]

In addition, da Silva *et al.*[55] investigated the influence of different sonication parameters (time, power level, and tip size) on the crystal size and C_a of oleogels from HOSO structured by CW, MGs, and HF. These authors showed that even though the power level also reduced the crystal size, this effect was not as significant as the one observed for sonication time. These findings are in contrast with those previously reported in fats,[6,14,28] where significant differences were observed in crystal size when sonication was used at different power levels. The authors suggested that the oleogelator's high melting point may cause the ultrasound amplitude (or power level) to be ineffective.

Regarding the effect of HIU on the hardness of these materials, no changes or reduction in the hardness due to HIU application were found, except for one sonication condition in which ultrasound was applied for 3 min using a 3.2 mm probe at a vibration amplitude of 216 μm. Under this condition, a significant increment (~18%) of the hardness of these oleogels was observed. Results from this research also showed that HIU application did not change the melting profile of these oleogels. It was found that the shape of the melting thermograms remained unchanged; no changes in ΔH_M melting onset temperature (T_o), or in melting peak temperature (T_p) were observed, suggesting that neither the amount of crystallized material in the sample nor the level of molecular organization of the crystalline network was significantly affected by HIU application. This result contrasts with earlier observations in oil/wax systems obtained by Jana and Martini,[61] who reported that the use of HIU in mixtures of 0.5 and 1.5 wt% of beeswax and different oils generated a less fractionated network, where stronger interactions between the gelator molecules were noticed. Within this context, differences among the chemical compositions of the gelators and the concentration used could lead to different trends in modifying the crystallization process by applying HIU and thus, in the melting thermograms.

Concerning the elastic nature of these oleogels, frequently quantified using the elastic modulus (G'),[63] a reduction and/or no changes in G' due to sonication was observed in these samples. In addition, the SFC is another parameter that must be analyzed in oleogels. The macroscopic properties of a fat crystal network are influenced by SFC, the microstructure (crystal size and shape) as well as by the polymorphism, and by chemical composition. No changes were observed in the SFC that were due to HIU application in these oleogel systems.

The main property that determines the functionality and applicability of oleogels is the oil binding capacity (OBC), which represents the ability of the oleogel matrix to physically entrap liquid oil.[64] A decrease or no change in OBC due to sonication was found in oleogels formed by structuring HOSO with CW/MGs/HF. da Silva *et al.*[55] explained that the decrease in crystal size achieved due to sonication was not enough to change the OBC of the network. This correlated with the reported hardness, as softer samples resulted in lower OBC. Moreover, the authors found that a sonication time of 3 min promoted the formation of smaller crystals in oleogels from MGs, CW, and

HF and thus increased their OBC, in comparison with the lower sonication (1 min) time. However, the maximum time tested in this work (3 min) was not enough to improve the oil retention of the samples in comparison with that of the non-sonicated oleogels as already discussed. Furthermore, they showed that OBC was reduced during 7 days of storage at 25 °C, and this decrease was greater in sonicated samples compared with the non-sonicated ones. This finding was related to the presence of spherulites in the oleogel's crystalline network, which are often associated with phase separation in mixtures of waxes and vegetable oils,[61] suggesting that the mechanical stability of the sonicated samples might be significantly affected over time.

Silva and Danthine[65] recently evaluated a system similar to the one described above but without the CW. In this case, the total amount of MGs/HF was kept as 6 wt% but its relative proportions were changed (0 : 6, 1 : 5, 2 : 4, 3 : 3, 4 : 2, 5 : 1, 6 : 0, MF : HF, respectively). HIU (20 kHz) was applied using 57 W and a 12.7 mm tip for a shorter time than in the previous study discussed [10 s, 30 s, or 10 s in three pulses (10sON/10sOFF)], and instead of applying HIU before nucleation, sonication was applied during cooling at the temperature where the first crystals appeared. This temperature was different for each blend, and as the HF content increased the temperature decreased. This was not expected since HF and MGs have close melting points (61 °C and 66.5 °C for HF and MGs, respectively). These results indicated that MGs/HF showed some incompatibility, and that HF solubility was increased in the presence of MGs. This behavior led to the formation of soft (MGs 3 wt% or higher) or liquid gels (MGs lower than 3 wt%). When HIU was applied in these systems, cavitation induced oleogelation and formed more crystals that entrapped the liquid phase resulting in a reduction of oil loss and improved viscosity, hardness, and elastics modulus of the blends. The use of only 10 s of sonication was the best condition and responsible for a 3-fold increase in hardness. By observing the pseudo-phase diagrams of the blends, the authors explained that sonication inhibited the formation of the eutectic mixture and improved the synergism between the gelators. Sonication was also able to reduce the dissolution effect of the HF in rapeseed oil. These are interesting results that showed the effectiveness of HIU in oleogels to improve the synergism of different gelators, which might result in reducing the amount of gelator used in the future. Surprisingly, even though HIU improved the physical properties of HF/MGs oleogels, it showed different effects in the HF and MGs alone. HF oleogels were very liquid and sonication did not improve their physical properties; while in the MGs oleogels, sonication worsened their physical properties.[65]

3.3.2 Mono-component Oleogels

3.3.2.1 Oleogels Structured by Propolis Wax

The effects of applying ultrasound with a 13 mm probe at different power levels (100, 200, and 300 W) and times (30–120 s) on oleogels obtained by structuring olive oil with 10 wt% of a propolis wax were studied by Sharifi *et al.*[56] These authors employed response surface methodology to maximize the hardness and OBC of these materials. They evaluated different ultrasonic power levels (100, 200, and 300 W), ultrasonic times (30–120 s), and cooling temperatures (5, 15, and 25 °C to achieve different cooling rates −1.35, 1.07, and 0.62 °C per min, respectively). Samples were analyzed after storage for 24 h. The optimized conditions for oleogel preparation were found by applying HIU for 120 s at 300 W after cooling samples at 5 °C. These sonicated oleogels were compared to non-sonicated ones. Different results regarding the effect of ultrasound on the physical properties of these materials were found compared to the previously reported results in oleogels from HOSO structured by CW, MGs, and HF. Optimized sonicated oleogels from olive oil and propolis wax showed a significant increase (~50%) in their hardness due to ultrasound application.

In contrast to what was found in oleogels from HOSO and MGs/CW/HF (Section 3.3.1), a significant increase in the melting enthalpy (ΔH_M) of oleogels from olive oil was reported as a consequence of sonication, which was attributed to more solid material. These different trends could be found according to the source of oleogels. While CW, MGs, and HF have very high melting points (~70 °C), propolis wax has a lower melting point (~60 °C). As previously discussed in this chapter, in general HIU is more efficient for fats that crystallize slowly and that have lower melting points.

With respect to the effect of HIU on the chemical composition of oleogels from olive oil and propolis wax, it was demonstrated that sonication did not modify their fatty acid profile, in accordance with the discussion above regarding fats (Section 3.2.2.4).

3.3.2.2 Oleogels Structured by Saturated MGs

The effects of MGs concentration (3, 4.5, and 6 wt%), cooling rate (0.1 and 10 °C per min), and HIU application on the physical properties of oleogels from saturated MGs and HOSO were studied by Giacomozzi *et al.*[57] Mixtures were stirred (200 rpm) and cooled from 80 to 25 °C. At the onset of crystallization, agitation was stopped and HIU (20 kHz) was applied for 30 s using 3 pulses (10 s ON/5 s OFF) and a tip with a 3.2 mm diameter (amplitude of vibration = 216 μm), resulting in a power level of 96 W. After that, the samples were statically cooled to 25 °C and stored for 24 h at 25 °C. It was found that HIU application led to a reduction in crystal size in all MG-based oleogels, similar to what was found in oleogels from HOSO and MGs/CW/HF (Section 3.3.1). The application of HIU did not allow materials harder than those prepared without the use of sonication to be obtained. Even though oleogels from MGs and HOSO showed a significant increase in G' because of ultrasound application immediately after the gelation process was completed, this positive effect was lost after 24 h of storage.

Similar to the results found in oleogels from HOSO structured by CW, MGs, and HF, no changes were reported in the melting profile of oleogels obtained with the same liquid oil structured by saturated MGs at a different gelator concentration due to HIU application. Another similarity between the effects of sonication in oleogels from CW/MGs/HF and from saturated MGs was found regarding the OBC results. Giacomozzi *et al.*[57] showed no improvements in the OBC by using ultrasound in oleogels from HOSO and MGs at 3 wt%, but a slight increase in this parameter was observed by working with higher gelator concentrations (4.5 or 6 wt%) only in the case of the slow cooling rate. In contrast to these findings, better results were found by Jana and Martini,[61] who demonstrated that HIU delayed phase separation in beeswax/oil systems from beeswax and canola, soybean, olive, and safflower oils, and thus, allowed improvement in the sample's ability to keep oil entrapped in its matrix. Moreover, they noticed that this effect was highly influenced by the SFA content of the oil used. A higher SFA content in oils led to an improvement in the physical stability of the crystalline network.

Based on the previously mentioned results, Giacomozzi *et al.*[66] proposed an alternative procedure for the preparation of these materials with the aim of reaching more highly effective sonication over the physical properties of oleogels. The authors hypothesized that in their first study, crystallization occurred too fast (10 °C per min) or too slow (0.1 °C per min) hindering the effect of sonication.[57,65] Similar results regarding the lack of effect on fat samples that crystallize slowly or quickly were reported by Martini *et al.*,[18] and Rincón-Cardona *et al.*[46] In this second work, these authors aimed to achieve a more controlled crystallization process by cooling the samples under agitation at a specific cooling rate (0.1 or 10 °C per min) to the crystallization temperature (T_c), where an intermediate degree of crystallization was observed.[66] HIU was used with the same conditions as in their previous work[57] but in a different moment during the crystallization process: when T_c was reached (not when the first crystals were observed as it was performed before). After that, the samples continued the cooling process under static conditions without controlling the cooling rate. This new methodology allowed the formation of more crystals before HIU application and promoted faster crystallization after sonication, especially in slowly cooled samples. Under these alternative conditions, the effect of applying HIU to reduce crystal size was greater in oleogels formed with the lowest MGs concentrations, an interesting result based on the need to minimize gelator concentrations in oleogel formulations. From the PLM analysis, it was shown that oleogels obtained without the use of ultrasound appeared to be more crowded, showing spherulitic clusters formed through the aggregation of needle-like crystals (Figure 3.5A, C, E, and G), in accordance with previously described observations for mixtures from waxes and vegetable oils.[61]

Furthermore, it was found that oleogels obtained with 3 wt% MGs without HIU application did not show a well-structured network and flowed after 24 h of storage at 25 °C, whereas sonicated samples presented a solid-like consistency (Figure 3.5B and D). Similarly, sonication contributes to obtaining a visibly stronger network in oleogels formulated with 6 wt% of MGs (Figure 3.5F and H).

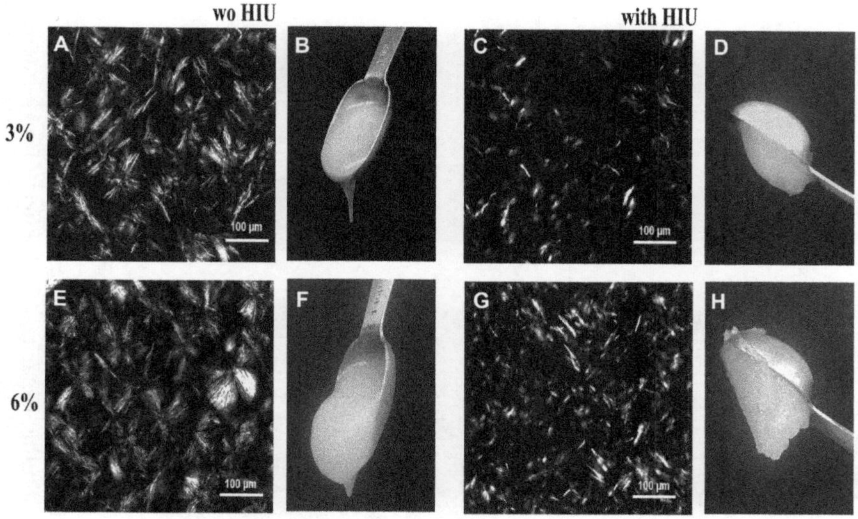

Figure 3.5 Polarized light microscopy images and visual appearance of oleogels obtained without high-intensity ultrasound (wo HIU) and with HIU: (A–D) 3 wt% of monoglycerides (MGs) and high oleic sunflower oil (HOSO); (E–H) 6 wt% of MGs and HOSO.
Adapted from ref. 66 with permission from Elsevier, Copyright 2020.

Another example of the effect of processing on HIU efficiency was found in the hardness of MG-based oleogels. By modifying their first approach, Giacomozzi *et al.*[66] demonstrated that HIU application could be used as a potential tool to generate harder materials even at lower gelator concentrations (3 wt%) (Figure 3.6A and B). Moreover, it was found that sonication in samples cooled to 10 °C per min was not as effective as sonication in samples cooled at 0.1 °C per min, probably because of the differences in crystallization behavior of MGs by changing the cooling rate. In slowly cooled samples, the gelator molecules might have more time to interact and thus form more crystals, compared with the fast-cooled samples.[66]

The most striking observation from these findings is that the same level of hardness can be obtained in oleogels with 3 wt% MGs using HIU compared to an oleogel with 6 wt% MGs without HIU at a rate of 0.1 °C per min (Figure 3.6A). Considering that the main goal of using oleogels in food formulations is to reduce the level of saturated fats while keeping the functional properties of a solid fat,[54] this is an important finding since similar physical properties can be obtained by using HIU in oleogels with half the amount of MG. Another interesting aspect of these results is that a harder structure was found in sonicated oleogels with a lower gelator concentration compared to the non-sonicated ones (as it can be seen from Figure 3.6A by lowering the concentration from 6 to 4.5 wt%).

The general agreement from all studies performed in oleogels until now was that HIU modified the gelator crystallization process during oleogel formation, generating smaller crystals which may have more contacts and

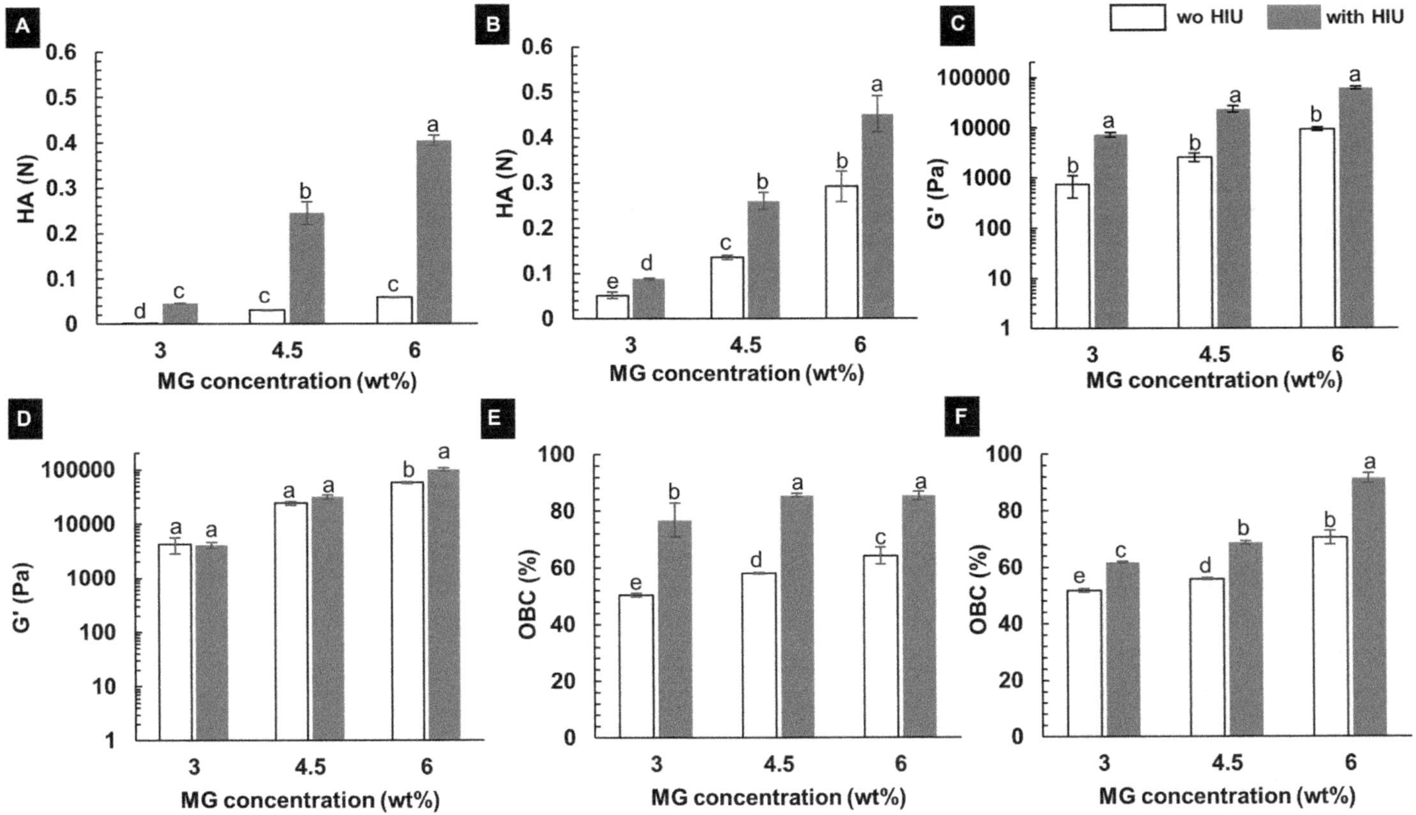

Figure 3.6 Physical properties of oleogels formulated with monoglycerides (MGs) and high oleic sunflower oil and stored for 24 h at 5 °C: (A and B) Hardness (HA), (C and D) elastic modulus (G'), and (E and F) oil binding capacity (OBC). Effect of high-intensity ultrasound (HIU) application: without HIU (wo HIU) and with HIU at a cooling rate of 0.1 °C per min (left) and 10 °C per min (right). Values with the same lower-case letter are not significantly different ($p > 0.05$).
Reproduced from ref. 66 with permission from Elsevier, Copyright 2020.

thus contribute to a harder structure compared to the network obtained without sonication, which is composed of larger crystals.

Moreover, contrary to the results obtained in previous studies where no or a negative effect of HIU was observed over the G' of MG-based oleogels,[57,65] Giacomozzi *et al.*[66] found a way that allowed the elastic properties of these materials to be improved by changing the moment at which sonication was applied during the preparation of oleogels, as it was explained before. These authors found a significant improvement in the elastic properties of these materials by applying HIU mostly for samples cooled at 0.1 °C per min in comparison with a faster cooling rate (10 °C per min) (Figure 3.6C and D). G' values showed approximately 10-fold increase after HIU was applied in oleogels formulated with 3 and 4.5 wt% of MGs, and close to a 7-fold rise in samples obtained with the highest MG concentration. Furthermore, higher crossover points were found in the sonicated samples, indicating that HIU application results in a stronger gel network[67] under this sonication condition.

On the other hand, Giacomozzi *et al.*[66] found that HIU helped to create a crystalline network with a better ability to entrap liquid oil in comparison with non-sonicated samples (Figure 3.6E and F). They demonstrated that the cooling rate established throughout the oleogel formation could be combined with HIU to produce different effects. In this work, improvements of the OBC by using HIU in the slowly cooled samples were higher than those of the rapidly cooled ones at the same gelator concentration. The authors suggested that a synergism exists between a low cooling rate and this condition of HIU application (10 s ON-5 s OFF, 3 pulses) to create gel networks that are more effective at structuring the liquid oil. The effectiveness of sonication toward the improvement of the OBC of oleogels becomes different depending not only on the ultrasound conditions (power levels, tip size, the time at which HIU is used, duration time, and application mode) but also on the composition of the samples and their crystallization behavior. The differences between results from da Silva *et al.*[55] and Giacomozzi *et al.*[66] provide a good example of that. Both studies evaluated the effect of sonication in oleogels, the former being focused in a mixture of three gelators (CW, MGs, and HF) and the second only in MGs. Moreover, in the first-mentioned work, oleogels were formed by quenching the sample and without controlling the cooling rate, in contrast with the second, where two cooling rates (0.1 and 10 °C per min) were studied. Giacomozzi *et al.*[66] applied HIU as three short pulses (10 s), in contrast with the single one tested in the previously mentioned work. The aim of using shorter pulses instead of only one is to allow enough sonication effect while minimizing the temperature increase associated with the process that could melt the crystals,[12] which is particularly important in the case of MGs as observed by da Silva and Danthine.[65] The positive effect of HIU on the OBC of oleogels formulated with 3 wt% of MGs reported by Giacomozzi *et al.*[66] was obtained when sonication was applied in three pulses in combination with a low cooling rate (0.1 °C per min). This might be related to the fast crystallization kinetics of MGs. A low cooling rate might slow down the gelator crystallization process before ultrasound waves are applied in comparison to a faster one and therefore, a greater effect of sonication would be achieved. This finding is in accordance with those previously reported in fats[48,51] where a higher efficiency of ultrasound treatment by either working at a low driving force, or by using slow cooling rates was demonstrated.

It seems that HIU application during the cooling process at a rate of 0.1 °C per min improves the packing of gelator crystals, resulting in a stronger network structure with no differences in the oil binding capacity when half the MG amount was used compared to non-sonicated oleogels. Additionally, higher oil retention was obtained at 0.1 °C per min with 4.5 wt% of MGs with HIU compared to oleogels formulated with 6 wt% without HIU. These findings highlight the possibility of tailoring oleogel properties by combining HIU with cooling rate to reduce the amount of monoglycerides in the oleogels (and hence, the amount of saturated fat used in the formulation) while maintaining, or even improving, the oil binding capacity of the network. These results correlate well with those obtained from microscopy and textural analysis and can be explained by the formation of comparatively smaller crystals in sonicated samples than those generated at the non-sonicated ones. As expected, as the size of the crystals decreases, more surface area is available to entrap the surrounding oil compared to a network composed of crystals of a larger size.[64]

Even though the SFC of MG-based oleogels was not affected by ultrasound application, a stronger network was developed as a consequence of sonication (higher G'). That is, an increase in the interactions that form the network could increase the elasticity without changing the amount of crystallized material. G' gives information about crystal–crystal interactions, which could be modified by the crystal size. The small crystals present in the crystalline network of sonicated MG-based oleogels may provide more surface area for crystal–crystal interactions compared to the larger crystals found in non-sonicated samples. Thus, samples with the same SFC can exhibit quite different elasticities. This type of behavior has been previously reported in sonicated TAGs.[15,19]

Concerning the effect of ultrasound on the fatty acid composition of oleogels from saturated MGs and HOSO, it was demonstrated that the use of HIU waves in the processing of oleogels did not lead to chemical modification of their fatty acid composition, in agreement with a previously reported study by Sharifi *et al.*[56] However, these changes may depend on the HIU conditions selected such as power level, duration, and mode (continuous *vs.* pulses). Based on these considerations, further research should be performed on this topic to explore how different sonication variables affect the chemical modification of oleogels.

3.4 Effect of HIU on Lipid Oxidation

Sonication is usually associated with temperature increases as previously described. This increase in temperature and cavitation events generates free radicals in aqueous systems as previously reported by various authors.[68] Therefore, concerns about lipid oxidation have been reported regarding sonocrystallized fats. This oxidation would be especially significant in oleogels and fats with low levels of SFA.

Oxidation due to sonication was evaluated in lipid systems. First, sunflower oil was sonicated using 20 kHz HIU at 150 W for 0.5–30 min. The authors reported a visual change in the sample and a metallic oxidized smell after 1 min of sonication, nevertheless no changes in peroxide value and free fatty acids due to sonication were found. The authors found that limonene and hexanal (products of secondary oxidation) could be related to the visual changes observed in the oil after sonication, and these visual changes were detected by sensory panelists.[69] Olive, sunflower, and sesame oils, and tallow olein were also evaluated according to their physicochemical parameters due to HIU application (24 kHz, 400 W, 5 min, 25–50%, and 100% amplitude) by Hosseini *et al.*[70] These authors observed an increase in peroxide value and conjugated diene, a reduction in the oil stability index measured by Rancimat, and a decrease in beta-carotene content for all oils. These effects were more pronounced as higher power levels were used.[70]

The studies reported above applied HIU for long periods of time and do not represent conditions commonly used to crystallize lipids (Tables 3.1 and 3.2). Therefore, the degree of oxidation in soybean oil and interesterified soybean oil was evaluated in sonicated and non-sonicated samples according to common processing conditions reported in the literature. Samples were sonicated (20 kHz, 101 W, 10 s) and stored at 25 and 40 °C and the peroxide value was measured as a function of storage time. No difference in peroxide value was found for either sample with or without HIU during 105 days at 25 °C (peroxide lower than 2 mEq kg^{-1}) which was attributed to the short time of sonication. When samples were transferred to 40 °C storage, sonicated and non-sonicated samples were not recommended for consumption after 14 days since samples had peroxide values higher than 10 mEq kg^{-1}. These high peroxide values were due to long storage at a high temperature.[71] The same interesterified soybean oil sample was evaluated by da Silva and Martini[8] using 5 s pulses of 11 or 57 W and no changes in peroxide value due to sonication were observed. Gregersen *et al.*[17] also evaluated oxidation in mixtures of anhydrous milk fat and rapeseed oil and they report that secondary products of oxidation were not found even after sonication for 10 min using 40 W.

In sonicated oleogels, the concern becomes even more worrisome since the preparation itself already consists of mixing oil and gelator molecules under continuous agitation at high temperatures ($\geq$80 °C), which may play a negative role on the quality of these materials.[72] In addition, as was shown in the previous sections, an increase in sample temperature was usually detected during sonication due to cavitation events. In fact, da Silva *et al.*[55] observed an extreme overheating with temperatures above 100 °C after 3 min of sonication at an amplitude of vibration of 108 μm (~93 W) during the formation of oleogels. Giacomozzi *et al.*[66] evaluated HIU application using short pulses of 10 s at 96 W, and shorter sonication times which lead to a lower temperature increase associated with the use of ultrasound. Therefore, we can hypothesize that oxidation might occur even under these moderate conditions. Nevertheless, literature regarding the effect of ultrasound on the oxidative stability of oleogels is scarce. In fact, to the best of our knowledge, the only contribution on this topic has been the study of Sharifi *et al.*[56] These authors found an increase (~28%) in the peroxide value of wax oleogels due to HIU application for 2 min at a power level of 300 W, in accordance with the results obtained in some edible oils.[69,70]

Overall, within the data available we can conclude that for fats with high or low levels of SFA, lipid oxidation due to sonication is not a problem if the sonication is used for a short time (a few seconds) as is used to induce fat crystallization (Table 3.2). If oils that have a higher amount of unsaturation are sonicated for longer times and at high-power levels, oxidation might occur due to the formation of free radicals and/or to the increase in temperature associated with sonication conditions. For this reason, the use of HIU on oleogels has to be carefully monitored and designed.

3.5 Scale Up to a Pilot Plant

As described in the previous sections, the effect of HIU on a lab scale has been well explored, mostly in batch systems. However, processing of fats at an industrial scale occurs under continuous flow and with agitation. For HIU to be used in an industrial setting, sonication must be viable, and cavitation should be able to occur under continuous flow when a much higher amount of sample is processed and in combination with other parameters such as shaved agitation and temperature.

Ye and Martini[15] and Gregersen *et al.*[17] have taken the first step in evaluating the effect of HIU in a flow system. These authors observed a reduction in induction time and acceleration in the crystallization rate, with more nucleation, smaller crystals, and a more connected and organized crystalline network when samples were processed with HIU. In addition, Ye and Martini[15] observed a higher elasticity and Gregersen *et al.*[17] observed no change in the hardness of sonicated samples. Moreover Gregersen *et al.*[17] emphasized the dependency of sonication efficiency due to the degree of saturation and, in this case, a higher amount of oil in the blend showed more effects due to sonication. However, neither of the studies mentioned above evaluated the effect of HIU under shear in a continuous system. Based on the previous discussion about ultrasound efficiency in the presence of agitation,[45] studies that evaluate the effect of sonication in a flow cell under shear are needed.

Thus, in order to better represent the use of HIU in an industrial setting the effect of sonication was further evaluated in a scraped surface heat exchanger (SSHE).[8,50,73–75] The fats evaluated and the conditions tested in the SSHE can be found in Table 3.4.

Table 3.4 Studies of sonication effect on a scraped surface heat exchanger.

Lipid material	What was evaluated	HIU conditions	Crystallization conditions	Main findings	References
Soybean base	HIU conditions	Amplitude 10–80% Time (continuous, 5 s P or 10 s P) HIU-3[a]	32 °C 11 L h^{-1}[b] 344/208 rpm[c]	↑ Physical properties HIU-3 using 50% 5 s P	72
	HIU position	HIU-0, HIU-1, HIU-2 and HIU-3 20 or 50% Amplitude Time 5 s P	32 °C 11 L h^{-1} 344/208 rpm	↑ Physical properties HIU-3 using 50% 5 s P	8
	SSHE conditions and HIU positions	HIU-1, HIU-2, and HIU-3 50%/5 s P	20 or 25 °C 11 L h^{-1} 344/208 rpm or 185/81 rpm	↑ Physical properties HIU-2 and HIU-3 using low agitation on both T_c.	73
Palm based	T_c batch *versus* SSHE	HIU-3 50% 10 s P	26.28.30 and 32 °C 11 L h^{-1} 344/208 rpm	In batch HIU improved physical properties on all T_c. In SSHE only at 26 °C. Higher properties were found for SSHE 26 °C	49
	HIU condition and position	20, 50, and 80% amplitude Time (continuous, 5 s P or 10 s P) HIU-1, HIU-2 and HIU-3	26 °C 11 L h^{-1} 344/208 rpm	↑ Physical properties HIU-3 using 50% 10 s P	74

[a]HIU positions HIU-0, HIU-1, HIU-2, and HIU-3 can be better visualized in Figure 4.7.
[b]barrels and pin worker agitation speed.
[c]flow rate.

Studies showed that HIU can be attached to a flow cell in different positions throughout a SSHE[8] to change the crystallization behavior of the fat. In addition, HIU was combined with other crystallization parameters in the SSHE using different HIU conditions such as power level and pulse duration.[73] A schematic of the system used can be seen in Figure 3.7.

The HIU-SSHE system showed effects like the ones observed in batch such as the generation of small crystals (Figure 3.2) and more organized and connected crystalline networks, lower oil migration, greater hardness and elasticity due to sonication. However, these results were dependent on many factors: (1) *Sonication conditions*, power level, and pulse duration. HIU generated excessive heat when applied at high power levels and melted the growing nuclei, damaging the crystal network instead of improving it. However, the heat promoted by HIU in the SSHE dissipates quickly due to the continuous flow and the cooling efficiency of the shaved wall.[73,74] (2) *Step of crystallization where HIU is applied*. As previously discussed for samples crystallized in batch, if HIU is applied in the presence of few crystals HIU is more efficient at inducing crystallization by promoting secondary nucleation. In the SSHE this was also true, and the HIU/flow cell had to be placed within the SSHE in a step where the nucleation has already started. Depending on specific crystallization conditions and sample composition this step might change.[74] (3) *Shaved agitation*. In addition to applying HIU where nucleation events are occurring, the rate of shaved agitation promoted by the SSHE can also influence the efficiency of cavitation. The positive effect of shaved agitation on inducing nucleation and on reducing crystal size is well known,[43,76] nonetheless, if agitation is applied in excess or at an inappropriate moment during the crystallization process it can favor dissolution of the bubbles generated during sonication reducing or annulling the effect of HIU, or can even break the crystalline structure already induced by HIU[50,74] (4) *Supercooling*. As previously discussed, supercooling has a significant effect on sonication when studied in batch. Supercooling also showed a significant effect in samples crystallized in the SSHE. The combination of appropriate supercooling with HIU position and optimized SSHE conditions showed the best improvement on physical properties of the fat.[50,74] It is important to note that contrary to the batch systems where no effect of HIU was observed under high supercooling, there is an effect of HIU on all supercooling if HIU is in the appropriate position (applied close to T_c).[74] For example, the effect of HIU in a soybean-based fat processed in a SSHE was optimized using a specific supercooling condition (32 °C, 50% amplitude in 5 s pulses at the HIU-3 position) which resulted in the best improvement of physical properties. The use of higher supercooling (25 or 20 °C) did not result in an improvement in physical properties when samples were processed with or without HIU.[8,74] (5) *Fat composition* also affects HIU efficiency in the SSHE. A soybean-based fat with a low amount of SFA was able to crystallize in the SSHE with HIU using low supercooling ($T_c = 32$ °C).[73] However, a palm-based sample with the same amount of SFA (33%) was able to crystallize with HIU only at higher supercooling

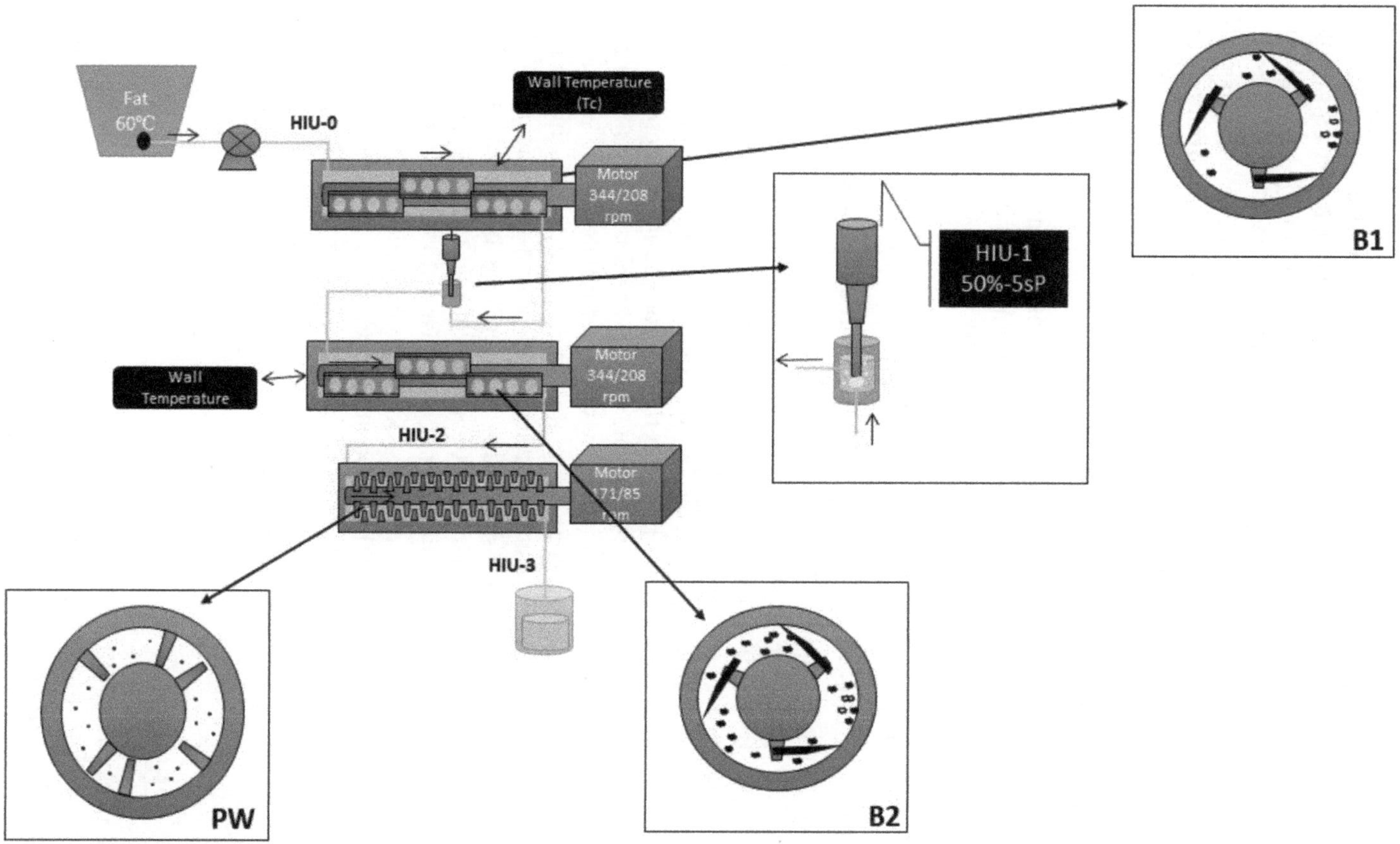

Figure 3.7 Crystallization system of high-intensity ultrasound (HIU) in line on a continuous pilot crystallization (scraped surface heat exchanger – SSHE). Where, B1 is barrel 1, B2 is barrel 2, and PW is the pin worker, and HIU-0, HIU-1, HIU-2, and HIU-3 are the different steps where HIU was applied.

($T_c = 26$ °C).[50] Moreover, the increase in supercooling ($T_c = 20$ °C) in the soybean-based fat in combination with HIU did not improve the physical properties as much as when the sample was crystallized at lower supercooling[74] (results shown in Figure 3.8). However, in the palm-based sample the increase in supercooling ($T_c = 20$ °C) in combination with HIU did improve the physical properties of the fat (data not published).

In this context, the scale up of the HIU followed the same effects as observed in a batch system (100 g), and the benefits of sonocrystallization can be obtained in line and on a large scale (20 L h^{-1}).

3.6 Application

Few studies have been performed on the use of sonicated fats in final products to evaluate if the effects of HIU on fat crystallization and on physical properties will stand in the final product.

3.6.1 Baked Products

Sonicated and non-sonicated soybean-based shortening with low levels of SFA were used to produce baked products such as cakes, cookies, and pie crust.[77] The quality of the cake was not affected by the sonication process, but the cookie and pie crust were. Taller cookies with lower spread were obtained in cookies formulated with sonicated shortening. These changes made the product more acceptable by consumers in terms of overall and textural acceptance, and more importantly did not change its appearance. Similar results were observed for the pie crusts where pie crusts were obtained that had higher break strength when formulated with sonicated shortening. However, appearance acceptability was reduced even though the overall and texture characteristics remained the same for consumers. In another study, a commercial shortening was replaced by a sonicated rice bran wax oleogel in pie crusts. Both treatments tested (1 or 2.5 min sonication) showed significantly similar force and work to break a pie crust, indicating similar amounts of flakiness in the final products. But the sonication for only 1 min showed the most similar product compared to a non-sonicated shortening. The hypothesis was that HIU improved the crystal structure of the oleogel and, as a consequence, the overall ability of the oleogel to withstand the mechanical agitation of preparing pie crust dough.[78]

3.6.2 Chocolate

The application of HIU in chocolate was evaluated to replace the tempering process. Authors found that the use of 9 s of sonication produced the desired stable polymorphic form for cocoa butter (β_V) with a melting profile ($T_p = 34.4$ °C) and hardness like the one obtained through a traditional tempering process that took 40 min. They also evaluated the sensory profile

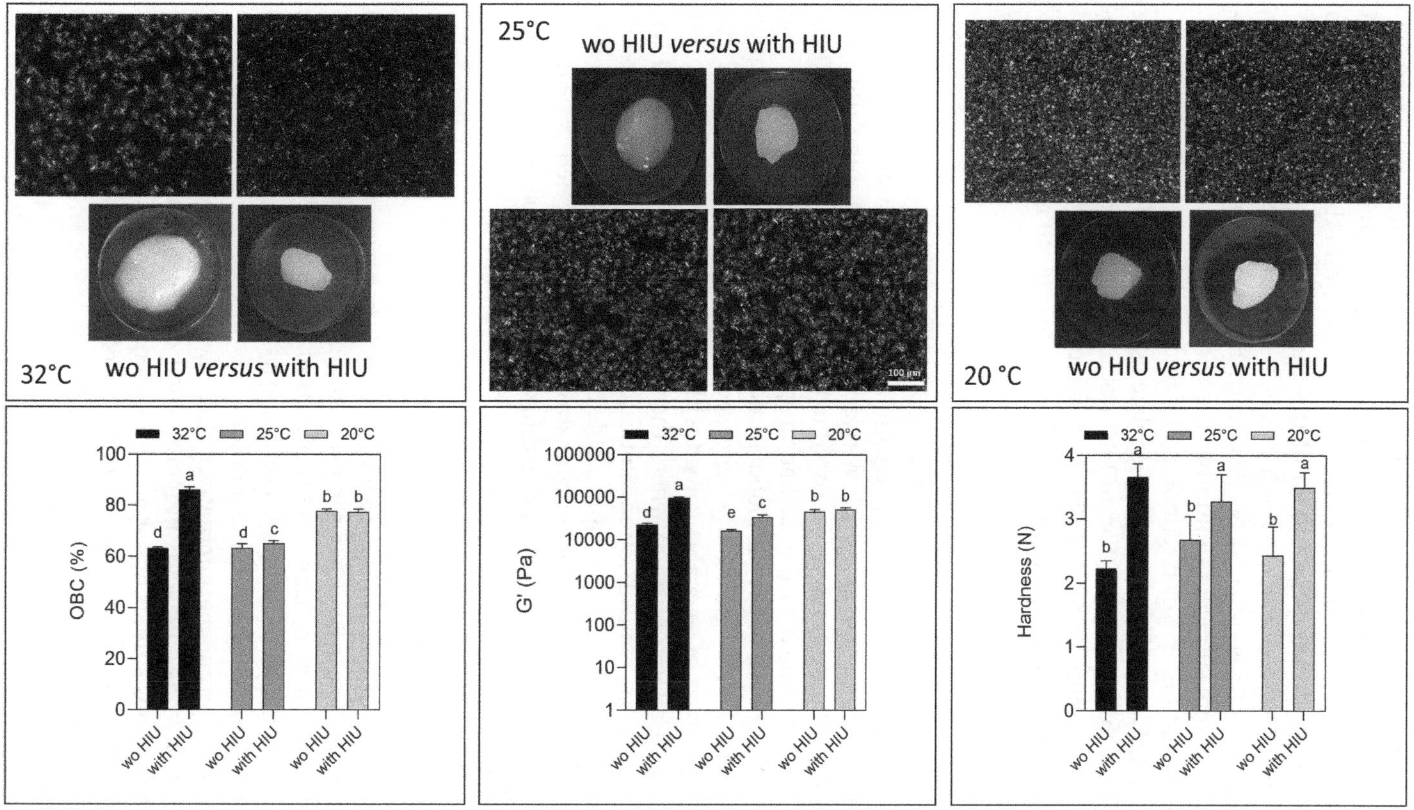

Figure 3.8 Effect of high-intensity ultrasound in a soybean-based fat under different supercoolings ($T_c = 20$, 25, and 32 °C) in a scraped surface heat exchanger. Effect on the microstructure, visual appearance, oil binding capacity (OBC), viscoelasticity (G'), and hardness after storage for 48 h at 25 °C.

of the product and a "metallic" off-flavor was detected only on over-sonicated samples (>15 s).[79] In this work, HIU showed the same fat bloom stability as a traditionally tempered chocolate, and both conditions were not resistant to temperature oscillation. Further studies from this group focused on the heat resistance to fat bloom of the chocolate and found that sonication by itself is not enough to produce a heat resistant chocolate. However, a combination of fractionated fat (5%) with higher melting point and sonication (9 s) were sufficient to improve the heat resistance of the chocolate. The combination of 5% of fractionated fat and sonication makes the chocolate more resistant than the same chocolate using the traditional tempering process.[80] In this case, the effect of HIU not only improved the chocolate's resistance to temperature oscillations, which is important in tropical countries, but also reduced the production process by 40 min.

3.6.3 Emulsions

HIU was recently tested in cream during the processing of butter. The physical properties of sonicated cream and butter obtained using the sonicated cream were evaluated.[81] A relatively shorter churning time was observed for sonicated cream compared to the non-sonicated sample. These results were attributed to the cavitation generated during sonication that reduced the strength of the fat globular membrane. In addition, sonication of the cream for 10 s significantly increased the hardness of the butter. The physical properties of vegetable spread with 5 and 25% of water in oil were also evaluated after sonication where the sonication process replaced traditional ultraturrax emulsification. Sonication in this case made the spreads with higher amounts of water more stable even after temperature oscillation due to smaller and more stable water droplets. The sonicated spread was also harder, more viscoelastic, and had lower oil loss compared to the non-sonicated ones.[55]

3.7 Future Trends

The effect of sonocrystallization in lipid materials has been well studied on a lab batch scale. Recently, it has been shown that it is possible to add HIU in line in a SSHE for a soybean- and palm-based fat. Further studies in other lipid matrices important for the food industry should be explored such as AMF, lauric fats, cocoa butter, and others.

Oleogels are promising alternatives that have been developed to replace saturated and *trans* fatty acids in foods. Their combination with HIU as reported in previous studies has shown improvement in physical properties of those gels. For this reason, more studies on oleogel systems should be explored. Further research is still necessary for this topic since many factors have to be considered when applying ultrasound in oleogel production, such as storage time and temperature and chemical composition. Moreover, it would be interesting to explore how the functionality and sensory properties of oleogels can be improved further by combining the use of various gelators, including healthier molecules, and ultrasound.

Fewer data are available related to the use of sonicated fats in lipid-based applications to evaluate if in fact the improvement in physical properties observed due to sonication of the fat would remain in the final product. Moreover, the effects of this technology in foods should also be evaluated according to its shelf-life and sensory acceptance.

Some studies have been performed to understand the cavitation events that occur during lipid sonocrystallization,[32,33,39,82] but their discussion falls outside the scope of this chapter. More information on this topic is needed to fully comprehend the physical events associated with cavitation in lipid systems and provide molecular mechanisms that explain changes in physical properties caused by sonication.

Abbreviations

AMF	anhydrous milk fat C_a: crystallized area
D_m	diameter mean
CW	candelilla wax
DSC	differential scanning calorimetry
FDA	Food and Drug Administration
GRAS	Generally Recognized as safe
G'	elastic modulus
HIU	high-intensity ultrasound
HF	fully hydrogenated crambe oil
HOSO	high oleic sunflower oil
ΔH_M	melting enthalpy
LaLaLa	trilaurin
MGs	monoglycerides
OBC	oil binding capacity
PLM	polarized light microscopy
PPP	tripalmitin
SFA	saturated fatty acids
SFC	solid fat content
SSHE	scraped surface heating exchange
StStSt	tristearin
TAGs	triacylglycerols
WHO	World Health Organization

Acknowledgements

This book chapter was approved by the Utah Agricultural Experiment Station as paper Number 9424. The authors would like to acknowledge the support provided by Utah State University, and Consejo Nacional de Investigaciones Científicas y Técnicas (CONICET, Argentina).

References

1. S. Martini, *Sonocrystallization of Fats*, Springer, London, 2013, First.
2. S. Martini, C. Bertoli, M. L. Herrera, I. Neeson and A. Marangoni, *J. Am. Oil Chem. Soc.*, 2005, **82**, 319–328.
3. S. Martini, C. Bertoli, M. L. Herrera, I. Neeson and A. Marangoni, *J. Am. Oil Chem. Soc.*, 2005, **82**, 305–312.
4. M. Häupler, F. Peyronel, I. Neeson, J. Weiss and A. G. Marangoni, *Food Bioprocess Technol.*, 2014, **7**, 3186–3196.
5. A. H. Suzuki, J. Lee, S. G. Padilla and S. Martini, *Food Eng. Phys. Prop.*, 2010, **75**, 208–214.
6. F. Chen, H. Zhang, X. Sun, X. Wang and X. Xu, *J. Am. Oil Chem. Soc.*, 2013, **90**, 941–949.
7. A. Patist and D. Bates, *Innov. Food Sci. Emerg. Technol.*, 2008, **9**, 147–154.
8. T. L. T. da Silva and S. Martini, *J. Food Eng.*, 2020, **274**, 109824.
9. K. Sato, *Chem. Eng. Sci.*, 2001, **56**, 2255–2265.
10. T. G. Leighton, *The Acoustic Bubble*, San Diego, 1994.
11. S. Martini, R. Tejeda-Pichardo, Y. Ye, S. G. Padilla, F. K. Shen and T. Doyle, *J. Am. Oil Chem. Soc.*, 2012, **89**, 1921–1928.
12. A. Wagh, P. Birkin and S. Martini, *Annu. Rev. Food Sci. Technol.*, 2016, **7**, 23–41.
13. K. Higaki, S. Ueno, T. Koyano and K. Sato, *J. Am. Oil Chem. Soc.*, 2001, **78**, 513–518.
14. Y. Ye, A. Wagh and S. Martini, *J. Agric. Food Chem.*, 2011, **59**, 10712–10722.
15. Y. Ye and S. Martini, *J. Agric. Food Chem.*, 2015, **63**, 319–327.
16. S. Ueno, R. I. Ristic, K. Higaki and K. Sato, *J. Phys. Chem. B*, 2003, **107**, 4927–4935.
17. S. B. Gregersen, R. P. Frydenberg, M. Hammershøj, T. K. Dalsgaard, U. Andersen and L. Wiking, *Eur. J. Lipid Sci. Technol.*, 2019, **121**, 1–9.
18. S. Martini, A. H. Suzuki and R. W. Hartel, *J. Am. Oil Chem. Soc.*, 2008, **85**, 621–628.
19. T. L. T. da Silva, Z. Cooper, J. Lee, V. Gibon and S. Martini, *J. Am. Oil Chem. Soc.*, 2020, **97**, 141–155.
20. R. P. Frydenberg, M. Hammershøj, U. Andersen and L. Wiking, *Cryst. Growth Des.*, 2013, **13**, 5375–5382.
21. WHO, Healthy Diet, https://www.who.int/news-room/fact-sheets/detail/healthy-diet.
22. R. Micha and D. Mozaffarian, *Nat. Rev. Endocrinol.*, 2009, **5**, 335–344.
23. W. C. Willett, *J. Intern. Med.*, 2012, **272**, 13–24.
24. F. A. D. A. FDA, *Final Determination Regarding Partially Hydrogenated Oils*, 2015, vol. 80.
25. R. W. Hartel, *Crystallization in Foods*, Springer, US, 2001.
26. S. Martini, M. L. Herrera and R. W. Hartel, *J. Am. Oil Chem. Soc.*, 2002, **79**, 1055–1062.
27. S. Sonwai, P. Ornla-ied, S. Martini, H. Hondoh and S. Ueno, *J. Am. Oil Chem. Soc.*, 2021, **98**, 43–52.

28. Y. Ye, C. Y. Tan, D. A. Kim and S. Martini, *J. Am. Oil Chem. Soc.*, 2014, **91**, 1155–1169.
29. J. Lee, M. Marsh and S. Martini, *J. Food Sci.*, 2020, **85**, 3391–3399.
30. A. A. Atchley and L. A. Crum, in *Ultrasound: Its Chemical, Physical and Biological Effects*, ed. K. S. Suslick, VCH, New York, 1988, pp. 1–64.
31. M. Patrick, R. Blindt and J. Janssen, *Ultrason. Sonochem.*, 2004, **11**, 251–255.
32. P. R. Birkin, T. M. Foley, T. T. Truscott, A. Merritt and S. Martini, *Phys. Chem. Chem. Phys.*, 2017, **19**, 6785–6791.
33. P. R. Birkin, H. L. Martin, J. J. Youngs, T. T. Truscott, A. S. Merritt, E. J. Elison and S. Martini, *J. Am. Oil Chem. Soc.*, 2019, **96**, 1197–1204.
34. L. Wu, J. Cao, X. Bai, H. Chen, Y. Zhang and Q. Wu, *J. Oleo Sci.*, 2016, **65**, 967–976.
35. J. Mayumi, A. Wagh, L. Antonio, R. Claro and S. Martini, *FRIN*, 2016, **86**, 54–63.
36. T. L. T. da Silva, M. Marsh, V. Gibon and S. Martini, *J. Food Sci.*, 2020, **85**, 964–971.
37. W. FAO, *Fats and Fatty Acids in Human Nutrition. Report of an Expert Consultation*, Rome, Italy, 2010.
38. U. S. D. of H. and H. S. HHS and U. S. D. of A. USDA, *2015–2020 Dietary Guidelines for Americans*, 2015.
39. J. Lee, J. Youngs, P. Birkin, T. Truscott and S. Martini, *J. Am. Oil Chem. Soc.*, 2020, **97**, 1105–1117.
40. R. C. Silva, J. Lee, V. Gibon and S. Martini, *J. Am. Oil Chem. Soc.*, 2017, **94**, 1063–1076.
41. V. De Graef, P. Van Puyvelde, B. Goderis and K. Dewettinck, *Eur. J. Lipid Sci. Technol.*, 2009, **111**, 290–302.
42. T. Tran and D. Rousseau, *Food Res. Int.*, 2016, **81**, 157–162.
43. N. C. Acevedo and A. G. Marangoni, *Food Bioprocess Technol.*, 2014, **7**, 575–587.
44. M. L. Herrera and R. W. Hartel, *J. Am. Oil Chem. Soc.*, 2000, **77**, 1177–1188.
45. J. V. Kadamne and S. Martini, *J. Am. Oil Chem. Soc.*, 2018, **95**, 571–582.
46. J. A. Rincón-Cardona, L. M. Agudelo-Laverde, M. L. Herrera and S. Martini, *J. Am. Oil Chem. Soc.*, 2015, **92**, 473–482.
47. J. Lee, R. Claro da Silva, V. Gibon and S. Martini, *J. Food Sci.*, 2018, **83**, 902–910.
48. J. V. Kadamne, E. A. Ifeduba, C. C. Akoh and S. Martini, *J. Am. Oil Chem. Soc.*, 2017, **94**, 3–18.
49. H. M. D. N. Lida and A. R. M. Ali, *J. Am. Oil Chem. Soc.*, 1998, **75**, 1625–1631.
50. T. L. T. da Silva, S. Danthine and S. Martini, *Lwt-Food Sci. Technol.*, 2021, **138**, 110593.
51. J. V. Kadamne, E. A. Ifeduba, C. C. Akoh and S. Martini, *J. Am. Oil Chem. Soc.*, 2017, **94**, 1045–1062.
52. J. V. Kadamne, M. Moore, C. C. Akoh and S. Martini, *J. Am. Oil Chem. Soc.*, 2018, **95**, 699–707.

53. E. D. Co and A. G. Marangoni, *J. Am. Oil Chem. Soc.*, 2012, **89**, 749–780.
54. A. R. Patel and K. Dewettinck, *Food Funct.*, 2016, **7**, 20–29.
55. T. L. T. da Silva, D. A. Barrera and S. Martini, *J. Am. Oil Chem. Soc.*, 2019, **96**, 681–691.
56. M. Sharifi, S. A. H. Goli and G. Fayaz, *Int. J. Food Sci. Technol.*, 2019, **54**, 509–515.
57. A. S. Giacomozzi, C. A. Palla, E. Carr and S. Martini, *J. Food Sci.*, 2019, **84**, 2549–2561.
58. H. Pehlivanoglu, M. Demirci and O. S. Toker, *Int. J. Food Prop.*, 2018, **20**, S2856–S2867.
59. A. R. Patel, *Edible Oil Structuring: Concepts,* Methods and Applications, 2017.
60. D. Tang and A. G. Marangoni, *J. Am. Oil Chem. Soc.*, 2006, **83**, 377–388.
61. S. Jana and S. Martini, *J. Agric. Food Chem.*, 2014, **62**, 10192–10202.
62. M. D. Bin Sintang, T. Rimaux, D. Van de Walle, K. Dewettinck and A. R. Patel, *Eur. J. Lipid Sci. Technol.*, 2017, **119**, 1–14.
63. J. A. Lopes da Silva and M. A. Rao, in *Rheology of Fluids and Semisolid Foods*, 2007, pp. 331–390.
64. A. I. Blake and A. G. Marangoni, *Food Biophys.*, 2015, **10**, 456–465.
65. T. L. T. da Silva and S. Danthine, *J. Food Sci.*, 2021, **86**, 343–356.
66. A. Giacomozzi, C. Palla, M. E. Carrin and S. Martini, *Food Res. Int.*, 2020, **134**, 109231.
67. C. D. Doan, V. D. W. Davy, A. R. Patel and K. Dewettinck, *J. Am. Oil Chem. Soc.*, 2015, **92**, 801–811.
68. P. Riesz and T. Kondo, *Free Radical Biol. Med.*, 1992, **13**, 247–270.
69. F. Chemat, I. Grondin, P. Costes, L. Moutoussamy, A. S. C. Sing and J. Smadja, *Ultrason. Sonochem.*, 2004, **11**, 281–285.
70. S. Hosseini, M. Gharachorloo, B. G. Tarzi, M. Ghavami and H. Bakhoda, *J. Am. Oil Chem. Soc.*, 2015, **92**, 1717–1724.
71. J. Lee, Y. Ye and S. Martini, *J. Am. Oil Chem. Soc.*, 2015, **92**, 305–308.
72. J. Lim, H. S. Hwang and S. Lee, *Appl. Biol. Chem.*, 2017, **60**, 17–22.
73. T. L. T. da Silva and S. Martini, *J. Food Eng.*, 2019, **263**, 341–347.
74. T. L. T. da Silva, S. Danthine and S. Martini, *J. Food Sci.*, 2020, **85**, 3380–3390.
75. T. L. T. da Silva and S. Martini, *J. Am. Oil Chem. Soc.*, 2020, **97**, 1253–1264.
76. J. D. Pérez-Martínez, J. Reyes-Hernández, E. Dibildox-Alvarado and J. F. Toro-Vazquez, *J. Am. Oil Chem. Soc.*, 2012, **89**, 199–209.
77. H. Zhong, K. Allen and S. Martini, *Food Res. Int.*, 2014, **55**, 239–246.
78. A. Y. Moon, *College of the University of Illinois at Urbana-Champaign*, 2017.
79. E. Rosales, *Graduate College of the University of Illinois at Urbana-Champaign*, 2014.
80. D. H. P. Padilla, *Graduate College of the University of Illinois at Urbana-Champaign*, 2018.
81. J. Lee and S. Martini, *J. Dairy Sci.*, 2019, **102**, 1918–1926.
82. P. R. Birkin, J. J. Youngs, T. T. Truscott and S. Martini, *Ultrason. Sonochem.*, 2020, **67**, 105168.

CHAPTER 4

Pulse and Oilseed Protein-based Oil Structuring for Baking Application

YAN RAN TANG, MANISHA SHARMA AND SUPRATIM GHOSH*

Department of Food and Bioproduct Sciences, University of Saskatchewan, 51 Campus Drive, Saskatoon, SK S7N 5A8, Canada
*Email: Supratim.ghosh@usask.ca

4.1 Oil Structuring and Its Importance in Food

High melting crystalline triacylglycerol (TAG) hardstocks, comprised of mainly saturated fatty acids, provide a significant role in structuring as well as texture and palatability of many food products, such as butter, cream, and a variety of baked goods.[1,2] The TAG hardstocks form a fat crystal network which comprises crystalline TAG lamella that stacks into crystalline nanoplatelets. The crystalline nanoplatelets aggregate *via* van der Waals forces to form spherulitic flocs, and finally weakly link between the flocs to create a 3D network that entraps liquid oil.[3,4] This oil structuring mechanism relies on the limited solubility of high-melting TAGs in liquid oil, which crystallizes upon cooling from a hot oil solution.[2,5] Butter, margarine, and shortening, widely used in many baked products, exhibit similar crystalline structures but differ in fat composition and water content.[6] However, there are two main drawbacks of this conventional way of oil structuring. First, a high fraction of the crystalline TAG phase (>20%) is usually required to achieve efficient oil structuring. Second, the crystalline TAG structure is generally built by saturated and *trans* fatty acids due to their long, straight

Food Chemistry, Function and Analysis No. 31
Development of *Trans*-free Lipid Systems and their Use in Food Products
Edited by Jorge F. Toro-Vazquez

Published by the Royal Society of Chemistry, www.rsc.org

hydrocarbon chains.[7] *Trans* fat is a by-product of partial hydrogenation to turn liquid vegetable oil into a solid and to enhance oxidative stability and crystalline structure formation.[8] Partial hydrogenation increases the shelf life of the product as well as enhances its suitability for making spread, margarine, shortening, and other baking applications by increasing the amount of crystalline fat.[9]

4.2 Potential Health Implication of Traditional Oil Structuring

High consumption of saturated and *trans* fat is associated with increased risk for cardiovascular disease.[10,11] A large body of data from the epidemiological, clinical trial, and *in vitro* studies has shown the deteriorative effect of industrially synthesized *trans* fatty acids on the risk of cardiovascular disease (CVD) and type 2 diabetes.[12–15] A high intake of *trans* fats increases the amount of harmful low-density lipoprotein (LDL) cholesterol in the bloodstream and reduces the amount of beneficial high-density lipoprotein (HDL) cholesterol. *Trans* fat also induces inflammation, which has a direct effect on coronary heart disease, stroke, diabetes and other chronic conditions.[15] Due to all the adverse effects of *trans* fat on human health, both FDA and Health Canada have officially banned the use of partially hydrogenated oil (PHO), since it is the primary source of dietary *trans* fat.[16,17]

On the other hand, a high intake of saturated fat also increases the total cholesterol: LDL cholesterol ratio in the bloodstream.[18] Therefore, the American Heart Association recommends 5 to 6% of calories from saturated fat. However, there is still a debate over the effect of saturated fat on CVD, but it was reported that substituting part of the saturated fat with polyunsaturated fat can reduce the risk of CVD and type-2 diabetes.[15,19] Therefore, there is an urgent need to develop fats without the use of *trans* fat and a significantly lower amount of saturated fat. However, the structure of fats still has to be retained due to its vital role in food texture, functionality, and flavor perception.[3]

4.2.1 Oleogel – A New Way of Oil Structuring

Oleogelation is an approach in which liquid oil is fabricated into a gel-like structure with the use of oleogelators.[7] In conventional approaches, food-grade triacylglycerol oil is structured with fat crystals rich in saturated or *trans* fatty acids. In oleogels, the conventional fat crystal network is replaced with small-molecules or (bio)polymers (commonly known as gelators), capable of forming a network structure and binding the oil.[20] This technique has been investigated extensively with many different approaches using different types of gelators.[21–23] The availability of gelators, their approval for food use in regard to health implications and toxicity, and the sustainability of the processing steps are some of the major factors that must be considered when selecting an oleogelator for food applications. In food, it is also desirable to achieve similar functional characteristics of texture, palatability, appearance, and stability compared to conventional *trans* and saturated fat-based structured oils.[24] Besides food applications, oleogels can also be applied for pharmaceutical and cosmetic purposes.[25] In the following sections, different types of oleogelation are discussed based on processing and types of oleogelators.

4.3 Different Types of Oleogelation Based on Processing – Direct and Indirect Approaches

Different approaches and gelator types have been tested to develop oleogels, which resulted in distinct rheological properties and thermal behavior.[26–28] The different methodologies for oleogelation can be divided into two main categories: direct and indirect approaches. The direct method involves dispersing the self-assembled molecules or crystalline particle gelators in liquid oil at a higher temperature to produce a gelator–oil solution, which upon cooling develops a 3D crystal network that physically entraps the liquid oil.[20] In the direct approach, an ideal gelator must have a suitable balance between their affinity or solubility in the edible oils *via* weak interaction, and sufficient insolubility in the oil. The gelators can self-assemble or crystalize to create a higher-ordered structure through non-covalent interactions such as hydrogen bonding, short-range van der Waals attractions, and π–π stacking.[29] In general, a weak solvent–gelator interaction results in dominant gelator–gelator interactions, which may lead to the formation of a robust continuous network, which on cooling, could eventually lead to the precipitation of crystalline or amorphous molecular clusters. The optimum balance of these interactions controls the formation of a continuous network of the self-assembled gelators and subsequent gelation of the liquid oil.[30]

The indirect approach of oleogelation uses biopolymers such as proteins and polysaccharides.[31] These molecules are mostly hydrophilic in nature, thus, exhibit low solubility in non-polar oil. However, an indirect oleogelator must exhibit amphiphilic behavior and a certain degree of surface activity so that the liquid oil can be entrapped physically as well as through weak interactions in the gel network. The polymers first expose their hydrophobic parts at the interface in foaming or emulsification, then the foam or emulsion needs to be dried to obtain a 3D porous microstructure that could adsorb liquid oil forming oleogels.[32,33] Another type of indirect approach involves the stepwise solvent exchange of a biopolymer hydrogel *via* a series of solvents with reduced polarity to fully replace the aqueous phase with liquid oil. However, the oleogel developed through solvent-exchange exhibited twice the stiffness compared with that of the corresponding hydrogel.[34] The use of acetone and other toxic solvents in the solvent-exchange process also becomes a concern as the requirement for clean-labeled food is in high demand among consumers.[35]

4.4 Different Types of Oleogelation Based on Oleogelators – Small Molecules and Biopolymers

The different types of gelators that have been studied over the years fall into three main categories: 1) self-assembled low molecular weight (LMW) molecules, 2) crystalline particles, and 3) biopolymers (Table 4.1). Low molecular weight surfactants such as sorbitan monostearate (also known as Span 60) (Table 4.1) self-assemble into toroidal inverse vesicles at gelation temperature (41–44 °C).[27] Upon cooling, the vesicles are turned into rod-shaped tubules that associate with each other to form three dimensional networks that immobilize the solvent, thus causing gelation. They formed an opaque and thermoreversible semisolid at critical gel concentration (10% gelator molecules).[27] Phytosterols such as β-sitosterol and γ-oryzanol have also been used to structure liquid oil where the molecules self-assembled into a tubular network that gelled the oil phase.[54] Guo *et al.*[39] recently reported water-induced self-assembly of mixed oleogelators (ceremide and lecithin). After the addition of water, the microscopic ceramide–lecithin gel changed from three-dimensional dendritic crystals into two-dimensional rod-like crystals. Their FTIR study revealed that water participated in the structure formation rather than just dispersing within the network and the primary driving force for structure formation was through hydrogen bonding and van der Waals interactions.

Fatty acids have also been used as crystalline particles to structure liquid vegetable oil (Table 4.1).[21,38] It was reported that the gelling of vegetable oils, essential oils, and hydrocarbon occurred at ambient temperature by using fatty acids with carbon chain lengths between 10 and 31. Daniel and Rajasekharan[21] demonstrated that in addition to the carboxyl group, the position of an additional hydroxyl group (hydroxy fatty acids) and acyl chain length played an essential role in gelation. They also showed gelation of vegetable oils using long-chain saturated fatty alcohols, wax esters, and dicarboxylic acids. Rogers *et al.*[22] investigated the effect of storage temperature on the oil binding capacity and crystallinity of 12-hydroxystearic acid-canola oil oleogels. As the gels were stored at 5 °C, they exhibited a highly branched crystalline structure with small uniform pores. At 30 °C, the gels showed longer fibers and less branched networks leading to a higher oil syneresis rate than that at 5 °C.

Recent studies have also explored biopolymers as oleogelators due to their sustainability, availability, and the consumer demand for 'clean label' products. Biopolymers such as modified cellulose (*e.g.*, methylcellulose, hydroxymethyl cellulose) and proteins (*e.g.* gelatin, soy protein, zein particle) have been used to indirectly structure liquid oil by using emulsion or foam-templated approaches (Table 4.1).[33,45,55] Ethylcellulose is the only biopolymer known so far to be able to disperse directly in liquid oil.[52] However, to disperse the polymer in the oil, the mixture must be heated to a temperature higher than the glass transition temperature (130–140 °C), which may not be ideal for unsaturated oil due to lipid oxidation. For the foam and emulsion-templated approaches, freeze-drying, vacuum drying, and oven drying (60–90 °C) of the emulsions have been employed to remove water and create

Table 4.1 Use of different oleogelators for vegetable oil gelation based on gelator types as reported in the literature.

Oleogelator type	Examples	Approach	References
Crystalline particles	Rice bran wax	Direct	Amoah *et al.*[36]
	Lecithin	Direct	Bodennec *et al.*[26]
	Fatty acids	Direct	Daniel and Rajasekharan[21]
	Fatty acid + alcohols	Direct	Schaink *et al.*[37]
	Derivatized fatty acids	Direct	Wright and Marangoni[38]
Low molecular weight molecules	Sorbitan monostearate	Direct	Murdan *et al.*[27]
	Ceramides	Direct	Guo *et al.*[39]
	β-sitosterol + γ-oryzanol	Direct	Sawalha *et al.*[20]
Biopolymers	Soy protein + κ carrageenan	Indirect, emulsion-templated	Tavernier *et al.*[33]
	Regenerated cellulose + CMC	Indirect, emulsion-templated	Jiang *et al.*[40]
	Hydrophobic cellulose derivatives (MC, HPMC)	Indirect, emulsion, and foam-templated	Lee[41]
	Sodium caseinate + alginate	Indirect, emulsion-templated	Wijaya *et al.*[42]
	Gelatin–tannic acid–flaxseed gum complexes	Indirect, emulsion-templated	Qiu *et al.*[43]
	Tea polyphenol–palmitate particles + citrus pectin	Indirect, emulsion-templated	Luo *et al.*[44]
	Xanthan gum + HPMC	Indirect, emulsion-templated	Meng *et al.*[45]
	Zein colloidal particles + sodium stearate	Indirect, emulsion-templated	Gao *et al.*[46]
	Quillaja saponin	Indirect, emulsion-templated	Chen and Yang[47]
	Pea and faba bean protein + Xanthan gum	Indirect, foam-templated	Mohanan *et al.*[48]
	Hydroxypropyl methylcellulose (HPMC)	Indirect, foam-templated	Patel *et al.*[31]
	HPMC and methylcellulose	Indirect, foam-templated	Tanti *et al.*[49]
	Whey protein	Indirect, hydrogel templated	de Vries *et al.*[34]
	κ-carrageenan	Indirect, hydrogel templated	Manzocco *et al.*[50]
	κ-carrageenan + lettuce homogenate filler	Indirect, hydrogel templated	Plazzotta *et al.*[51]
	Ethylcellulose	Direct	Davidovich-Pinhas *et al.*[52]
Inorganic particles	Fumed silica	Direct	Patel *et al.*[53]

structured oil.[33,45] The present chapter will focus on indirect oleogelation, especially the utilization of pulse and oilseed proteins in the development, characterization, and application of emulsion-templated oleogels.

4.5 Indirect Oleogelation

4.5.1 Foam-templated Indirect Oleogelation

Foam-templated oleogelation is a novel, environment-friendly, and simple approach to develop oleogels at room temperature without using harsh chemicals or crosslinkers. It involves the formation of aqueous foam from a polymer solution and then freeze-drying the aqueous foam to obtain a porous dried foam.[31,53] The porous dried foam was able to absorb a large amount of oil, leading to the formation of oleogels.[31] During foam-templated oleogelation, the foamability and foam stability are crucial in the development of well-structured oleogels. Factors such as pH, surface hydrophobicity, the presence of salt, sugar and lipids, temperature, and protein source affect the surface properties, therefore, the foaming ability of the protein.[56] Film drainage, bubble coalescence, and Ostwald ripening (disproportionation) contribute to foam destabilization.[57] Among them, Ostwald ripening is a major factor that causes destabilization due to the diffusivity of gas molecules through the foam stabilizers and the polydisperse bubble size that accelerate disproportionation.[57] Therefore, thick and elastic lamellae are desirable to reduce air diffusivity and lamellae rupture to ensure stable foam formation.[58] There are only a few studies that have reported successful foam-templated oleogels. The oleogels made with hydroxypropyl methylcellulose (HPMC)-based foam were reported to show strong shear-thinning behavior with a 40-fold drop in viscosity by a two-fold increase in shear rate and negligible shear recovery.[31] However, the oleogel maintained a gel-like behavior at high temperatures (100 °C) under dynamic shear. Mohanan *et al.*[48] reported foam-templated oleogels stabilized with pulse proteins from faba beans and peas. Xanthan gum was added to the protein solution to enhance foam formation and stability. The effect of pH on the oil binding capacity and rheological properties of the oleogels was investigated. It was found that pea protein concentrate oleogels displayed a higher oil binding capacity at pH 5 than faba bean protein oleogels, while faba bean protein concentrate oleogels exhibited a higher oil binding capacity than pea protein oleogel at pH 9, although about 30–40% of the liquid oil leaked out of the foam structure leading to poor properties of the oleogels and greater hardness and chewiness of the cakes baked with the oleogel.

4.5.2 Emulsion-templated Indirect Oleogelation

In emulsion-templated oleogelation, the continuous aqueous phase is removed *via* vacuum or freeze-drying, resulting in a solid-like structured system with a high concentration of liquid oil droplets.[59] A stable emulsion is crucial prior to the formation of oleogels *via* this process. Surface-active biopolymer stabilized O/W emulsions, and particle-stabilized Pickering emulsions were developed to fabricate emulsion-templated oleogels.[33,46] Proteins are widely used as an emulsifier to stabilize emulsion droplets through electrostatic repulsion and steric hindrance[60] thereby providing a strong interfacial barrier, which is essential for oil droplet stability during emulsion-templated oleogelation. Biopolymer complexes made of proteins and polysaccharides were also introduced to increase the emulsion stability due to the accumulation of the complexes around the oil droplets, which thickens the interfacial layers and retards coalescence among the oil droplets.[33] In general, it was found that the gel strength of an oleogel increased to a high value (storage modulus, $G' > 40\,000$ Pa) as the biopolymer concentration increased. This property was attributed to a stronger biopolymer network at the interface at higher concentrations.[33] Factors such as the viscosity of the emulsion continuous phase and oil droplet size are also important for oleogelation. Therefore, non-surface-active polysaccharides, such as xanthan gum (XG), have also been used as thickeners to increase the viscosity of the continuous phase to improve emulsion stability during drying.[45]

4.5.3 Hydrogel-templated Indirect Oleogelation

In the hydrogel-templated approach, the polymer is hydrated to form a hydrogel, and then the hydrogel is freeze-dried to form a porous polymer network, which takes up the oil.[61] Conventional air or oven drying limits the oil adsorption capacity due to the collapsed structure resulting from capillary pressure when water evaporates.[62] Other than freeze-drying, approaches such as stepwise solvent exchange and supercritical drying were adopted so that the polymer network of the hydrogel is well maintained.[34,50] Since these two approaches do not involve vapor transition or intense capillary force, they can form low density and highly porous structures.[63,64] de Vries *et al.*[34] reported that up to 91% of oil could be incorporated into a whey protein isolate gel matrix through a solvent exchange procedure without affecting the gel network noticeably.

4.6 Characterization of Biopolymer-based Oleogels for Food Application

4.6.1 Oleogel Stability in Terms of Oil Binding Capacity and Oil Loss

The stability of oleogels in terms of oil loss can be expressed as the oil holding and binding capacity or the oil loss value. This property is one of the key factors used to determine whether an oleogel is feasible for food applications. A lower or slower oil loss is desirable in food manufacturing.[65] In a foam-templated oleogel, the oil holding capacity is calculated by dividing the final oil content of the oleogel with the weight of the dried foam after the initial addition of liquid oil to the dry foam. The oil binding capacity of the foam-templated oleogels, however, is determined by removing the loosely bound oil *via* centrifugation and measuring the amount of oil remaining in the oleogel.[48] In an emulsion-templated oleogel, the oil holding capacity is determined by dividing the oil loss with the initial weight of the sample or the initial oil content in the sample. In contrast, oil loss is obtained by measuring the weight of the oil released after a certain period of time[65] or through centrifugation to accelerate the process.[42] The oil binding capacity reported by Mohanan *et al.*[48] ranged from about 12 up to 28 g of oil bound per g of dried foam, where faba bean concentrates with 0.25% xanthan gum at pH 9 exhibited the highest oil binding capacity. Patel *et al.*[31] fabricated foam with 1% HPMC, which was reported to absorb oil to an amount of 98 times its own weight. However, a question remains on whether centrifugation was applied to remove any loosely bound oil, so the values of these two studies are not comparable. To improve the oil binding capacity of foam-templated oleogels, Mohanan *et al.*[66] dissolved candelilla wax (CW) or high-melting monoglyceride (MAG) to canola oil before adding to the pulse protein-stabilized dried foam. Compared to the control oleogel, the CW or MAG-added foam-templated oleogels improved the oil binding capacity decreasing the oil loss from ∼34% for the control to ∼4% in the presence of 2% CW or 3% MAG in the oleogel. Interestingly, the presence of a similar amount of CW or MAG alone in the canola oil did not result in a good oil binding capacity (100% oil loss), hence, hydrophobic, van der Waals interactions and hydrogen bonding contributed by CW, MAG, and pulse proteins improved the oil binding capacity of the whole system with both crystal and foam networks.[48,67] For the emulsion-templated oleogel, the centrifugation method is commonly used to determine oil loss.[42,44,45] Due to a variety of factors such as emulsifier types, concentration, pH, emulsion drying methods, and shearing methods, the oil loss (ranges from about 5–35%) among the oleogels varies between the different studies. However, a common trend is observed in these studies, namely, as the emulsifier concentration increased, the oil binding capacity increased (oil loss decreased). These could be ascribed to the thick interfacial network at higher emulsifier concentrations that protect the oil droplets from coalescence or leakage from the network. Meng *et al.*[45] investigated the effect of different thickening agents such as carboxymethyl cellulose (CMC), xanthan gum, sodium alginate, gum Arabic, guar gum, flaxseed gum, and locust bean gum on the oil binding capacity of the HPMC-stabilized emulsion oleogels. Among them, xanthan gum exhibited the lowest oil loss value, at about 12%. Although the mechanism of how the interactions of different thickening agents with HPMC contribute to the improved oil binding capacity is not clear, the authors ascribed the decreased oil loss to the higher mechanical strength of the corresponding emulsions.

4.6.2 Rheology of Biopolymer-based Oleogels

The rheological properties of oleogels are evaluated to study their flow behavior under shear stress and gelation behavior as a function of oscillatory strain and frequency. This is an important aspect to consider in food applications since the rheology of oleogels is responsible for the final product's texture, appearance, and handling work. In general, the storage modulus (G') of oleogels, which indicates their solid-like elastic properties, ranges from about 10^3 Pa to 10^6 Pa depending on the oleogelation approach, emulsifier types, and concentrations.[43,44,47] The gel strength of the foam-templated oleogels reported ranged from 10^4 to 10^5 Pa within the linear viscoelastic region (LVR).[31,48] For the emulsion-templated oleogels, most of the reported values fall within the range of about 10^4 to 10^6 Pa.[33,40,42,44,45,47] Qiu *et al.*[43] studied the effect of drying on the gelatin–polyphenol–polysaccharide nanocomplex stabilized oleogels. They reported a lower G' within the LVR ($\sim 10^3$ Pa) of the oven-dried oleogels as compared to the freeze-dried oleogels ($\sim 10^4$ Pa), which could be ascribed to lower deformation of the network due to water sublimation during freeze-drying. In the studies of hydrogel templated oleogels, small deformation oscillatory tests were not reported, which could be due to the high stiffness of the oleogels.[34,50]

Shear-thinning behavior is also common among oleogels, for instance, emulsion-templated cellulose-rich oleogels, HPMC, and xanthan gum-stabilized oleogels and foam-templated oleogels using HPMC.[31,40,45] The thixotropy recovery of oleogels is commonly investigated, which gives us an insight into the repeated applicability of oleogels as structured oil in food applications.[68] Emulsion-templated oleogels structured with HPMC, cellulose, and gelatin have shown significant thixotropic recovery properties with the highest recovery at 90% for HPMC and methylcellulose stabilized-emulsion-templated oleogels.[40,43,69] Emulsion-templated oleogels structured with gelatin and xanthan gum exhibited an interesting property with partial thixotropic recovery even at high temperature (80 °C).[59] In contrast, foam-templated oleogels showed very weak thixotropic recovery, with negligible structure recovery as the shear rate decreased from 10 s^{-1} to 0.1 s^{-1}.[31] The excellent thixotropic recovery properties of emulsion-templated oleogels can be attributed to the elasticity of the interfacial network that allows the clusters of highly packed droplets to return to a more random distribution, resulting in resistance to flow.[70] On the other hand, both emulsion and foam-templated oleogels structured with HPMC exhibited thermal stability. It was also proposed that the retention of a gel-like structure ($G'' < G'$) when temperature cycled from 5 to 80 °C and back indicated their potential for non-food applications as bio-lubricants.[31,69]

4.6.3 Textural Analysis of Biopolymer-based Oleogels

Since oleogels can be applied as a spreadable or butter substitute, the texture analysis of oleogels was carried out using a spreadability test. In the force–distance curve, the positive peak was taken as firmness, while the negative peak was taken as stickiness. The firmness of the foam-templated oleogels ranged from 2–60 N depending on the presence of additives, such as CW or MAG.[48,66] For emulsion-templated oleogels structured with citrus fibers, the firmness was about 10 N.[71] There are very limited spreadability studies reported on emulsion-templated oleogels; rather, the firmness of the dried emulsions was measured through a uniaxial compression test. The firmness of the dried emulsion products also varies within a huge range (~3 to 100 N), which could be due to the drying methods, emulsifier type, and concentrations.[44,47] In general, increasing the concentration of oleogelators was reported to result in an enhancement in the fracture force, indicating a harder gel.[59,72] A high concentration of polymer forms a strong interfacial layer as well as a sturdy network in the bulk phase that contributes to the hardness of the gel. These thick interfacial layers can resist the capillary pressure induced by the evaporation of water during drying, therefore also enhance the oil holding ability of the gel network.[59] Qiu *et al.*[43] reported a higher firmness of freeze-dried emulsions (~100 N) than that of oven-dried emulsions (~40–60 N). A similar finding was also observed by Chen and Yang.[47] The authors have ascribed the high mechanical strength in the freeze-dried product to the inter-crosslinked networks of the oil droplets. Among the three indirect oleogelation approaches, hydrogel-templated oleogels have shown the highest firmness in uniaxial compression tests. In the study of oleogel fabrication through a stepwise solvent exchange, de Vries *et al.*[34] reported an increase in oleogel stiffness (from 10^6 to 10^7 Pa) as the protein concentration increased from 10% to 16% and also as compared to the corresponding hydrogels (10^3–10^5 Pa). In a stepwise solvent exchange approach, protein–protein interactions became more favorable compared to protein–solvent interactions as the solvent polarity decreased. Hence, the strength of the protein networks increased tremendously, leading to the very high stiffness of the oleogels. In the hydrogel-templated approach, Manzocco *et al.*[50] converted a κ carrageenan hydrogel to an alcoholgel and dried it using supercritical CO_2 to obtain an aerogel, followed by the addition of liquid oil to form oleogels. The authors reported firmness of the oleogel ranging from ~158 N to 312 N (~10^6 Pa, with 6.25 mm probe diameter), which is similar to the values reported for freeze-dried oleogels by Qiu *et al.*[43] and Chen and Yang.[47]

4.6.4 Food Applications of Biopolymer-based Oleogels

Food application of oleogels was tested in a variety of products such as ice cream, cake, short dough products (cookies, pastry *etc.*), muffins, and sausages.[36,41,73–75] A detailed listing of research papers on the food applications of biopolymer-based oleogels is shown in Table 4.2. Of these, applications in cake have been the most widely studied.[36,41,76,77] Prior to cake baking, the properties of the cake batter were studied using rheology and air incorporation, indicated by density and microstructure.[41] The methylcellulose and xanthan gum structured oleogel cake batter was compared with cake batter made from margarine, shortening, and liquid oil.[76] Oleogel cake batter showed more air incorporation (density 0.92 $g\,mL^{-1}$) than liquid oil cake batter (1.08 $g\,mL^{-1}$), but less than margarine and shortening cake batters (0.78 and 0.84 $g\,mL^{-1}$). Similar observations were also reported by

Mohanan *et al.*[48] in their study of foam-templated oleogels stabilized with pea or faba bean proteins as a fat replacer in cakes. It was proposed that the lower air incorporation in the oleogel cake batter was due to the absence of emulsifiers and oil crystals in the matrix, compared with the margarine and shortening samples. The addition of high melting MAG to foam-templated oleogels has been proven to decrease the batter specific gravity indicating higher air incorporation.[66] The addition of CW to the foam network, however, didn't improve the air incorporation of cake batter. It was proposed that although both MAG and CW formed crystals in the oleogels, the surface activity of the MAG was responsible for the incorporation and stabilization of the air bubbles in the cake batters.[66] The cake batter made with a 4 wt% HPMC oleogel as a partial shortening replacer showed a decrease in air incorporation when the replacement level of the oleogel was more than 25%.[41] Micro-CT images of the cake batter also showed that shortening replacement with HPMC oleogels up to 50% by weight produced muffins with a similar total porosity compared to the control with 100% shortening[41]

The final cake products were tested by using a texture analyzer based on textual parameters such as hardness, chewiness, cohesiveness, adhesiveness, and springiness index.[36,76,77] The hardness of oleogel cake was shown to be higher than that of shortening cake (4.56 ± 0.15 N). However, among the wax-based oleogel cakes (beeswax, rice bran wax, and candelilla wax), the beeswax oleogel cake exhibited the softest texture at 8.14 ± 0.14 N.[36] Mohanan *et al.*[66] compared the cake texture of the foam-templated pulse protein-stabilized oleogels with and without the addition of MAG and CW. Though all the oleogel cakes exhibited greater hardness compared to the shortening cakes, the MAG added oleogel cake showed a significant reduction in hardness as compared to the oleogel cake without MAG, while the addition of CW didn't result in an improvement of the cake texture. Since the cake structure was set through the protein–protein and protein-gelatinized starch interactions, the greater hardness of the pulse protein oleogel-based cakes could be due to more protein and starch interactions.[78] The presence of MAG might have interrupted the interactions and staling process, hence, lowering the overall cake hardness.[79] Although not many studies have been carried out to determine the storage stability of oleogel cakes, Patel *et al.*[76] reported improved hardness and chewiness of oleogel-added cakes compared to liquid oil-added cakes during 3 day storage. This was attributed to the structured oleogel in the cake matrix that prevented the cross-linking of the gluten network over a certain time period, thereby reducing the hardness and chewiness during storage.

4.6.5 Sensory Evaluation of Food Made From Biopolymer-based Oleogels

Sensory studies of food products give us insight into consumer acceptability and the subsequent improvements based on the evaluation report. Several sensory evaluations of food products prepared with oleogels have been

Table 4.2 Summary of research work on food applications of oleogels.

References	Oleogels	Application	General observation
Amoah *et al.*[36]	Direct; rice bran wax, beeswax, candelilla wax structured sunflower oil oleogel	Cake (baked with cake flour)	Beeswax oleogels cakes showed the lowest hardness (8.1 N) among the three wax-based oleogel cakes. Comparable cake volume compared to the shortening cakes.
Demirkesen and Mert[83]	Direct; beeswax structured sunflower oil	Gluten-free cake (baked with rice flour)	15, 30, and 45 wt% replacement of shortening with oleogels showed comparable cake attributes
Willett and Akoh[84]	Direct; phytosterol blends (β sitosterol + γ oryzanol) or SSAP (sucrose stearate/ascorbyl palmitate) + structured menhaden oil with caprylic acid/stearic acid	Yellowcake (baked with cake flour)	All structured lipid cakes and corresponding oleogel cakes showed lower hardness (5–8 N) and chewiness (4–6 N) than the shortening cake (12 N, 9N).
Pehlivanoglu *et al.*[77]	Direct; blend oil (30% palm oil, 25% cottonseed oil, 25% palm kernel, and 20% palm stearin) and carnauba wax structured high oleic sunflower oil + cottonseed oil	Cake (baked with wheat flour)	The overall acceptability of oleogels cake was comparable to the shortening cakes
Kim *et al.*[85]	Direct; carnauba wax structured canola oil	Cake (baked with cake flour)	25 wt% replacement of shortening with oleogels showed comparable attributes. However, 100% replacement showed an almost 3-times increase in hardness.
Alvarez-Ramirez *et al.*,[86]	Direct; candelilla wax structured canola oil	Wheat sponge cake	With increased shortening replacement, the specific cake volume increased, hardness decreased, but other textural properties increased.
Jung *et al.*[87]	Direct; candelilla wax structured rice bran oil	Sweet pan bread	No significant changes of bread volume until 75% replacement of butter with oleogel. The hardness of the oleogel-based breads increased with % replacement.
Tabibiazar *et al.*[81]	Direct; carnauba wax and adipic acid structured soybean oil	Cake and beef burger	Overall acceptance of carnauba oleogel products was the lowest. The addition of adipic acid improved overall acceptance compared to the control.

Adili *et al.*[80]	Direct; ethylcellulose and adipic acid structured soybean oil	Cake and beef burger	Overall, acceptance of ethylcellulose oleogel products was the lowest. The addition of adipic acid improved overall acceptance compared to the control.
Patel *et al.*[73]	Direct; shellac structured sunflower oil	Chocolate paste & cake	Chocolate paste prepared with oleogels showed no oiling off during several weeks of storage. Shellac oleogel cakes showed a comparable score in texture, moistness, and volume but a lower score in cell uniformity and higher crumbliness
Yılmaz and Öğütcü[82]	Direct; sunflower wax and beeswax structured hazelnuts oil	Cookies	The hedonic score of the cookies was not significantly different than the control. Beewax cookies exhibited similar storage stability to the control, compared to the sunflower wax cookies.
Mert and Demirkesen[88]	Direct; carnauba and candelilla wax structured sunflower oil	Cookies	The moisture content of all cookies was not significantly different. The shortening cookies showed the lowest hardness. The wax type did not affect the texture of the cookie, but the increased wax content decreased the hardness of the cookie.
Zulim Botega *et al.*[74]	Direct; rice bran wax, CW, carnauba wax structured high oleic sunflower oil	Ice cream	15% rice bran wax in oleogels led to a lower meltdown rate and better shape retention compared to the candelilla wax and carnauba wax.
Bemer *et al.*[89]	Direct; rice bran wax or ethyl cellulose structured soybean oil	Cream cheese	Rice bran wax oleogel cream cheese showed no significant difference in texture and sensory mouthfeel compared to the control. Ethylcellulose oleogels showed a similar texture except for a lower adhesiveness. Rice bran wax oleogel cream cheese had a higher bitterness.
Koumakis and Petekidis[90]	Direct; monoglyceride and phytosterol structured sunflower oil	Sausages	The overall sensory score of the oleogel sausage was comparable to the control sausage. The oleogel sausage scored lower in juiciness but higher in appearance, color, and smell.
Lim *et al.*[91]	Direct; carnauba wax structured soybean oil	Frying media for instant noodles	Oleogel deep-fried noodles absorbed 16% less oil compared to those with soybean oil and palm oil. The peroxide value of oleogel fried noodles was lower than the soybean oil fried noodles.

Table 4.2 (*Continued*)

References	Oleogels	Application	General observation
Luo *et al.*[44]	Indirect, emulsion-templated; tea polyphenol-palmitate particles and citrus pectin structured camellia oil	Cake (baked with wheat flour)	All oleogel cakes showed a lower overall quality score than butter cakes.
Patel *et al.*[76]	Indirect, emulsion-templated; methylcellulose + xanthan gum structured sunflower oil	Cake (baked with wheat flour)	Textural properties of oleogel cakes were closer to the margarine and shortening cakes compared to those of the canola oil cakes
Tavernier *et al.*[65]	Indirect, emulsion-templated; soybean protein + CW structured sunflower oil	Shortcrust pastry cookies	No differences of hardness among cookies with a different fat phase.
Mohanan *et al.*[48]	Indirect, foam-templated; faba bean and pea protein + xanthan gum structured canola oil	Cake (baked with all-purpose wheat flour)	Cake specific volumes were not significantly different among the shortening, canola oil, and oleogel cakes. However, the hardness and springiness, and chewiness of the canola oil and oleogel cakes were higher than that of shortening cake.
Mohanan *et al.*[66]	Indirect, foam-templated; faba bean and pea protein + xanthan gum foam + monoglycerides or candelilla wax structured canola oil	Cake (baked with all-purpose wheat flour)	All cake specific volumes were similar. The addition of MAG significantly improved the hardness of oleogel cakes, making it the closest to the shortening cake.
Tanti *et al.*[49]	Indirect, foam-templated; HPMC and methylcellulose structured canola oil	Sandwich cookies cream	50 and 75% shortening replacement with oleogels showed acceptable texture attributes within the range of commercial cookie creams. Full replacement with oleogels showed the texture fell out of the range.
Lee[41]	Indirect, foam-templated; HPMC structured sunflower oil	Muffin (baked with wheat flour)	25 and 50 wt% replacement of shortening with oleogels didn't sacrifice the original texture, volume, and total porosity.

reported, such as cakes,[44,73,77,80,81] cookies,[82] and beef burgers.[80,81] The sensory acceptability of the products varied among the different studies. In the shellac oleogel cake (direct oleogelation), the volume, sponginess, stickiness, and moistness of the oleogel cakes were comparable to the shortening cakes, while the crumbliness was higher and the uniformity in cell size score was lower for the shellac oleogel cakes ($p < 0.05$).[73] In the texture-based sensory analysis, parameters such as firmness, cohesiveness, and adhesiveness were not significantly different, while springiness and chewiness of the oleogel cake were significantly higher than the shortening cake ($p < 0.05$).[73] Pehlivanoglu *et al.*[77] developed four direct oleogel formulations with a combination of various oil, water, shortening, cottonseed oil, and high oleic sunflower oil contents. Though the hardness from the texture analysis was higher (~2.3–7.6 N), the oleogel cakes without any shortening (oil + water + cotton seeds oil + high oleic sunflower oil) have shown to have comparable overall acceptability compared to the control shortening cake (hardness 1.7 N). In contrast, both carnauba wax (CW) and ethylcellulose oleogel cakes (direct oleogelation, 50% shortening replacement) showed inferior acceptability compared to the control shortening cake.[80,81] However, the addition of adipic acid in the CW oleogels had improved overall acceptability, though the control shortening cake remained the highest in acceptability.[80] Adipic acid is a linear dicarboxylic acid that is found naturally in sugar cane and beets, which helps in nucleation and crystal formation. The hedonic score of the emulsion-templated oleogel (tea polyphenol-palmitate particles and citrus pectin structured camellia oil) cake reported by Luo *et al.*[44] was lower (21.5–27.6) than that of the control butter cake (32.0). In this study, the flavor and odor scores were comparable to the butter cake, while the texture, crust, and crumb color were less acceptable. The reported oleogel cakes showed a wide range of variation in acceptability from the sensory panelists, which might come from a variety of oleogelators and formulations. More effort and research need to be done to improve the cake sensory properties when made with different oleogelators using different oleogelation approaches.

4.7 Utilization of Pulse and Oilseed Proteins for the Development of Emulsion-templated Oleogels

Plant proteins are a major ingredient in delivering the world food supply-demand. Their production is sustainable, and they have been promoted as a possible replacement for animal proteins in various food applications. Apart from their nutritional benefit, plant proteins are also used as functional ingredients in food and beverages. Most proteins are also surface-active, meaning they can adsorb at the oil–water interface, and they be used to form highly stable oil-in-water emulsions. The global plant-based protein market is at about $9 billion US and is expected to grow at a compound annual growth rate (CAGR) of about 8% to $14.3 billion by 2025 (Plant-based protein market by type and application, www.researchandmarkets.com, 2019).

This rapid growth is attributed to increasing consumer awareness and demand for a protein-rich, plant-based diet, and the innovation in ingredient technologies, such as emulsification and microencapsulation (www.researchandmarkets.com). Leading the plant-based protein market is soy, followed by wheat, pea, and other pulses. Most recently, pulse proteins have seen a remarkable growth in market size owing to the numerous research and development studies on peas, faba beans, and other pulses. Oilseed proteins, such as those extracted from canola, however, have not seen such growth in demand. For example, only a mere 0.14% of the plant protein market is taken up by canola protein, which is expected to grow at a CAGR of 4.2% to only $16.7 million by 2025 (Global canola protein market, www.researchandmarkets.com, 2018). This is despite the fact that canola is the second largest cultivated oilseed crop in the world after soybean and produces a protein-rich meal that would add tremendous value to the crop. However, at present most of the canola meal produced is used in animal feed, which undermines the huge potential of such a high-quality, nutritious, economical, and sustainable source of protein for human consumption. Keeping this in mind, we have investigated the potential of canola protein isolates (CPIs) in the development of emulsion-templated oleogels. CPIs were extracted from cold-pressed meal (donated by Pleasant valley Oil meal, Alberta, Canada) according to Tang and Ghosh[92] and had a protein content of 94.5% (dry basis). The results were also compared with similar emulsion-templated oleogels made from faba bean protein isolates (FBPIs, donated by AGT Foods and Ingredients, Saskatchewan, Canada) that were extracted *via* isoelectric precipitation and had a protein content of 90.5% (dry basis).

The first step in the oleogel development was the preparation of 50 wt% canola oil-in-water emulsions with 4 wt% protein dispersions using a high-pressure homogenizer at 20 000 psi for 6 cycles. Both the CPI and FBPI emulsions were developed with a similar volume average droplet diameter of 9.3 ± 0.8 μm and 10.9 ± 1.0 μm, respectively. To improve the protein interfacial strength and hence emulsion stability, both the emulsions were heat-treated to induce protein denaturation, followed by cooling to room temperature to initiate protein aggregation at the oil droplet surface. For complete protein denaturation, CPI emulsions were heated at 90 °C, while the FBPI emulsions were heated at 100 °C for 30 mins. Oleogels were obtained by vacuum drying both the heated (HE) and unheated (UE) emulsions at 60 °C until reaching a constant weight followed by shearing using a hand blender to obtain a uniform texture. Figure 4.1 shows a visual representation of the oleogels. In general, CPI oleogels were darker in color compared to the FBPI oleogels due to the darker color of the canola proteins compared to faba bean proteins. Both the UE oleogels were liquid-like, while between the HE oleogels, CPIs showed a strong gel-like material that could hold its own weight against gravity. The HE FBPI oleogel was semi-transparent and flowed slowly under gravity. The oleogels were then studied for their oil binding capacity, viscoelasticity, and cake baking properties.

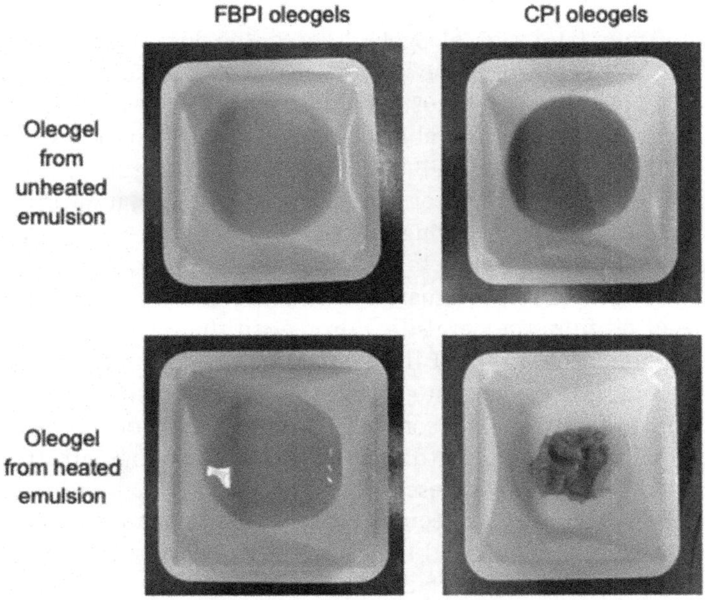

Figure 4.1 Visual observation of FBPI and CPI oleogels made from unheated and heated oil-in-water emulsions.

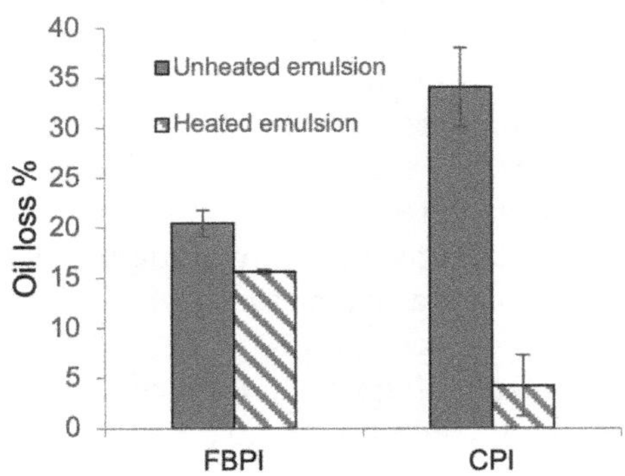

Figure 4.2 Oil biding capacity of the FBPI and CPI oleogels expressed by their oil loss percentage under centrifugal force. Higher oil loss is an indication of lower oil binding capacity.

4.7.1 Characterization of Faba Bean and Canola Protein Isolate-based Oleogels

The ability of the oleogels to retain oil within their structure was measured by the amount of oil loss under centrifugal force (Figure 4.2). The percent oil loss of the oleogels was calculated from the amount of weakly bound oil separated from the oleogels after centrifugation. A lower oil loss indicates a better oil binding capacity, according to Blake and Marangoni.[93] The unheated emulsion (UE) CPI oleogel showed the highest percentage of oil loss, while the heated emulsion (HE) CPI oleogels showed the lowest oil loss. Without heating, the UE FBPI oleogel showed a lower oil loss ($20.5 \pm 1.3\%$) compared to the UE CPI oleogel ($34.1 \pm 4.0\%$), indicating the better ability of the pulse proteins to provide interfacial stability to the oil droplets in the oleogels. Upon heating, both the HE CPI and FBPI oleogels exhibited significantly ($p < 0.05$) lower oil loss compared to the corresponding UE oleogels, indicating that the heating of emulsions before drying helped to improve the oil binding capacity of the oleogels. During the process of oleogel development, there are two steps that affect the structural conformation of the oil droplets, which might lead to the release of oil from the emulsified droplets if the interfacial network was not strong enough to withhold the oil droplets. First, upon drying, the oil droplets were tightly packed into a compact polygonal structure due to the evaporation of water from the continuous phase.[42] Second, the structure of the dried emulsions might get disrupted during shearing with the kitchen blender, which leads to the coalescence of oil droplets. Therefore, the lower oil loss from the heated oleogels could be attributed to the formation of thick interfacial protein layers around the oil droplets as a result of heat-induced protein denaturation and interfacial aggregation as well as gelation of proteins in the continuous phase. Interestingly, after heat treatment, the oil loss from CPI oleogels was significantly lower than from the FBPI oleogels, indicating the ability of canola proteins to undergo more heat-induced denaturation and subsequent protein aggregation leading to a better ability to protect the oil droplets from coalescence and disruption.

The viscoelastic properties of the oleogels were characterized by the dynamic oscillatory measurements (strain and frequency sweep) (Figure 4.3). In the strain sweep study, all oleogels showed higher G' than G'' before crossover. A linear viscoelastic region (LVR) was also observed between 0.01 to about 0.6% strain, indicating a strong gel structure. Beyond LVR, both G' and G'' dropped, went through a strain region where the values of G' and G'' overlapped, and then the G'' became higher than G', indicating gel breakdown. For the UE CPI oleogel (Figure 4.3B), a peak in G'' was observed at around 250% strain, which could be due to a two-step yielding leading to a gel breakdown and release of oil droplets from the oleogel. Such a peak in G'' at higher strain was much less prominent in the HE CPI oleogels (Figure 4.3B), indicating improved structural stability of the oleogels made from the heated emulsions. Unlike the CPI oleogels, no peaks in G'' were observed after the crossover in both UE and HE FBPI oleogels (Figure 4.3A). The UE FBPI oleogels also exhibited longer LVR and higher G' within the LVR (~10 000 Pa) than the UE CPI oleogels (~1000 Pa), which led to the lower oil loss reported in Figure 4.2 for the pulse protein oleogels compared to canola proteins without any heat treatment.

In the frequency sweep study, all oleogels showed a higher G' than G'' at all frequencies up to 100 Hz (Figure 4.3C and D). At 100 Hz, a crossover of G' and G'' was observed in the UE CPI oleogels, indicating a gel breakdown at high-frequency shear. The HE CPI oleogels exhibited stronger gel-like

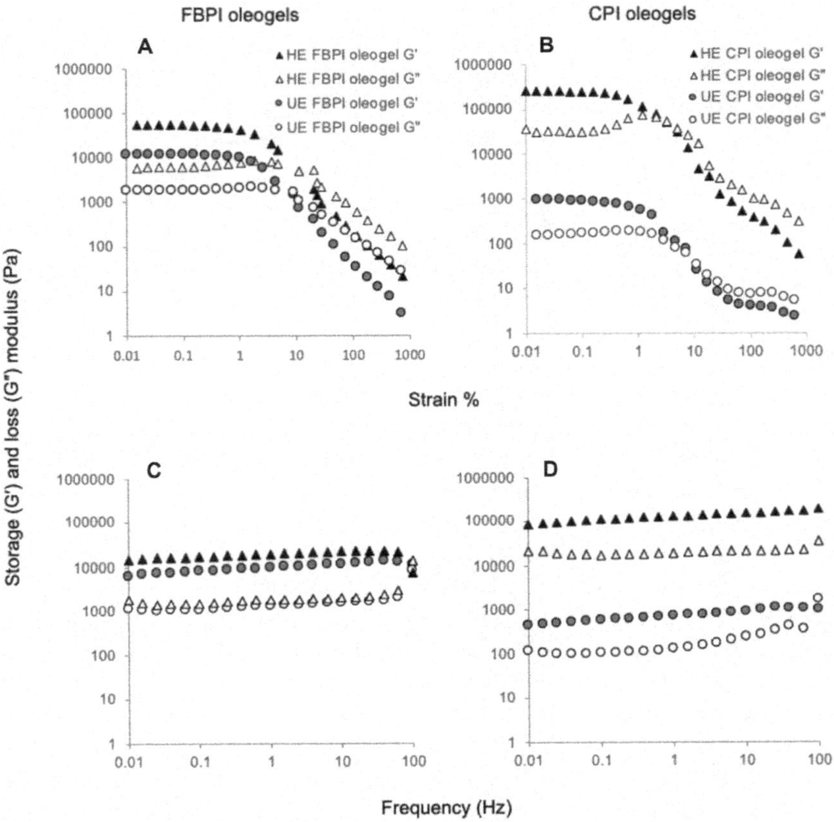

Figure 4.3 Oleogel viscoelasticity. (A and B) Strain sweep viscoelasticity of (A) FBPI and (B) CPI oleogels made from heated (triangle) and unheated (circle) emulsions. (C and D) Frequency sweep viscoelasticity of (C) FBPI and (D) CPI oleogels made from heated (triangle) and unheated (circle) emulsions. Storage moduli (G') are denoted with closed symbols, while loss moduli (G'') are shown with open symbols.

behavior with higher G' and G'' than the UE CPI oleogels at all frequencies without any crossover. In contrast, the crossover was observed both in the UE and HE FBPI oleogels. Moreover, the difference of G' in the UE and HE oleogels was not as large for the FBPI oleogels as in the CPI oleogels. Overall, the HE CPI oleogels showed the highest G' at about 100 000 Pa with a LVR in the frequency range studied, indicating the strongest oleogels of the four.

4.7.2 Cake Batter and Cake Properties Prepared with Faba Bean and Canola Protein Isolate-based Oleogels

Cakes were prepared according to the AACC[94] method# 10–90. Each cake contained 175 g water, 140 g sugar, 100 g all-purpose flour, 50 g fat, 12 g non-fat dried milk, 9 g dried egg white powder, 6.25 g baking powder, and 3 g NaCl. Different oleogels were used as a full replacement for the fat phase of the cake. As a control, conventional high-saturated fat-containing shortening and liquid canola oil were used as the fat phase. The cake batter was prepared by mixing the ingredients using a KitchenAid Ultra Power Mixer (KitchenAid, Whirlpool Canada LP, Mississauga ON) with a 4.5 qt (4.3 L) stationary bowl and rotating stirrers. The batter was transferred to a muffin pan (1.5×7.24×12.5″) and baked at 375 °F (190.6 °C) for 15 min followed by cooling down to room temperature. The cakes were appropriately wrapped and stored at room temperature before further characterization 24 h after baking.

The specific gravity of the cake batters (ratio of the weight of equal volume cake batter to water) is an indication of the ability of the fats to entrap and stabilize air during the batter mixing process. Though the shortening cake batters showed the lowest average specific gravity (higher air incorporation), all the cake batters were not significantly different, except for the shortening and HE CPI oleogel cake batter (Table 4.3). The cake specific volume determined using a rapeseed displacement method[48] is the measure of cake volume per unit weight. The cake made with shortening showed the lowest specific volume but was not statistically different than the UE and HE FBPI oleogel cakes. Among all the cakes, the HE CPI oleogel cakes showed the highest specific cake volume indicating the ability of heated CPI to better hold air bubbles in the cake structure.

The effect of the fat types on the cake appearance was also studied through color analysis. The color of the cake crumbs was determined using a colorimeter and reported as values such as L^* (lightness/darkness), a^*

Table 4.3 Cake batter specific gravity, cake specific volume, and color analysis of cake crumbs made with different fat types. Different letters in the same row indicates significant differences ($P<0.05$).

Fat type	Shortening	Canola oil	Unheated CPI oleogel	Heated CPI oleogel	Unheated FBPI oleogel	Heated FBPI oleogel
Batter Specific gravity	1.0 ± 0.0^{a}	1.1 ± 0.0^{ab}	1.1 ± 0.0^{ab}	1.1 ± 0.0^{b}	1.1 ± 0.0^{ab}	1.1 ± 0.0^{ab}
Cake specific volume	1.5 ± 0.1^{a}	1.7 ± 0.1^{b}	1.7 ± 0.1^{bcd}	1.8 ± 0.1^{d}	1.6 ± 0.1^{abc}	1.6 ± 0.1^{ab}
Color analysis L^*	73.9 ± 0.8^{a}	80.3 ± 0.7^{b}	73.5 ± 1.3^{a}	71.7 ± 0.8^{c}	78.1 ± 0.9^{d}	77.9 ± 0.8^{d}
Color analysis a^*	-0.4 ± 0.2^{a}	-0.6 ± 0.0^{b}	1.1 ± 0.4^{c}	1.1 ± 0.2^{c}	0.1 ± 0.1^{d}	0.2 ± 0.0^{d}
Color analysis b^*	19.8 ± 1.6^{a}	17.8 ± 0.2^{b}	21.6 ± 0.4^{c}	21.3 ± 0.3^{c}	18.9 ± 0.3^{d}	19.1 ± 0.3^{ad}

(redness/greenness), and b^* (blueness/yellowness) (Table 4.3). The L^* values were the highest for the canola oil cakes, followed by both the UE and HE FBPI oleogel cakes and the shortening and UE CPI oleogel cakes. The HE CPI oleogel cakes showed the lowest L^* value, indicating a darker crumb color. For the a^* values, both UE and HE CPI oleogels exhibited the highest value, indicating a higher redness than the rest of the cakes. Overall, the lower L^* (darker), higher b^* (higher yellowness), and higher a^* (higher redness) of the CPI oleogel cakes could be attributed to the presence of the CPIs, which are dark brown due to the oxidation from the phenolic compounds.[95] The a^* values for UE and HE FBPI oleogels were not significantly different from each other ($p > 0.05$), while the shortening and canola oil cakes showed a negative a^* value, indicating a higher greenness than the rest of the cakes ($p < 0.05$). For the b^* values, both UE and HE CPI oleogels had the highest value, indicating a higher yellowness in the crumbs, while the canola oil cake crumbs exhibited the lowest b^*. The b^* of the HE FBPI oleogel cakes was not significantly different than those for the UE FBPI and shortening cakes.

4.7.3 Texture Analysis of Cakes Prepared with Different Fat Types

Figure 4.4 shows the results of the texture analysis of cakes from the uniaxial two-step compression test. From the peak and area of the force *vs* time graph, the hardness, springiness, cohesiveness, and chewiness can be calculated. Hardness is defined as the maximum force obtained during the first compression curve (Figure 4.4A). In this study, the shortening cake exhibited the highest hardness (∼ 10 N), while the UE FBPI cakes showed the lowest hardness (∼6 N). Springiness (Figure 4.4B) represents the extent that the cake recovers to its initial height after the first compression. The shortening, canola oil, CPI oleogel cakes showed fairly close springiness values (∼0.96–0.97), but they were significantly higher for the shortening cake ($p < 0.05$). Both the UE and HE FBPI oleogel cakes displayed the lowest springiness (∼0.93), indicating a smaller recovery of cake height after the compression. The cohesiveness (Figure 4.4C) of the cake was calculated by taking the ratio of the first peak area and the second peak area. This indicates the crumbliness or perceptions of denseness in the sensory studies.[96] The shortening cake had the lowest cohesiveness, while the oleogel cakes showed no significant difference in cohesiveness among each other ($p > 0.05$). Lastly, chewiness (Figure 4.4D) is a product of the primary parameters such as hardness, springiness, and cohesiveness. The shortening cake showed the highest chewiness; however, it was not significantly different than the canola oil and UE CPI oleogel cakes ($p > 0.05$). The UE FBPI cakes had the lowest chewiness, but it was not significantly different than the HE CPI and FBPI oleogel cakes ($p > 0.05$). Overall, CPI and FBPI emulsion-templated oleogels showed potential as a fat replacer in cake baking. Though the texture profile of the cakes varies, sensory tests on the cakes could be a better indicator of consumer acceptability of oleogel-based cakes.

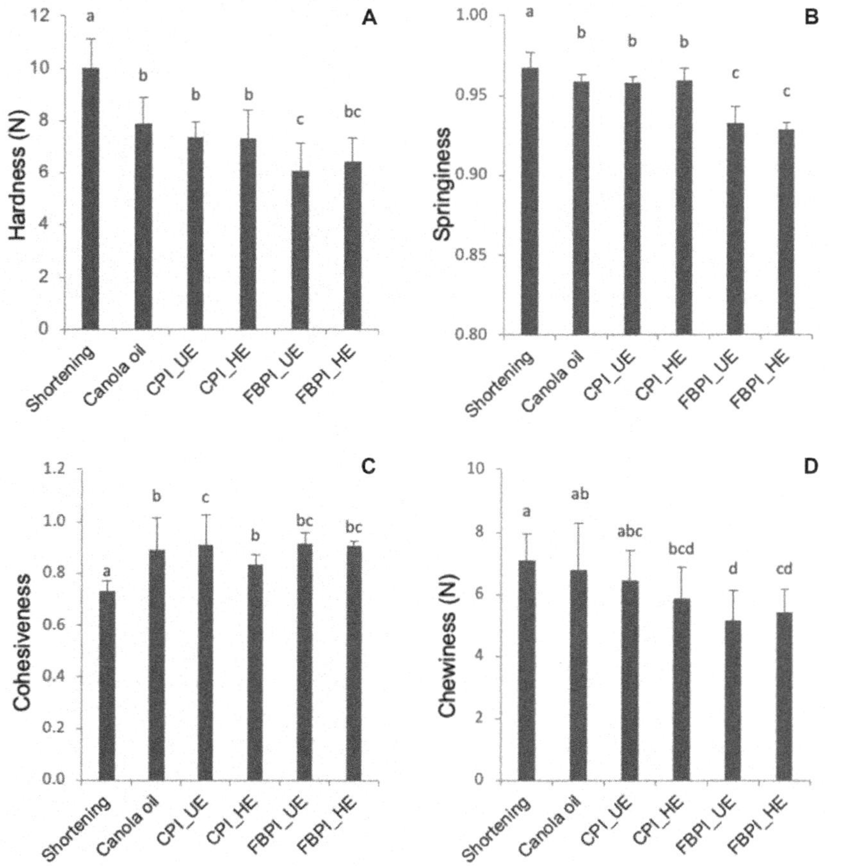

Figure 4.4 Texture parameters (A) hardness, (B) springiness, (C) cohesiveness, and (D) chewiness of cakes prepared with different fat types. Different letters indicate significant difference ($p<0.05$).

4.8 Summary

The current chapter focused on the utilization of biopolymers to create structure in liquid oil, thereby developing the so-called oleogels *via* an indirect approach. It first discussed different types of oleogelators based on their direct or indirect approach to oleogelation. The crystalline organic or inorganic particles and low-molecular-weight molecules were mostly used for direct oleogelation, while biopolymers, such as plant proteins and polysaccharides, formed oleogels *via* an indirect approach. So far emulsion, foam, and hydrogel templates as starting materials have been used for indirect oleogelation. Since the biopolymers tend to have more hydrophilic moieties on the outside and hydrophobic moieties buried inside the structure, these templates help to expose the hydrophobic moieties first, and then by removal of the water and addition of liquid oil, the system is transformed into an oleogel.

Most of the biopolymer-based oleogels have been characterized by their oil binding capacity, rheology, and texture. Drying methods, namely oven-drying *vs.* freeze-drying, also proved to be an important factor of oleogel stability, rheology, and texture. However, only very few studies have directly compared the effect of these two drying methods, where freeze-drying led to a higher firmness and mechanical strength of the oleogels. With an increase in biopolymer concentration, an increase in interfacial and network strength was observed, which led to a higher oil binding capacity and firmness of the oleogels. Some authors showed that the addition of low molecular weight emulsifier or crystalline additive in the oil phase further improved the oil binding capacity. All oleogels showed strong viscoelastic behavior with typically higher gel strength for emulsion than foam-templated oleogels. The emulsion-templated oleogels also showed better thixotropic recovery compared to foam-templated oleogels. Although not studied in detail, some oleogels also showed thermal stability, which could be important to understand their behavior under baking conditions or in various non-food applications, such as bio-lubricants. The hydrogel-templated oleogels showed higher firmness compared to the foam and emulsion-templated oleogels.

A review of food applications of biopolymer-based oleogels showed a range of products from cakes, cookies, ice cream, and sausages, although cakes were studied the most. Usually, the hardness of protein-stabilized oleogel-based cakes is higher than conventional shortening-based cakes, which could be ascribed to enhanced interactions between proteins and wheat starch. It was also shown that the addition of lipophilic emulsifiers, such as MAG, could prevent such interactions, leading to a softer cake texture. Although biopolymer-based oleogels have been investigated by many, only a handful of studies have looked into the sensory acceptability of the foods developed with them. Even then, a wide range of variation in acceptability can be seen in sensory analysis. This is one area where much research is needed before oleogels can be successfully incorporated into food systems.

Finally, this chapter also reported some of our recent work on the pulse and oilseed protein-stabilized emulsion-templated oleogels and their food applications. Concentrated oil-in-water emulsions, made with 50 wt% canola oil and stabilized with 4 wt% faba bean protein isolates (FBPIs) or canola protein isolates (CPIs), were dried in a vacuum oven and subsequently sheared to obtain the oleogels. It was found that heat treatment of the emulsions prior to drying was essential to improve protein interfacial strength leading to a higher oil binding capacity, and better rheology of the oleogel. Overall, the heat-treated CPI-based oleogels were more stable with significantly lower oil loss and gel strength than the heat-treated FBPI-based oleogels, indicating that the protein's ability of heat-denaturation and aggregation are essential factors in oleogel development. The stability of the oleogels, however, did not affect the hardness of the cakes baked with them, and both FBPI and CPI oleogel cakes were softer than the conventional shortening-based cakes. CPI-oleogel-based cakes were, however, darker in color due to the dark color of the canola proteins.

4.9 Future Prospects

Compared to the application of other oleogelators, such as crystalline particles and LMW self-assembled molecules, which might have some restrictions with food safety regulations, plant proteins pose fewer food safety concerns. Plant proteins also have higher consumer acceptability, and the demand for plant-based food has been rising in recent years. Moreover, the utilization of plant proteins as oleogelators could enhance its value in the food ingredients market. Various studies have also shown the nutritional benefits of the application of oleogels in food, for example, a reduction of lipid digestibility (compared to canola oil), serum triacylglycerol (compared to margarine and butter), and insulin levels (compared to liquid oil).[97–99] However, the nutritional benefits of plant-protein-based oleogels have not been reported yet. In terms of functionality, although several studies on the characteristics and application of plant protein-based oleogels have been reported, the underlying mechanisms of protein–oil interaction and its action during baking and similar food processing remain unknown, which is critical to extending its consistent application in food. Sensory studies on the food products developed with plant protein-oleogels are also lacking. Hence, extensive research is still needed for the application of plant-protein-based oleogels as saturated fat replacers due to the variables such as protein source, properties, food formulations, processing, ingredient interactions, and production costs.

References

1. A. Gravelle, M. Davidovich-Pinhas, A. Zetzl, S. Barbut and A. Marangoni, *Carbohydr. Polym.*, 2016, **135**, 169–179.
2. F. D. Gunstone, *Modifying Lipids for Use in Food*, Woodhead Publishing, 2006.
3. A. G. Marangoni, N. Acevedo, F. Maleky, F. Peyronel, G. Mazzanti, B. Quinn and D. Pink, *Soft Matter*, 2012, **8**, 1275–1300.
4. M. A. Rogers, *Food Res. Int.*, 2009, **42**, 747–753.
5. M. Pernetti, K. F. van Malssen, E. Flöter and A. Bot, *Curr. Opin. Colloid Interface Sci.*, 2007, **12**, 221–231.
6. A. Juriaanse and I. Heertje, *Food Struct.*, 1988, **7**, 8.
7. A. G. Marangoni, *J. Am. Oil Chem. Soc.*, 2012, **89**, 749–780.
8. J. W. Coenen, *J. Am. Oil Chem. Soc.*, 1976, **53**, 382–389.
9. D. Singh, M. Rezac and P. Pfromm, *J. Am. Oil Chem. Soc.*, 2009, **86**, 93.
10. A. Keys, *Circulation*, 1970, **41**, 186–195.
11. W. Willett, *Atheroscler. Suppl.*, 2006, **7**, 5–8.
12. A. Erkkilä, V. D. de Mello, U. Risérus and D. E. Laaksonen, *Prog. Lipid Res.*, 2008, **47**, 172–187.
13. F. B. Hu and W. C. Willett, *JAMA*, 2002, **288**, 2569–2578.
14. R. P. Mensink, P. L. Zock, A. D. Kester and M. B. Katan, *Am. J. Clin. Nutr.*, 2003, **77**, 1146–1155.
15. U. Risérus, W. C. Willett and F. B. Hu, *Prog. Lipid Res.*, 2009, **48**, 44–51.

16. Health Canada, 2018, https://www.canada.ca/en/health-canada/news/2018/09/canadian-ban-on-trans-fatscomes-into-force-today.html (accessed August 2019).
17. FDA, 2018, https://www.fda.gov/food/food-additivespetitions/final-determination-regarding-partially-hydrogenated-oils-removing-trans-fat (accessed August 2019).
18. W. C. Willett, *Science*, 1994, **264**, 532–537.
19. P. W. Siri-Tarino, Q. Sun, F. B. Hu and R. M. Krauss, *Am. J. Clin. Nutr.*, 2010, **91**, 535–546.
20. H. Sawalha, R. den Adel, P. Venema, A. Bot, E. Flöter and E. van der Linden, *J. Agric. Food Chem.*, 2012, **60**, 3462–3470.
21. J. Daniel and R. Rajasekharan, *J. Am. Oil Chem. Soc.*, 2003, **80**, 417–421.
22. M. A. Rogers, A. J. Wright and A. G. Marangoni, *Soft Matter*, 2008, **4**, 1483–1490.
23. X. Yan, Y. Cui, Q. He, K. Wang and J. Li, *Chem. Mater.*, 2008, **20**, 1522–1526.
24. G. List and R. Reeves, *Trans Fat Alternatives*, AOCS Press, Champaign, 2005, pp. 71–86.
25. A. G. Marangoni and N. Garti, *Edible oleogels: structure and health implications*, Elsevier, 2018.
26. M. Bodennec, Q. Guo and D. Rousseau, *RSC Adv.*, 2016, **6**, 47373–47381.
27. S. Murdan, G. Gregoriadis and A. T. Florence, *J. Pharm. Sci.*, 1999, **88**, 608–614.
28. A. R. Patel and K. Dewettinck, *Eur. J. Lipid Sci. Technol.*, 2015, **117**, 1772–1781.
29. B. O. Okesola, V. M. Vieira, D. J. Cornwell, N. K. Whitelaw and D. K. Smith, *Soft Matter*, 2015, **11**, 4768–4787.
30. A. R. Patel, *Curr. Opin. Food Sci.*, 2017, **15**, 1–7.
31. A. R. Patel, D. Schatteman, A. Lesaffer and K. Dewettinck, *Rsc Adv.*, 2013, **3**, 22900–22903.
32. I. Schmidt, B. Novales, F. Boué and M. A. Axelos, *J. Colloid Interface Sci.*, 2010, **345**, 316–324.
33. I. Tavernier, A. R. Patel, P. Van der Meeren and K. Dewettinck, *Food Hydrocolloids*, 2017, **65**, 107–120.
34. A. de Vries, J. Hendriks, E. van der Linden and E. Scholten, *Langmuir*, 2015, **31**, 13850–13859.
35. D. Asioli, J. Aschemann-Witzel, V. Caputo, R. Vecchio, A. Annunziata, T. Næs and P. Varela, *Food Res. Int.*, 2017, **99**, 58–71.
36. C. Amoah, J. Lim, S. Jeong and S. Lee, *LWT*, 2017, **86**, 430–437.
37. H. Schaink, K. Van Malssen, S. Morgado-Alves, D. Kalnin and E. Van der Linden, *Food Res. Int.*, 2007, **40**, 1185–1193.
38. A. J. Wright and A. G. Marangoni, *J. Am. Oil Chem. Soc.*, 2006, **83**, 497–503.
39. S. Guo, M. Lv, Y. Chen, T. Hou, Y. Zhang, Z. Huang, Y. Cao, M. Rogers and Y. Lan, *Food Funct.*, 2020, **11**, 2048–2057.

40. Y. Jiang, L. Liu, B. Wang, X. Sui, Y. Zhong, L. Zhang, Z. Mao and H. Xu, *Food Hydrocolloids*, 2018, **77**, 460–464.
41. S. Lee, *Food Hydrocolloids*, 2018, 77, 796–802.
42. W. Wijaya, Q.-Q. Sun, L. Vermeir, K. Dewettinck, A. R. Patel and P. Van der Meeren, *Food Struct.*, 2019, **21**, 100112.
43. C. Qiu, Y. Huang, A. Li, D. Ma and Y. Wang, *J. Agric. Food Chem.*, 2018, **66**, 13243–13252.
44. S.-Z. Luo, X.-F. Hu, Y.-J. Jia, L.-H. Pan, Z. Zheng, Y.-Y. Zhao, D.-D. Mu, X.-Y. Zhong and S.-T. Jiang, *Food Hydrocolloids*, 2019, **95**, 76–87.
45. Z. Meng, K. Qi, Y. Guo, Y. Wang and Y. Liu, *Food Chem.*, 2018, **246**, 137–149.
46. Z.-M. Gao, X.-Q. Yang, N.-N. Wu, L.-J. Wang, J.-M. Wang, J. Guo and S.-W. Yin, *J. Agric. Food Chem.*, 2014, **62**, 2672–2678.
47. X.-W. Chen and X.-Q. Yang, *J. Agric. Food Chem.*, 2019, **67**, 2637–2646.
48. A. Mohanan, Y. R. Tang, M. T. Nickerson and S. Ghosh, *RSC Adv.*, 2020, **10**, 14892–14905.
49. R. Tanti, S. Barbut and A. G. Marangoni, *Food Hydrocolloids*, 2016, **61**, 329–337.
50. L. Manzocco, F. Valoppi, S. Calligaris, F. Andreatta, S. Spilimbergo and M. C. Nicoli, *Food Hydrocolloids*, 2017, **71**, 68–75.
51. S. Plazzotta, S. Calligaris and L. Manzocco, *Food Hydrocolloids*, 2019, **96**, 1–10.
52. M. Davidovich-Pinhas, S. Barbut and A. Marangoni, *Carbohydr. Polym.*, 2015, **127**, 355–362.
53. A. Patel, B. Mankoč, M. B. Sintang, A. Lesaffer and K. Dewettinck, *RSC Adv.*, 2015, **5**, 9703–9708.
54. A. Bot and W. G. Agterof, *J. Am. Oil Chem. Soc.*, 2006, **83**, 513–521.
55. Y. Zhu, X. Chen, D. J. McClements, L. Zou and W. Liu, *Food Hydrocolloids*, 2018, **77**, 870–878.
56. E. A. Foegeding, P. Luck and J. P. Davis, *Food Hydrocolloids*, 2006, **20**, 284–292.
57. B. S. Murray and R. Ettelaie, *Curr. Opin. Colloid Interface Sci.*, 2004, **9**, 314–320.
58. W. Kloek, T. Van Vliet and M. Meinders, *J. Colloid Interface Sci.*, 2001, **237**, 158–166.
59. A. R. Patel, P. S. Rajarethinem, N. Cludts, B. Lewille, W. H. De Vos, A. Lesaffer and K. Dewettinck, *Langmuir*, 2015, **31**, 2065–2073.
60. C. Chang, S. Tu, S. Ghosh and M. Nickerson, *Food Res. Int.*, 2015, 77, 360–367.
61. K. Chen and H. Zhang, *ACS Appl. Mater. Interfaces*, 2020, **12**, 7795–7804.
62. G. W. Scherer and D. M. Smith, *J. Non-Cryst. Solids*, 1995, **189**, 197–211.
63. H. Gesser and P. Goswami, *Chem. Rev.*, 1989, **89**, 765–788.
64. L. W. Hrubesh and J. F. Poco, *J. Non-Cryst. Solids*, 1995, **188**, 46–53.
65. I. Tavernier, C. D. Doan, P. Van der Meeren, B. Heyman and K. Dewettinck, *Eur. J. Lipid Sci. Technol.*, 2018, **120**, 1700393.

66. A. Mohanan, M. T. Nickerson and S. Ghosh, *J. Am. Oil Chem. Soc.*, 2020, **97**, 1319–1333.
67. F. R. Lupi, V. Greco, N. Baldino, B. de Cindio, P. Fischer and D. Gabriele, *J. Colloid Interface Sci.*, 2016, **483**, 154–164.
68. A. R. Patel, *Mater. Chem. Phys.*, 2017, **195**, 268–274.
69. A. R. Patel, N. Cludts, M. D. Bin Sintang, B. Lewille, A. Lesaffer and K. Dewettinck, *Chem. Phys. Chem.*, 2014, **15**, 3435–3439.
70. A. Y. Malkin and A. I. Isayev, *Rheology: Concepts, Methods, and Applications*, Elsevier, 2017.
71. P. Y. Phoon and C. J. Henry, *Food Funct.*, 2020, **11**, 1349–1361.
72. Z. Meng, K. Qi, Y. Guo, Y. Wang and Y. Liu, *Food Hydrocolloids*, 2018, **77**, 17–29.
73. A. R. Patel, P. S. Rajarethinem, A. Grędowska, O. Turhan, A. Lesaffer, W. H. De Vos, D. Van de Walle and K. Dewettinck, *Food Funct.*, 2014, **5**, 645–652.
74. D. C. Zulim Botega, A. G. Marangoni, A. K. Smith and H. D. Goff, *J. Food Sci.*, 2013, **78**, C1334–C1339.
75. D. Kouzounis, A. Lazaridou and E. Katsanidis, *Meat Sci.*, 2017, **130**, 38–46.
76. A. R. Patel, N. Cludts, M. D. B. Sintang, A. Lesaffer and K. Dewettinck, *Food Funct.*, 2014, **5**, 2833–2841.
77. H. Pehlivanoglu, G. Ozulku, R. M. Yildirim, M. Demirci, O. S. Toker and O. Sagdic, *J. Food Process. Preserv.*, 2018, **42**, e13621.
78. A. Paraskevopoulou and V. Kiosseoglou, *J. Food Sci.*, 1997, **62**, 208–211.
79. S. Cauvain, *Trends Food Sci. Technol.*, 1998, **9**, 56–61.
80. L. Adili, L. Roufegarinejad, M. Tabibiazar, H. Hamishehkar and A. Alizadeh, *LWT*, 2020, 109277.
81. M. Tabibiazar, L. Roufegarinejad, H. Hamishehkar and A. Alizadeh, *Food Chem.*, 2020, **333**, 127446.
82. E. Yılmaz and M. Öğütcü, *Food Funct.*, 2015, **6**, 1194–1204.
83. I. Demirkesen and B. Mert, *J. Am. Oil Chem. Soc.*, 2019, **96**, 545–554.
84. S. A. Willett and C. C. Akoh, *J. Food Sci.*, 2019, **84**, 1390–1399.
85. J. Y. Kim, J. Lim, J. Lee, H. S. Hwang and S. Lee, *J. Food Sci.*, 2017, **82**, 445–452.
86. J. Alvarez-Ramirez, E. Vernon-Carter, Y. Carrera-Tarela, A. Garcia and C. Roldan-Cruz, *LWT*, 2020, 109701.
87. D. Jung, I. Oh, J. Lee and S. Lee, *LWT*, 2020, 109194.
88. B. Mert and I. Demirkesen, *LWT – Food Sci. Technol.*, 2016, **68**, 477–484.
89. H. L. Bemer, M. Limbaugh, E. D. Cramer, W. J. Harper and F. Maleky, *Food Res. Int.*, 2016, **85**, 67–75.
90. N. Koumakis and G. Petekidis, *Soft Matter*, 2011, **7**, 2456–2470.
91. J. Lim, S. Jeong and S. Lee, *LWT*, 2017, **84**, 788–794.
92. Y. R. Tang and S. Ghosh, *Food Hydrocolloids*, 2020, **113**, 106399.
93. A. I. Blake and A. G. Marangoni, *J. Am. Oil Chem. Soc.*, 2014, **91**, 885–903.

94. AACC, *AACC International. Approved Methods of Analysis, 11th Ed. Method 90-10.01. Baking Quality of Cake Flour. Approved October 08, 1976. Re-approval November 3, 1999*, AACC International, St. Paul, MN, USA, 1999, https://methods.aaccnet.org/methods/10-90.pdf.
95. L. Xu and L. Diosady, *Food Res. Int.*, 2002, **35**, 23–30.
96. M. A. Brandt, E. Z. Skinner and J. A. Coleman, *J. Food Sci.*, 1963, **28**, 404–409.
97. A. G. Marangoni, S. H. Idziak, C. Vega, H. Batte, M. Ollivon, P. S. Jantzi and J. W. Rush, *Soft Matter*, 2007, **3**, 183–187.
98. T. A. Stortz, A. K. Zetzl, S. Barbut, A. Cattaruzza and A. G. Marangoni, *Lipid Technol.*, 2012, **24**, 151–154.
99. M. Chloe, M. Davidovich-Pinhas, A. J. Wright, S. Barbut and A. G. Marangoni, *Food Funct.*, 2017, **8**, 1438–1451.

CHAPTER 5

Physicochemical Properties and Polymorphic Behavior of Tropical Fats: Their Potential and Practical Use in Trans-*free Food Systems*

M. R. RAMOS-RAMOS, V. A. GARCIA-LONDOÑO, V. BORRONI AND M. L. HERRERA*

Institute of Polymer Technology and Nanotechnology, Facultad de Arquitectura Diseño y Urbanismo, University of Buenos Aires-CONICET, Intendente Güiraldes, 2160, (C1428EGA) Ciudad de Buenos Aires, Argentina
*Email: lherrera@fi.uba.ar

5.1 Introduction

Hydrogenated oils have been widely used in the past for their technological properties and low cost. However, *trans*-fats legislation forced the food industry to reformulate products.[1] In this context, tropical fats gained increasing interest. They are natural sources of symmetrical stearic-rich triacylglycerols (TAGS) useful for chocolate and confection production. In addition, as tropical fats are semi-solid edible vegetable fats, they provide levels of solid fat content (SFC) as a function of temperature that might be suitable for margarine and shortening manufacture. The replacement of

Food Chemistry, Function and Analysis No. 31
Development of *Trans*-free Lipid Systems and their Use in Food Products
Edited by Jorge F. Toro-Vazquez

Published by the Royal Society of Chemistry, www.rsc.org

partially hydrogenated vegetable oils with tropical fats could greatly reduce the incidence of *trans*-fats in fatty food products and the health risks associated with *trans*-fats consumption.[2]

In addition to *trans*-fat concerns, saturated fat content is another subject of controversy in the food industry. Several associations around the world advise reducing the amount of saturated fat in the diet because of its negative effect on blood cholesterol levels. The web page of the American Heart Association recommends aiming for a dietary pattern that achieves 5% to 6% of calories from saturated fat. However, the World Health Organization (WHO) still maintains its recommendation that the intake of saturated fats should be less than 10% of total energy intake. For *trans*-fats, the recommendation is less than 1% of total energy intake. The WHO advises people to switch their fat intake from saturated fat and *trans*-fat to unsaturated fat. Blending tropical fats with vegetable oils is an alternative way to obtain a shortening with appropriate SFC but a better fatty acid profile from a nutritional point of view. These improved formulations would widen tropical fat applications in food systems. Nowadays, tropical fats are used to prepare several products such as margarines, non-dairy creams, shortenings for baking and frying, chocolate, and confectionery. In chocolate and confections, new cocoa butter substitutes or equivalents are being developed to formulate *trans*-fat-free coatings and fillings.[3]

Tropical fats may be categorized into three groups according to the fatty acid composition: those having medium chain fatty acids, usually called lauric fats, like coconut or palm kernel oils, those rich in palmitic acid, like palm oil and its fractions, and those rich in stearic acid, like cocoa butter, shea, and cupuassu fats. Palm oil and coconut oil are two of the most important oil crops worldwide. These oils have good shelf stability, a creamy texture, and other features desired by food manufacturers. Palm kernel oil extracted from palm kernels has become the most important lauric oil nowadays. Recently, other non-traditional fats such as mango kernel fat and winged bean oil have attracted great attention for their good physicochemical properties.[4–6] These fats, not exploited commercially yet, could be good alternatives to *trans*-fat.

The functional and nutritional properties of fats and oils depend on the fatty acid distribution in the glycerol backbone. Therefore, the TAGS composition of a fat system is very important as it will determine its physical properties. Tropical fats usually have a saturated TAGS content higher than required for plastic fats. Thus, they are not used as extracted but after fractionation, full hydrogenation, interesterification, mixing with other fats, or a combination of these processes to reach the desired chemical composition. When developing a structured lipid for this purpose, it is important to consider the novel fat physical properties and functionality in comparison with those properties provided by the *trans*-fat that will be replaced. In this way, the final fat product will have the required appearance and shelf life.[7] The removal of *trans*-fat from food products is still challenging since some beneficial functional characteristics of products such as margarines are attributable to them. *Trans*-isomers influence physical properties such as thermal and polymorphic behaviors, and texture of margarines or shortenings. To describe the physical behavior of a fat system the selected properties measured are usually melting and crystallization behavior, SFC, polymorphism, rheological properties, and microstructure.

5.2 Chemical Composition

5.2.1 Fatty Acid Composition

The fatty acid composition of commercial hydrogenated oils used to formulate margarines or shortenings before and after *trans*-fat legislation has been reported by several authors (see Table 5.1). The examples shown in Table 5.1 were intended for different applications and therefore have different *trans*-fatty acid contents.

The first two examples indicated in Table 5.1 correspond to hydrogenated shortening used for laminated bakery products.[8] The notable difference between partial-hydrogenated-oil-based shortening and tropical fats is the presence of *trans*-fatty acids, which are in significant amounts in the examples shown in Table 5.1. The method used for analysis only determined C18:1*t* isomers (elaidic acid, octadec-9*t*-enoic acid). However, C18;2*t* (linolelaidic acid, octadecadienoic acid) isomers may also be present, indicating that total *trans*-fat content could be higher in the first two examples in Table 5.1. In those hydrogenated shortenings the *trans*-fatty acid contents are high considering the recommendations provided by the WHO (limit *trans*-fat to 1% of total energy intake).

Table 5.1 Fatty acid composition of hydrogenated oils produced before and after *trans*-fat legislation. The composition is reported in percentage (%).

Fat type[a]	C12:0	C14:0	C16:0	C18:0	C18:1*c*	C18:1*t*	C18:2*c*	C18:2*t*
Hydrogenated vegetable and palm oils[8]	—	—	19.1	12.0	31.0	13.0	11.7	—
Hydrogenated soybean and cottonseed oils[8]	—	—	13.3	16.0	26.4	16.0	11.1	—
Hydrogenated sunflower oil[9]	—	—	8.6	12.0	42.0	13.4	12.3	—
Hydrogenated canola oil[9]	—	—	8.7	9.9	54.5	13.4	5.6	—
Hydrogenated sunflower oil[10]	—	0.1	6.7	29.5	38.8	9.2	2.2	12.6
Hydrogenated sunflower oil[10]	—	0.1	6.6	31.2	41.0	7.6	0.4	12.4
Hydrogenated soybean oil[11]	0.1	0.2	14.1	4.43	32.7	15.7	18.5	5.4
Hydrogenated vegetable fat for margarine[12]	—	0.1	11.8	9.1	22.3	35.6	21.0	—

[a]The fat type column indicates the reference where the corresponding fatty acid composition was originally reported. In all cases the standard deviation was less than 1%.

Because of the increased health concerns about *trans*-fatty acids in the last 20 years, new hydrogenation processes have been studied to seek ways for a substantial reduction of *trans*-fatty acids in hydrogenated products. Among the new hydrogenation processes a catalyst of diatomite activated by acid to increase the surface was investigated.[9] A low percentage palladium catalyst containing 0.2% of Pd/Diatomite was synthesized to hydrogenate sunflower and canola oils, at the laboratory and industrial scale, respectively. The catalyst significantly reduced the formation of *trans*-isomers in hydrogenated fat compared to a typical commercial Ni catalyst when performing the process under the same conditions of temperature and hydrogen pressure.[9] The classical hydrogenation process employed before *trans*-fat legislation gave very different results. For example, the hydrogenation process of sunflower oil, performed in a batch slurry reactor and with a supported nickel catalyst at a concentration of 0.05%, either under a high temperature of reaction and low pressure of hydrogen gas (190 °C, 1 $kg\,cm^{-2}$) or under low temperature and high pressure (170 °C, 2.5 $kg\,cm^{-2}$), led to not only a higher *trans*-fat content than with a Pd/Diatomite catalyst but also to a fat with a different *trans* isomer profile. C18:2*t* was the mayor *trans*-fatty acid instead of C18:1*t*.[10] The *trans* isomer content reported for sunflower and canola oils hydrogenated with Pd/Diatomite were similar to the ones found for commercial hydrogenated soybean oil.[11] The last example in Table 5.1 is a hydrogenated vegetable fat that was used for margarine before the *trans*-fat legislation.[12] In conclusion, with the reaction conditions the *trans*-fats resulted in contents that were at least twice the content achieved today in hydrogenated oils.

The fatty acid composition of different tropical fats, used or with the potential of being used as *trans*-fat alternatives, have also been reported by several authors (see Table 5.2). There is a wide range of fatty acid compositions indicating that these fats will have different physical properties, some suitable for food applications.

Table 5.2 Fatty acid composition of tropical fats used to formulate *trans*-fat-free products. The composition is reported in percentage (%).

Fat type[a]	C10:0	C12:0	C14:0	C16:0	C18:0	C18:1c	C18:2c	C20:0	C22:0
Palm kernel oil[14]	3.45	50.63	16.33	8.36	1.88	13.35	2.02	—	—
Coconut oil[15,16]	5.71	45.23	19.32	10.37	3.17	7.25	1.76	0.14	—
Palm oil[17]	0.07	0.87	1.17	45.74	4.49	40.59	5.54	—	—
Palm stearin[18–20]	0.07	0.90	1.13	46.50	4.58	38.70	8.25	0.18	—
Cocoa butter[21]	0.09	0.32	0.21	27.71	32.82	34.51	2.31	0.31	—
Cupuassu fat[20,22]	0.02	0.61	0.22	7.57	31.60	41.61	5.20	11.10	2.03
Shea butter[23]	—	—	—	3.5	42.1	48.6	3.2	1.4	—
Mango kernel fat[4]	—	—	0.2	6.7	40.1	45.3	5.8	—	—
Winged bean oil[6]	—	—	0.1	8.8	6.0	54.5	45.0	1.7	14.9

[a]The fat type column indicates the reference where the corresponding fatty acid composition was originally reported. When more than one reference is cited the mean values were calculated and reported.

Coconut oil has a saturated fatty acid content of around 84%, while palm kernel oil and palm oil contain 81 and 52%, respectively, indicating that palm kernel oil is more saturated than palm oil and comparable to coconut oil.[13] In both oils, palm kernel and coconut, the main fatty acid is lauric acid (C12:0, dodecanoic acid, see Table 5.2).[14–16] In addition, they contain larger amounts of C8:0 (caprylic acid or octanoic acid), C10:0 (capric or decanoic acid), and especially C14:0 (myristic or tetradecanoic acid) than the most used seed oils like soybean oil, rapeseed oil, and sunflower seed oil. Palm kernel and coconut oils have melting temperatures close to 30 °C and 25 °C, respectively. For spreadable products, formulations may also contain other vegetable oils with less short-length saturated fatty acid content that will improve the nutritional properties. The major fatty acid in palm oil and palm stearin is C16:0 (palmitic or hexadecanoic acid).[17–20] The melting temperature of palm oil and palm stearin are close to 50 °C. This melting temperature is too high to use these fats as the only ones in shortening formulations. Cocoa butter, cupuassu fat, and shea butter have high contents of C18:0 (stearic or octadecanoic) and C18:1 (oleic or octadec-9c-enoic) acids.[20–23] Cocoa butter also contains high levels of C16:0 acid. Cupuassu and shea butter, however, have a significantly lower amount of C16:0 acid than cocoa butter. The melting points of cocoa butter, cupuassu fat, and shea fat are around 36 °C, 34 °C, and 33 °C, respectively. They are in the same range of melting temperature as partially hydrogenated vegetable oils used in the past for margarine production (34–40 °C).[10] Mango kernel fat has an attractive lipid profile from a nutritional point of view. It has high contents of C18:1 (40–46%) and C18:0 (40–45%) fatty acids. Its physicochemical characteristics have been reported to be very similar to those of commercial cocoa butter.[4] Winged bean (*Psophocarpus tetragonolobus*) is a non-conventional leguminous crop that shares similar nutritional value with soybean crop. It has similar lipid and protein contents as the soybean crop and the advantage of being able to grow where soybean fails to grow. The oil contains approximately 70% unsaturated and 30% saturated fatty acids, with C18:1 (34%) and C18:2 (linoleic or octadecadienoic, 31%) as the major unsaturated fatty acids along with other minor components.[6] The disadvantage is that it is a liquid at ambient temperature. Lipids with a very low melting point, as in the case of oils, or a very high melting point, such as some fats like palm kernel and coconut, may have restricted industrial applications.[24] Combining oils and fats might lead to a melting behavior that guarantees broader applications.

5.2.2 Triacylglycerol Composition

TAGS containing elaidic fatty acids were used by Herrera *et al.* (1998) to follow the effect of two hydrogenation processes in sunflower oil (Table 5.3).[10] Overall, there is agreement among authors in the literature that high reaction temperatures promote the formation of *trans* isomers. This hydrogenation process is often called selective. On the contrary, a high concentration of hydrogen gas

Table 5.3 Main triacylglycerols (TAGS) containing elaidic fatty acid present in hydrogenated sunflower oils used to formulate margarines before *trans*-fat legislation.[a]

TAGS	High temperature/low pressure			Low temperature/high pressure		
	120 min	135 min	145 min	120 min	135 min	150 min
C18:1*c* C18:0 C18.1*t*	2.1	2.7	5.2	13.5	17.7	21.7
C18:1*t* C18:1*t* C18:1*t*	2.9	4.9	4.3	8.0	12.3	11.4
C18:1*t* C18:1*t* C18:1*c*	31.9	39.2	40.1	28.3	22.1	14.4

[a]Data reported in ref. 10.

results in low selectivity and low *trans*-isomer content. These conditions are usually referred to as non-selective. To follow both hydrogenation processes (selective and non-selective), samples were collected every 15 min during hydrogenation reactions, beginning 45 min after the start of the reaction. Hydrogenation processes were stopped after 145 (selective hydrogenation) and 150 min (non-selective hydrogenation). Changes in the main TAGS of the three last samples collected in each process clearly show hydrogenation evolution (see Table 5.3).

The TAGS compositions of both hydrogenated oils, processed under selective or non-selective conditions, were different. The TAGS C18:1*c* C18:0 C18:1*t* (OSE), C18:1*t* C18:1*t* C18:1*t* (EEE), and C18:1*t* C18:1*t* C18:1*c* (EEO) contents, especially, were different. OSE and EEE were present in much higher percentages in samples obtained in non-selective conditions compared to the percentages found for samples obtained in selective conditions. EEO was the main TAGS in samples collected during the selective hydrogenation process. The sample obtained at 120 min during the non-selective hydrogenation had higher percentages of OSE and EEE than the sample collected at 145 min (selective process), but lower than the samples obtained at 135 and 150 min (non-selective process) and the highest percentage of EEO of the three last samples taken during the non-selective hydrogenation process. Even though their fatty acid composition was similar (Table 5.1), the TAGS structures were different and from the TAGS composition different crystallization and polymorphic behaviors would be expected. The Mettler Dropping Point (MDP) of EEE and EEO as reported by Adlof and List (2007) were 43.5 and 14.4 °C, respectively.[25] For OOO, MDP was 6.1 °C while for SSS it was 73.3 °C.[25,26] *Trans*-isomer TAGS have a molecular geometry more similar to the saturated TAGS and melting point intermediate between unsaturated and saturated TAGS. When thermal behavior was studied by differential scanning calorimetry (DSC), it was found that the difference between the onset of melt (start of melt) and the melting points was small indicating that EEE, EOE, and EEO TAGS are sharply-melting materials, since their melting points were about 4–7 °C above the onset of melt.[25] According to Adlof and List (2007), EEE and SSS have similar heats of fusion, and the introduction of oleic acid into the molecule lowers melting enthalpies by approximately a factor of 2 (*i.e.* from 36.1 cal g^{-1} for EEE to 22.5 and 26.8 cal g^{-1} for EOE and EEO, respectively).[25]

The TAGS compositions of several tropical fats used to formulate *trans*-fat free products were selected to show the differences with hydrogenated oils TAGS compositions (see Table 5.4). As can be observed, they do not contain elaidic TAGS or C18:2t isomers. The hydrogenated oil samples shown in Table 5.3 had a very different TAGS profile than the fats reported in Table 5.4, where capric (Ca), lauric (La), myristic (M), palmitic (P), oleic (O), and stearic (S) TAGS have the highest contents.

Palm kernel and coconut oils main TAGS have capric, lauric, and myristic fatty acids bonded to a glycerol molecule. These oils contain a higher concentration of shorter carbon number TAGS than the other fats shown in Table 5.4 and the hydrogenated oils shown in Table 5.3. The main TAGS in palm oil and palm stearin are POP and POO. Palm oil also contains high amounts of OOO while palm stearin has a large content of PPP. Cocoa butter has three main TAGS: POP, POS, and SOS, which form a solid solution and are responsible for its unique properties. The main TAGS in cupuassu fat and shea butter are SOS, SOA, SOO, and POS. They also contain OOO, which is a liquid at ambient temperature (6.1 °C).[26] The SOO content in both cupuassu fat and shea butter is significantly higher than the one of cocoa

Table 5.4 Triacylglyceride (TAGS) composition of tropical fats used to formulate *trans*-fat-free products. The composition is reported in percentage (%).[a]

TAGS[b]	Palm kernel oil[14]	Coconut oil[15,16]	Palm oil[29]	Palm stearin[18–20]	Cocoa butter[21,22]	Cupuassu fat[20,22]	Shea butter[23]
CaCaCa	1.4	1.6	—	—	—	—	—
CaCaLa	14.2	13.3	—	—	—	—	—
CaCaM	18.5	20.2	—	—	—	—	—
LaLaLa	28.7	26.3	—	—	—	—	—
LaLaM	21.5	20.0	0.4	—	—	—	—
LaMM	11.3	12.2	1.2	—	—	—	—
MMM	4.4	3.9	2.5	—	—	—	—
PPP	—	—	5.6	20.3	—	—	—
MOP	—	—	2.2	—	—	—	—
PPS	—	—	1.0	2.8	—	1.3	0.5
POP	—	—	29.8	49.3	20.4	2.3	0.1
PLP	—	—	3.3	1.0	1.8	1.6	0.7
POS	—	—	5.2	6.9	41.4	13.2	5.2
POO	—	—	23.4	17.1	3.5	7.4	0.2
PLO	—	—	8.8	0.2	—	—	—
PSS	—	—	—	—	—	2.3	—
SOS	—	—	—	—	27.0	27.5	38.4
SOA	—	—	—	—	0.9	12.5	8.3
SOO	—	—	3.2	1.1	4.0	18.1	32.3
OOO	—	—	17.7	1.3	—	3.6	11.2
OLO	—	—	5.6	—	—	—	—
OOA	—	—	—	—	—	8.6	2.4

[a]When more than one reference is cited, the mean value was calculated.
[b]Ca: capric acid; La: lauric acid; M: myristic acid; P: palmitic acid; O: oleic acid; S: stearic acid; L: linoleic acid.

butter. Compared to the melting points of fully tri-saturated TAGS, including tristearin (SSS, 73.3 °C), tripalmitin (PPP, 65.5 °C), trimyristin (MMM, 57 °C), and trilaurin (LaLaLa, 46.4 °C), the introduction of oleic acid into position 2 at the glycerol molecule, forming symmetrical TAGS, lowers the melting point about 30 °C.[26] The melting point of the most stable polymorphic form of SOO is 23 °C.[27] The MDPs of POP, SOS, SLS, and SSL are 35.2, 44.1, 37.9, and 42.3 °C, respectively.[26] The different TAGS have different physical properties, therefore, depending on the types of fat used, the crystallization and polymorphic behaviors will be different. These behaviors will be determined by the types and contents of TAGS and by the interactions that occur among them. Mixtures of TAGS similar in melting points may form solid solutions over extensive, but not usually complete, ranges of composition. This leads to several types of phase behavior, eutectic formation being the most common, although peritectics and monotectics are also formed. Therefore, the mixtures may behave in a different way than the pure TAGS they contain.[28]

5.3 Solid Fat Content

Solid fat content (SFC) is a physical property usually employed to give an indication of fat functionality for example rolling ability at a given temperature or performance as a plastic fat in spreadable products. In a margarine formulation, the shape of the curve "SFC as a function of temperature" is used to predict the melting profile of the final product and performance of the margarine during packaging and storage as well as eating quality.[7] This is because the SFC is considered to be responsible for many quality parameters of plastic fats such as appearance, handling, plasticity, spreadability, aeration properties, oiling off, and sensorial attributes.[29,30] Regarding sensorial properties, a sharp SFC *vs.* temperature curve indicates that the melting rate is high in the temperature interval where the decrease is notorious. This sharp decrease is usually associated with the organoleptic attribute of "coolness", a pleasant sensation in the mouth that produces some products such as chocolate when they melt. The SFC values at higher temperatures (35 to 37 °C) are related to the capability of releasing aromas during melting and the sandy or waxy sensation in the mouth during consumption. SFC values higher than 5% at 35 °C may lead to unpleasant sensations during eating.[30] SFC values are also used to evaluate fat compatibility in a blend through building an iso-solid phase diagram where negative interactions such as dilution and eutectic behavior are evident. SFC *vs.* temperature curves for hydrogenated sunflower oil and three different tropical fats, shea, cupuassu, and cocoa butter have different shapes (see Figure 5.1).

The hydrogenated sunflower oil (HSFO) selected was obtained at high temperature and low hydrogen pressure, under processing conditions used before *trans*-fat legislation. For HSFO, the SFC at low temperature (10 to 20 °C) is lower than the ones of tropical fats, indicating that HSFO is more suitable for spreadable products than tropical fats. However, SFC values at

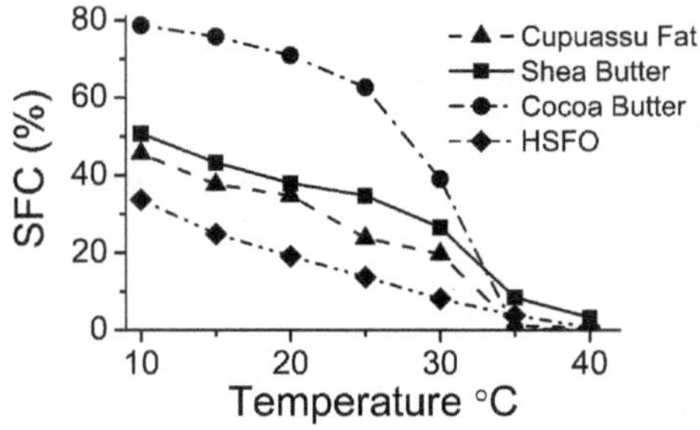

Figure 5.1 Solid fat content (SFC) *vs.* temperature profiles for hydrogenated sunflower oil (HSFO) and three different tropical fats: cocoa butter, cupuassu, and shea butter.

35 °C are higher than recommended for chocolate products leading to waxy feelings in chocolate toppings, for example. Depending on the application, margarine fats are formulated to reach a SFC ranging from 5–8% to 40% over a range of temperature from 10 to 30 °C. In addition to changing with temperature, the SFC also changes with the polymorphic transition of crystals from an intermediate-stability state to a more stable state, and of course, it is strongly influenced by TAGS composition and processing conditions. SFC values were reported to be higher for fast cooling than for slow cooling.[31] In foods such as ice cream or emulsion-based products, a minimum SFC is also required to entrap and stabilize air bubbles or water droplets in order to achieve the desired texture.

Several authors had evaluated the suitability of different systems containing tropical fats as alternatives to *trans*-fats. The physicochemical properties of ternary blends containing coconut oil, palm stearin, and partially hydrogenated soybean oil (0 to 100%) were studied.[10] These studies proved that the higher the concentration of partially hydrogenated soybean oil in the blend's, formulation the larger the difference in SFC between the system with and without *trans*-fats. For evaluating the suitability of a non-hydrogenated shortening for bakery products, its SFC with temperature profile was compared to the profiles of two commercial hydrogenated shortenings, whose chemical compositions are reported in Table 5.1.[8] At temperatures close to ambient temperature, the non-hydrogenated shortening had lower SFC values than the two hydrogenated shortenings evaluated, indicating that it will melt at a faster rate at eating temperature. These results are in agreement with the ones reported for ternary blends containing coconut oil, palm stearin, and partially hydrogenated soybean oil.[10] The profiles of hydrogenated shortenings had a flatter shape than the profile of the non-hydrogenated shortening, not showing a sharp decrease with temperature, which indicated a better functionality for warmer regions. This behavior was also reported for other hydrogenated oils, which had a profile similar to the one shown in Figure 5.1.[12] SFC profiles for hydrogenated oils obtained with different catalysts were studied.[9] According to these authors, the increase in SFC during hydrogenation was the result of an increase in saturated fat as well as an increase in *trans*-fat isomer contents. Although the use of 0.2% of Pd/Diatomite catalyst improved the chemical composition by reducing *trans*-fat content, the SFC was not modified with the type of hydrogenation process. New and traditional catalysts gave similar values of SFC with temperature.

Several studies analyzed formulations containing tropical fats for applications in margarine and shortenings. When hydrogenated coconut and cottonseed oils were studied by nuclear magnetic resonance, the profiles had an S-shape, similar to the one shown for cocoa butter in Figure 5.1. The SFC at 10 °C was high (85 and 45%, respectively), with a rapid reduction in SFC from 15 to 25 °C.[32] The decrease in SFC at those temperatures was most likely related to the high percentage of short- and intermediate-chain (≤C12 and C14) saturated fatty acids in these oils and in the TAGS they formed. The differences found in SFC profiles at low temperature suggest that there were some differences in the proportions of *trans*-isomers present in these hydrogenated oils. Since *trans*-isomers have a higher SFC than *cis*-isomers, the hydrogenated oil curves will be very much affected by *trans*-isomer content.[33]

Lauric fat blends grouping the various TAGS species according to their melting points into four groups were prepared: chilled (5 °C), room (5 to 25 °C), body (25 to 45 °C), and heated (above 45 °C).[30] The desired SFC profiles were built by formulating the TAGS distribution using different contents of selected groups. This approach allowed improving the suitability of a lauric fat blend for a specific product. Reports of palm kernel oil showed that the SFC profile significantly varied with temperature.[34] At low temperatures, it has an almost flat SFC with temperature profile, *i.e.*, from 5 °C to 15 °C, the SFC varies from 75% to 65%. At temperatures above 15 °C, the SFC declines sharply and it is 0% at 33 °C. Palm kernel oil is able to maintain the consistency of margarine at lower temperatures (5–15 °C), but the margarine texture softens at serving temperature (25–30 °C).[34] Since SFC is 0% at body temperature, consumers will not experience a waxy mouthfeel. To improve its performance as a plastic fat, high oleic acid *Moringa oleifera* seed oil was interesterified with palm kernel oil.[14] The interesterified system was not suitable for margarine application. However, ice cream may be formulated with this shortening. To obtain desirable chemical and physical characteristics for margarines, shortenings, and bakery fats, blends of semi-solid fats formulated with fully hydrogenated soybean oil and rice bran oil were interesterified with coconut oil using a lipozyme 1,3-specific lipase.[15] For a selected temperature, the interesterified samples showed lower SFC values than the physical blends with the same composition. This result could be expected because the physical blend had a higher melting point (from 58.0 to 65.5 °C depending on the different oil contents) than the interesterified sample with the same formulations (from 39.8 to 46.6 °C). The decreases in melting point and SFC were most likely due to the decrease in the proportion of high-melting TAGS (*i.e.*, SSS) that occurred for the rearrangement of saturated fatty acids within TAGS molecules during interesterification. The physical blends had a flat curve with a SFC content of around 60% at 35 °C. These SFC values are too high for spreadable products. SFC results showed that interesterified systems will be more suitable for margarine or shortening applications than physical blends.[15] In agreement with these findings, interesterified blends of *Moringa oleifera* seed oil and palm stearin had a wide range of plasticity and showed enhancements in spreadability, mouthfeel, and cooling effect.[14] Chemical interesterification of fully hydrogenated palm oil stearin, palm oil stearin, canola oil, and cottonseed oil mixed in various ratios was studied and data lead to the same conclusions.[35] The SFC profiles of interesterified blends were different from the non-interesterified blends. Chemical interesterification caused a decrease in SFC making blends more suitable for shortenings. Similar results were obtained by chemical interesterification of palm stearin and patawa oil, where SFC was similar to that of some margarine types.[18] The 50/50 shortening showed the same plastic and spreadable consistency at refrigeration temperature as soft table margarine, thus combining the physical and nutritional properties desirable for the food industry. Although interesterification led to better results in many cases, some successful examples of blending were also reported. Margarines were produced from rice bran oil and fractionated palm stearin.[19] By blending an oil with a semi-solid natural fat in ratios of 10 : 90, 20 : 80, and 30 : 70, shortenings with solid fat contents and crystal morphology similar to commercial margarine fats were obtained. These shortenings had appropriate physical properties to replace *trans*-fat and better nutritional properties. Other fat systems with interesting properties as a *trans*-fat replacement in table and functional margarine are pracaxi oil, passion fruit oil, and cupuassu fat that have interesting fatty acid profiles from a nutritional standpoint, displaying low atherogenicity and thrombogenicity indices. Their blends showed suitability for different industrial applications.[20] On the other hand, shea butter proved to be a good cocoa butter extender and it is increasingly popular in chocolate production and as a *trans*-fat replacement in many formulations.[23]

5.4 Crystallization Behavior

Knowledge of the crystallization kinetics of fats is highly important since it helps in the optimization of their usage in industrial processing, especially when the technology requires cooling steps. Crystallization kinetics deeply influences the final structure of fats and is strongly related to their polymorphic behavior and rheological properties.[16] When the isothermal crystallization behavior of sunflower seed oil hydrogenated under two different conditions was studied by means of pulse nuclear magnetic resonance, the SFC profiles obtained showed in most cases hyperbolic shapes. Crystallization initially was fast and there was no induction time of nucleation (defined as the time interval between when the crystallization temperature was reached and the start of crystallization) for either sample and therefore no step was observed in solid fat *vs.* time curves.[36] Crystallization behaviors of both hydrogenated sunflower oils were in agreement with the chemical composition of the samples (Tables 5.1 and 5.3). Samples which had the highest content of high-melting TAGs (especially trielaidin) showed only hyperbolic curves. In most cases, hydrogenated oils crystallize faster than semi-solid natural fats such as tropical fats. The low activation free energy of nucleation values (energy barrier for crystallization), calculated from the Fisher–Turnbull equation, reported for hydrogenated sunflower oil indicated that a very low supercooling was needed for crystallization since even at crystallization temperatures close to the melting point, crystallization occurred at shorter times than the ones reported for other fats such as palm oil.[10] This is highly relevant from a practical point of view since tropical fats will crystallize to more extent during storage than during production processes, which means that storage conditions must be controlled more carefully in order to obtain products with the desired physical properties. To improve the crystallization behavior of blends with tropical fats, the effect of adding fully hydrogenated coconut oil to non-hydrogenated coconut oil was investigated by measuring the SFC as a function of time.[16] The results showed that addition of fully hydrogenated fat to non-hydrogenated fat accelerated the crystallization process by increasing the crystallization rate and the maximum SFC (plateau value in SFC with time curve). Similar results were obtained for other fat blends such as fully hydrogenated cottonseed oil and canola oil, and high oleic sunflower oil and fully hydrogenated soybean oil, where the hard-fat accelerated the crystallization kinetics.[37,38]

5.5 Microstructure

Production of fats that are low in *trans* fatty acids requires modification of the base natural fats or oils by simple blends, fractionation, interesterification, or through a combination of these processes. A possible alternative is the replacement of *trans*-fat by solid tropical lauric fats in products that require solid fats for their texture or firmness. *Irvingia gabonensis* is a tropical crop that grows in central and western Africa. Its seed fat is a source of "lauric fat" with high proportions of myristic (~50%) and lauric (~40%) acids. Due to its high saturated fatty acid content, *I. gabonensis* seed fat presents a sharp melting profile with a high SFC at temperatures up to 30 °C and complete melting around 40 °C. Thanks to this, *I. gabonensis* seed fat can be used in blends, as hard stock, for structured product formulation.[39] To prepare zero *trans* margarine, it was blended with four vegetable oils in order to reduce the saturated fatty acid content and to improve nutritional quality: rapeseed oil, groundnut oil, palm super olein, and *Dacryodes edulis* pulp oil. The microstructure of the solidified fat showed some differences among different blends most likely due either to the nature of liquid oil used for blending or to the crystallization process. However, some blends were suitable for baking or table margarine with physical properties similar to commercial products. *Trans*-free plastic fats were also prepared by enzymatic interesterifications using a magnetic biocatalyst starting from solid palm stearin and liquid rice bran oil blends. After the interesterifications, the total fatty acid compositions of the binary blends were almost unchanged, whereas the fatty acid positional distribution and triacylglycerol species were significantly varied. As compared with the physical blends, the interesterified products had a lower slip melting point. The interesterification led in some cases to an obvious improvement in the microstructure of the studied systems. The blends of palm stearin and rice bran oil showed loosely-packed and low-density aggregates of spherulitic crystals when crystallized. After the enzymatic interesterification, the interesterified products exhibited smaller aggregates of fat crystals with a more uniform microstructure. Generally, larger sized fat crystals can result in a sandy mouthfeel of the finished product. Small compact crystals are desired for use in margarine fats and commercial shortenings since they can stabilize air bubbles and water droplets in the final food products, giving a fine and smooth texture. Similar results were reported for *trans*-free shortening formulated with coix seed oil, fully hydrogenated palm oil, and *Cinnamomum camphora* seed oil.[40] According to this report, in the shortenings studied, the crystal structure showed a great difference between physical blends and interesterified fats.[40] The fully hydrogenated palm oil they used had leaf-like crystals combined tightly together. When it was blended, the crystal morphology was rough and needle-shaped, whereas when it was interesterified discrete crystals were present. Interesterified samples had fewer and smaller crystals than physical blends. When sunflower seed oil was hydrogenated under two different conditions to obtain typical shortenings used before *trans*-fat legislation, the microstructure obtained correlated to the shape of crystallization kinetics curves. Hydrogenated samples with a SFC *vs.* time behavior described by hyperbolic curves showed needle-like crystals, while samples with kinetics behavior represented by sigmoidal curves showed spherical crystals.[36] In all cases the crystals were small (<3 μm) and the structures were very homogeneous, in agreement with the fact that interesterified systems, which had similar structural characteristics to these high-*trans* hydrogenated sunflower oils, have physical properties suitable for margarine.

As a different approach to replace *trans*-fats, the formation of molecular compounds in a multicomponent triglyceride system with particular focus on the ratios of SOS:OSO/SSO, where "O" and "S" denote oleic and specific saturated fatty acids, respectively, was proposed.[41] To obtain the molecular compound, specific fat fractions were mixed (a fraction rich in 1-palmitoyl-2-oleyl-3-palmitin and a fraction rich in a mix of 1-oleyl-2-palmitoyl-3-olein and 1,2-dipalmitoyl-3-olein-glycerol). The microstructure of crystalline mixtures was characterized by two distinct phases, one of them consisting of spherulites dispersed in a second continuous phase of small granular crystals associated with molecular compound formation. The implications of this work are that it may be possible to incorporate molecular compounds into fatty foods and thus lower the *trans* fatty acid content of these foods by reducing the need for hydrogenation to create solid crystalline materials, and also to increase the relative amount of healthier unsaturated fatty acids, such as oleic acid.

5.6 Rheological Properties

The texture of a fat-based product is related to the rheological properties of the fat resulting from its hierarchical structures. These structures are complex and vary in length from the molecular to the microscopic scale. They give fats a solid viscoelastic behavior ($G' > G''$) at small deformations, and their diversity reflects the adaptation of fats to their mechanical function. The rheological properties of fats as related to the crystal network have been a major field of study. Significant advances have been made in this area, enabling partial understanding of linear viscoelasticity in terms of fat microstructure. Recent studies proved that the rheological behavior of a fat can be used for product reformulation and design of alternative lipid structuring materials.[42] These studies described successful examples in which shortenings containing tropical fats replaced *trans*-fat in different food types, with good functionality. With the aim of using a healthier shortening, the effects of interesterification with sodium methoxide on physicochemical and rheological properties of *trans*-free fats made from palm stearin and soybean oil were evaluated.[43] Strain sweep, frequency sweep, and temperature sweep tests were used to evaluate the rheology of fat systems. Due to the high melting point and low elasticity, initial blends were not suitable for direct application in plastic fat production. Their results showed that interesterification led to an improvement in the elastic character of fat blends. The interesterified blends of palm stearin and soybean oil showed desirable characteristics suited for production of *trans*-free soft tub, stick, and Baker's margarine, all-purpose shortening, and Iranian vanaspati.

Puff pastry is a bakery product with a high-fat content, of up to 40%, and is known for its light and flaky, layered structure. Fat is important for the development of the characteristic puff pastry properties such as the flaky shape, layered effect, texture, appearance, volume, and lift. The suitability for application in puff pastry production of four vegetable fat blends of palm stearin and rapeseed oil with low-*trans* fatty acid content ($\leq$0.6%) and reduced saturated fatty acid content was investigated.[44] Rheological measurements of four blends revealed a more elastic than viscous behavior. Puff pastry fat blends containing 48.2% and 40.9% of total saturated fatty acids achieved excellent baking results for full-fat and fat-reduced puff pastry; hence these interesterified blends showed adequate shortening properties. To further analyze the role of shortenings in bakery, nonlinear viscoelastic properties under large deformations of laminating shortenings were reported and the results were correlated to the shortening structure described using ultra small angle X-ray scattering (USAXS) and scanning electron microscopy (SEM).[45] According to this study, a shortening must have the ability to remain elastic to facilitate dough layering and withstand high stresses without rupturing. A laminating shortening that is too soft will be absorbed into the dough or squeezed out. On the other hand, a shortening that is too hard will break and rupture the dough. They studied four commercial shortenings with a TAGS composition with a trisaturated content from

13% to 16% and unsaturated content from 43% to 46%. The laminated shortenings described showed a homogenous network made up of layered crystal aggregates with a rheological behavior desired for the shortenings used in laminated doughs. In this way, healthier and functional shortenings, with the required rheological properties were designed.[45] In agreement with this study, the addition of interesterified blend-based fast-frozen special fat to frozen dough had a significant influence on its physical properties and microstructure.[46] The frozen dough possessed better texture and rheological properties as shown by the enhanced extensibility, decreased hardness, increased cohesiveness and resilience, and improved flexibility. The selected fat, free of *trans*-fat, performed as a high-quality special fat for pre-frozen fast food preparation.

As mentioned in Section 5.4, tropical fats can be an alternative to partially hydrogenated vegetable oils for fat-based products given their semi-solid consistency and zero *trans*-fat content. However, their slow crystallization behavior results in hardening during storage. For this reason, the final texture obtained for a product may not have the desired values to make it accepted by the consumers. To improve rheological properties, the effect of adding confectioner's sugar on the crystallization and rheological behaviors of two commercial palm oils over four weeks of storage was investigated.[47] The results showed that the presence of sugar reduced the solid fat content of both palm oils, but increased their solid-like character. The effect was explained evaluating the role of the dispersed sugar on oil viscosity.[47] These findings established a foundation for the optimization and improvement of palm-based confections.

5.7 Polymorphic Behavior

The SFC profile is important in fat-based product formulation. However, despite its extensive usage, the SFC profile should not be used as the only parameter to define the applicability of a fat in obtaining the desired final product attributes. A plastic fat is developed by crystalline networks in a continuous oily matrix. The strength of the crystal network also depends on the polymorphic behavior and size of the fat crystals. The overall physical properties of the product are the result of complex interactions among SFC, polymorphism, and the microstructure of the solid fat. A balance of these properties is required to produce high-quality products.[8]

Fats used as raw materials in food manufacture have a characteristic polymorphic behavior, strongly influenced by heat and mass transfer during the crystallization process. Each polymorphic form is defined by a unique melting point and characteristic crystalline structural properties.[1] According to the literature, for a TAGS, three main forms are observed: α, β', and β. Melting point, density, and stability increased in this order. These polymorphic forms present two levels of organization: the lamellar stacking described by long spacings and the lateral chain packing (subcell packing) characterized by short spacings. Both structural levels can be easily described by X-ray scattering at small (SAXS) and wide (WAXS) angles, respectively.[1] Fat polymorphism is closely related to fat functionality in the food industry. For margarine and spreadable fats, optimum texture has been correlated to the presence of β' crystals. These crystals tend to be smaller in size and form a strong network able to include droplets and air bubbles giving plastic shortenings. The replacement of oleic acid with elaidic acid had a drastic effect on the polymorphism of TAGS. For example, the most stable polymorphic form for PEP is the β' while for EPE it is the β form. POP and OPO have a more complex polymorphism with forms not suitable for spreadable products.[48] The β crystals are often larger and associated with hard and brittle textures making them suitable for chocolate and confections. Several tropical fats are polymorphic in nature. Depending on the processing conditions, cocoa butter may crystallize in six polymorphic forms γ (I), α (II), β_2' (III), β_1' (IV), β_2 (V), and β_1 (VI). Cupuassu fat, however, crystallizes in four polymorphic forms: γ, α, β', and β, in increasing order of melting temperature and stability. For natural fat, only one β' form was described.[22] Although the existence of two β forms (β_2 and β_1) was described, the transition $\beta_2 \rightarrow \beta_1$ was not proved. This behavior was explained by the fact that the polymorphic transformation $\beta_2 \rightarrow \beta_1$ could not be followed due to the structural similarity of the corresponding X-ray patterns obtained in those experiments.[22] Most recent works proved that when cupuassu fat crystallized without agitation, the β_2-form was only obtained during storage and no transformation to the β_1-form occurred over a long time. This is very important from a technological point of view since it is not likely that cupuassu fat is involved in blooming. Several types of blooming were described but in all of them, the β_1-form was involved. The fact that cupuassu fat has a very stable β_2-form makes it suitable for chocolate products. When cupuassu fat was fractionated, for some fractions, and in selected conditions, β'_1 and β_1 forms were the most important polymorphs obtained as evidenced by their thermal, microstructural, and polymorphic behaviors.[49] The X-ray patterns and their corresponding microstructures for five polymorphic forms of cupuassu fractions are summarized (see Figure 5.2). The corresponding X-ray long and short spaces are reported (see Table 5.5). The γ-form was not included because it is very unstable and not very important from a technological point of view.

The five polymorphic forms shown have characteristic signals in SAXS (long spacing) and WAXS (short spacing) regions and a distinctive microstructure. They also have different melting points. The polymorphic form in which a fat crystallizes will define its suitability as alternative to *trans*-fat in a product since each product needs the fat phase in a specific polymorphic form to reach the desired macroscopical properties such as texture and appearance. Croissants with two different hydrogenated roll-in shortenings (Table 5.1) and a non-hydrogenated shortening formulated with canola oil, modified palm oil, and palm kernel oil were prepared to evaluate the effect of replacing *trans*-fat.[8] The non-hydrogenated shortening did not contain *trans* TAGS and contained a low percentage of saturated TAGS, making it healthier from a nutritional point of view. Both hydrogenated shortenings suffered only minor changes in

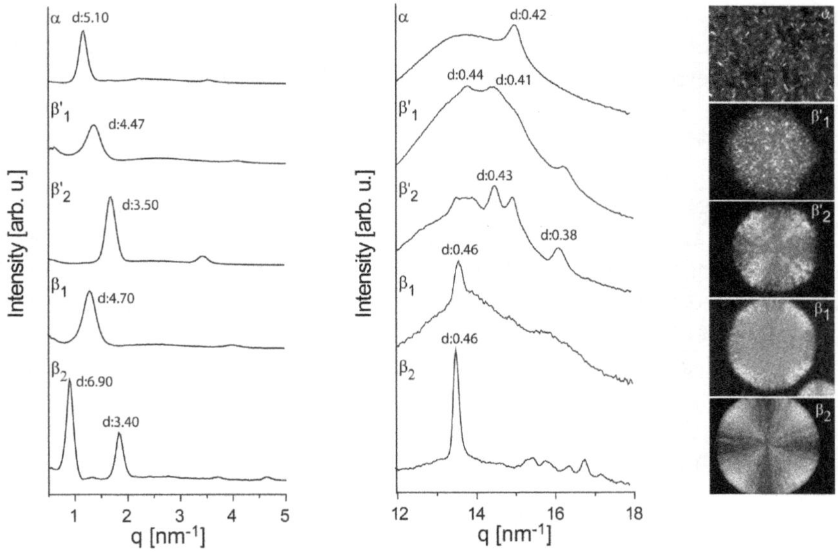

Figure 5.2 X-ray patterns in the SAXS and WAXS regions together with their corresponding microstructures for five polymorphic forms of cupuassu fat.

Table 5.5 Short and long spacings and melting temperature of five polymorphic forms of cupuassu fat.

Polymorphic form	Short spacing (nm)[a]	Long spacing (nm)	Melting temperature (°C)
α	0.42	5.1	17.7
β′2	0.38 (S) 0.43 (W)	3.5	24.4
β′1	0.41 (S) 0.44 (W)	4.47	27.1
β2	0.46	6.91 (S) 3.40 (M)	31.6
β1	0.46	4.70	43.6

[a]Intensity of the diffraction peaks: S, strong; M, medium; W, weak.

polymorphic behavior with time. They remained in the β′ polymorph during storage. On the contrary, the non-hydrogenated shortening experienced a significant conversion from β′ to the β form. According to the results reported, this conversion did not take place immediately upon cooling, but after approximately 24 h of storage time.[8] Although the non-hydrogenated shortening has improved nutritional properties compared to hydrogenated shortenings, its polymorphic behavior might lead to products with a shorter shelf life. Other healthier shortenings were prepared by interesterification of soybean oil and fully hydrogenated palm oil,[50] or of rice bran stearin, fully hydrogenated soybean oil, and coconut oil,[51] by blending palm and canola oils with fully hydrogenated palm oil,[52,53] or palm oil and anhydrous milk fat.[54] In all these systems, the β′ form was the main polymorph present and the fats remained in this form in a wide variety of processing conditions, showing their suitability for spreadable products.

Blends with winged bean oil, palm stearin, and palm olein were prepared to replace *trans*-fat in margarines. The blends were tested for textural, thermal properties and crystal growth pattern, and polymorphic behavior.[6] One of the blends contained 48.5% of palm stearin, 1.5% of palm olein, and 50% of winged bean oil, while the other contained 1.5% of palm stearin, 48.5% of palm olein, and 50% of winged bean oil. The blend with 48.5% palm stearin exhibited a harder texture than the one formulated with 1.5%. It had a crystal morphology with sizes ranging from 1.0 to 5.2 μm, evenly distributed, that corresponded to β′ polymorphs; the blend with 1.5% palm stearin showed a soft texture, and polymorphs with protruding needle-like crystals, unevenly distributed. The 48.5% palm stearin blend showed good physicochemical properties as multi-purpose for use as table and baking margarine, especially in tropical climates. This study described the potential use of underutilized oilseed crop resources for making specialty and structured fats with better nutritional properties for food formulation.

5.8 Conclusions

Tropical fats are *trans*-free fats. Although they have high saturated fatty acid contents, they may be blended or interesterified with liquid oils rich in unsaturated fatty acids to improve their nutritional values. Formulations containing tropical fats or its fractions and unsaturated oils have the desired SFC, microstructure, rheological properties, and polymorphic behavior for applications such as in margarines, non-dairy creams, shortenings for baking and frying, chocolate, and confectionery, allowing successful replacement of *trans*-fat in foods leading to healthier products.

Acknowledgements

This work was supported by the National Agency for the Promotion of Science and Technology (ANPCyT), (grant number PICT-2018-00868) and the University of Buenos Aires (grant number UBA-20020170100039BA).

References

1. K. Sato, Fundamentals and Applications, in *Food, Cosmetics, and Pharmaceuticals*, John Wiley & Sons, Hoboken, NJ, USA, 2018.
2. Z. Zhang, X. Ma, H. Huang and Y. Wang, *LWT – Food Sci. Technol.*, 2017, **86**, 492.
3. F. Peyronel, R. Campos and A. G. Marangoni, *Food Res. Int.*, 2016, **88**, 52.
4. C. Torres-León, R. Rojas, J. C. Contreras-Esquivel, L. Serna-Cock, R. E. Belmares-Cerda and C. N. Aguilar, *Trends Food Sci. Technol.*, 2016, **55**, 109.
5. J. Jin, Q. Jin, C. C. Akoh and X. Wang, *Crit. Rev. Food Sci. Nutr.*, 2018, **59**, 1794.

6. M. Makeri, M. M. Sahric, H. M. Ghazali, K. Ahmad and K. Muhammad, *LWT – Food Sci. Technol.*, 2019, **100**, 158.
7. D. R. Kodali, *Trans Fats Alternatives*, AOCS Press, Champaign, IL, USA, 2005.
8. K. D. Mattice and A. G. Marangoni, *Food Res. Int.*, 2017, **96**, 54.
9. K. Toshtay, A. B. Auyezov, Z. A. Bizhanov, A. T. Yeraliyeva, S. K. Toktasinov, B. Kudaibergen and A. Nurakyshev, *Eurasian Chem.-Technol. J.*, 2015, **17**, 33.
10. M. L. Herrera, C. Falabella, M. Melgarejo and M. C. Añón, *J. Am. Oil Chem. Soc.*, 1998, **75**, 1273.
11. J. Reyes-Hernández, E. Dibildox-Alvarado, M. A. Charo-Alonso and J. F. Toro-Vázquez, *J. Am. Oil Chem. Soc.*, 2007, **84**, 1081.
12. R. Grimaldi, R. Tieko Nassu, L. A. Guaraldo Gonçalves and R. Nogueira Cavaletti Moreira, *Grasas Aceites*, 1998, **49**, 1.
13. A. J. Dijkstra, *Encyclopedia of Food and Health*, Elsevier, Amsterdam, The Netherlands, 2016.
14. S. Dollah, S. M. Abdulkarim, S. H. Ahmad, A. Khoramnia and H. M. Ghazali, *J. Sci. Food Agric.*, 2016, **96**, 3321.
15. P. Adhikari, J. A. Shin, J. H. Lee, H. R. Kim, I. H. Kim, S. T. Hong and K. T. Lee, *Food Bioprocess Technol.*, 2012, **5**, 2474.
16. V. Dhaygude, A. Soós and L. Somogyi, *Period. Polytech.: Chem. Eng.*, 2018, **62**, 123.
17. V. Nguyen, T. Rimaux, V. Truong, K. Dewettinck and F. Van Bockstaele, *Food Res. Int.*, 2020, **137**, 109412.
18. P. D. Oliveira, A. M. C. Rodrigues, C. V. Bezerra and L. H. M. Silva, *Food Chem.*, 2017, **215**, 369.
19. P. Podchong, S. Sonwai and D. Rousseau, *J. Am. Oil Chem. Soc.*, 2018, **95**, 433.
20. C. Vieira Bezerra, A. M. da Cruz Rodrigues, P. D. de Oliveira, D. Albuquerque da Silva and L. H. Meller da Silva, *Food Chem.*, 2017, **221**, 1466.
21. C. H. Jia, J. A. Shin and K. T. Lee, *Food Sci. Biotechnol.*, 2019, **28**, 1649.
22. J. C. Silva, T. S. Plivelic, M. L. Herrera, N. Ruscheinsky, T. G. Kieckbusch, V. Luccas and I. L. Torriani, *Cryst. Growth Des.*, 2009, **9**, 5155.
23. A. C. Rodriguez Negrette, C. Huck Iriart and M. L. Herrera, *J. Am. Oil Chem. Soc.*, 2019, **96**, 239.
24. R. C. Silva and L. A. Gioielli, *Braz. J. Pharm. Sci.*, 2006, **42**, 235.
25. R. O. Adlof and G. R. List, *J. Am. Oil Chem. Soc.*, 2007, **84**, 427.
26. G. R. List, R. O. Adlof, C. J. Carrierre and R. O. Dunn, *Grasas Aceites*, 2004, **55**, 135.
27. P. Walstra, *Physical Chemistry of Foods*, Marcel Dekker, New York, USA, 2003.
28. M. L. Herrera, *J. Am. Oil Chem. Soc.*, 1994, **71**, 1255.
29. N. A. Rashid, N. A. Kamarulzaman and Z. Omar, *Food Res. Int.*, 2018, **106**, 982.

30. B. P. Nusantoro, N. De Clercq, K. Anthierens and K. Dewettinck, *J. Am. Oil Chem. Soc.*, 2013, **90**, 1607.
31. R. Campos, S. S. Narine and A. G. Marangoni, *Food Res. Int.*, 2002, **35**, 971.
32. Z. Shen, A. Birkett, M. A. Augustin, S. Dungey and C. Versteeg, *J. Am. Oil Chem. Soc.*, 2001, **78**, 387.
33. M. Cizmeci, A. Musavi, M. Kayahan and A. Tekin, *J. Am. Oil Chem. Soc.*, 2005, **82**, 925.
34. M. M. Sahri and N. L. H. M. Dian, *J. Oil Palm Res.*, 2011, **23**, 958.
35. I. Karabulut, S. Turan and G. Ergin, *Eur. Food Res. Technol.*, 2004, **218**, 224.
36. M. L. Herrera, C. Falabella, M. Melgarejo and M. C. Añón, *J. Am. Oil Chem. Soc.*, 1999, **76**, 1.
37. A. P. B. Ribeiro, R. Grimaldi, L. A. Gioielli and L. A. Gonçalves, *Food Res. Int.*, 2009, **42**, 401.
38. D. Li, P. Adhikari, J. A. Shin, J. H. Lee, Y. J. Kim, X. M. Zhu, J. N. Hu, J. Jin, C. C. Akoh and K. T. Lee, *LWT – Food Sci. Technol.*, 2010, **43**, 458.
39. J. Yamoneka, P. Malumba, G. Lognay, C. Blecker and S. Danthine, *LWT – Food Sci. Technol.*, 2019, **101**, 747.
40. Y. Xu, X. Zhu, X. Ma, H. Xiong, Z. Zeng, H. Peng and J. Hu, *Food Biosci.*, 2018, **22**, 1.
41. A. N. Sibbald, J. R. Carney and A. G. Marangoni, *J. Am. Oil Chem. Soc.*, 2016, **93**, 1441.
42. B. A. Macias-Rodriguez and A. G. Marangoni, *Crit. Rev. Food Sci. Nutr.*, 2018, **58**, 2398.
43. M. H. Naeli, J. Farmani and A. Zargaraan, *J. Food Process Eng.*, 2017, **40**, e12409.
44. C. Silow, E. Zannini and E. K. Arendt, *J. Food Sci. Technol.*, 2016, **53**, 2117.
45. B. A. Macias-Rodriguez, F. Peyronel and A. G. Marangoni, *J. Food Eng.*, 2017, **212**, 87.
46. T. Zhu, X. Zhang, B. Li and H. Wu, *Food Chem.*, 2019, **272**, 76.
47. R. West and D. Rousseau, *Food Res. Int.*, 2016, **85**, 224.
48. L. Zhang, K. J. Wei, J. C. Chen, M. Xiong, X. Li, H. Hondoh and S. Ueno, *Cryst. Growth Des.*, 2020, **20**, 1655.
49. A. C. Rodríguez-Negrette, M. J. Rodríguez-Batiller, V. A. García-Londoño, V. Borroni, R. J. Candal and M. L. Herrera, *J. Am. Oil Chem. Soc.*, 2020, **97**, 1215.
50. Y. Li, J. Zhao, X. Xie, Z. Zhang, N. Zhang and Y. Wang, *Food Chem.*, 2018, **255**, 405.
51. C. Shi, M. Chang, R. Liu, Q. Jin and X. Wang, *Int. J. Food Eng.*, 2015, **11**, 467.
52. G. Marangoni de Oliveira, M. A. Stahl, A. P. Badan Ribeiro, R. Grimaldi, L. Pavie, Cardoso and T. Guenter Kieckbusch, *Eur. J. Lipid Sci. Technol.*, 2015, **117**, 1762.
53. K. Martins Barbosa, L. Pavie Cardoso, A. P. Badan Ribeiro, T. Guenter Kieckbusch and M. H. Masuchi Buscato, *J. Food Sci. Technol.*, 2018, **55**, 1104.
54. V. Nguyen, T. Rimaux, V. Truong, K. Dewettinck and F. Van Bockstaele, *J. Food Eng.*, 2021, **291**, 110245.

CHAPTER 6

Edible Oleogels Produced with Fatty Alcohols: The Use of Policosanol as an Oleogelator

F. R. LUPI,*[a] A. SHAKEEL,[b,c] U. FAROOQ,[d] N. BALDINO[a] AND D. GABRIELE[a]

[a] Department of Information, Modeling, Electronics and System Engineering, (D.I.M.E.S.) University of Calabria, Via P. Bucci, Cubo 39C, I-87036 Rende (CS), Italy; [b] Faculty of Civil Engineering and Geosciences, Department of Hydraulic Engineering, Delft University of Technology, Stevinweg 1, 2628 CN, Delft, The Netherlands; [c] Department of Chemical, Polymer & Composite Materials Engineering, University of Engineering & Technology, New Campus, Lahore 54890, Pakistan; [d] Faculty of Aerospace Engineering, Aerospace Manufacturing Technologies, Delft University of Technology, Kluyverweg 1, 2629 HS Delft, The Netherlands

*Email: francesca.lupi@unical.it

6.1 Introduction

Organogels are structured systems where a large amount of liquid phase, an organic solvent, is entrapped within a three-dimensional network built by gelator molecules of low molecular weight. When the solvent is an edible oil, the term oleogel is used frequently instead of organogel.[1] Gelation can be promoted either by high molecular weight compounds, *i.e.*, polymers, or by low molecular weight agents.[2,3] Even though the studies on these systems are growing and more information is available, the discovery of new edible

Food Chemistry, Function and Analysis No. 31
Development of *Trans*-free Lipid Systems and their Use in Food Products
Edited by Jorge F. Toro-Vazquez

Published by the Royal Society of Chemistry, www.rsc.org

gelators is still the result of a trial and error activity, therefore, food-grade molecules are only available in limited number.[3] Only one edible polymeric gelator is currently used, *i.e.*, ethylcellulose, whereas a larger number of low molecular weight gelators (LMWGs) are available.[3]

6.1.1 LMW Oleogels: A General Overview

LMWGs are compounds with a molecular weight lower than 3000 Da,[2] that often have hydrophobic and hydrophilic groups that make them soluble enough in organic solvents to allow the formation of gelator–solvent interactions, but not so soluble as to yield a solution instead of a gel;[3] nevertheless, apolar compounds can also be excellent gelators. Among the most studied LMWGs, there are monoacylglycerols (or monoglycerides, MGs),[4–8] plant sterol derivatives,[9–11] fatty acids,[12–15] fatty alcohols,[14–17] sorbitan esters,[18,19] and lecithin.[20] LMWGs might be used alone resulting in mono-component oleogels[2] or in mixtures taking advantage of the potential synergistic effects among different gelators (*i.e.*, multi-component oleogels).[2,21]

When a mixture of gelators is used, they can organize an assembled structure together (co-assembly) or independently of each other (self-sorting).[2,22] For instance, a mixture of fatty acids and fatty alcohols can be considered as an example of a co-assembled structure where the mixed crystals are characterized by an altered morphology with respect to single gelators (needle-like for a mixed system instead of platelets for monocomponent gels), yielding a bi-component oleogel with different properties when compared to the monocomponent gels.[22] On the other hand, mixtures of phytosterols and saturated monoglycerides can be considered as self-sorting systems because microscopy evidenced that mixed gelators crystallize independently of each other, yielding the same structures observed in monocomponent oleogels; in this case the synergistic effect is attributed to the presence of different crystalline structures, which leads to a better spatial distribution preventing the aggregation phenomena.[22]

The use of gelator mixtures might improve the oleogel physical properties achieving technological and functional properties closer to those of traditional structured fats using unsaturated vegetable oils. Thus, oleogels might help to satisfy consumer requests for healthy food systems.[2,21] Mixtures can be based on the use of different molecules belonging to the same chemical "family"; for instance, sterol derivatives such as γ-oryzanol and β-sitosterol exhibit interesting gelation properties only when they are used in mixtures.[9,11,23,24] In a similar way, commercial MGs are usually formulated as mixtures of different esters *i.e.*, monopalmitate and monostearate, where the use of different proportions allows the oleogel's physical properties to be tailored.[7]

On the other hand, different types of gelators can be mixed to enhance the properties of one of them or to achieve a gelling capability. For instance, Pernetti *et al.*[19] observed that lecithin or sorbitan esters individually are not able to gel vegetable oils and only viscous systems were obtained using high concentrations. In contrast, a mixture of both of them showed gelling properties of the same vegetable oils when using concentrations above 4%. In the same way, Gandolfo *et al.*[15] observed synergistic effects by mixing fatty acids and fatty alcohols with the same chain length. These results were confirmed by Schaink *et al.*[14] who developed crystals of different shapes and sizes by changing the mixture composition. Different gelator mixtures have been investigated, and several of them are included in the reviews by Pakseresht *et al.*,[2] Puscas *et al.*,[21] and Patel.[22]

Among the different LMWGs studied, fatty alcohols have received limited attention and, in most cases, their tentative gelling properties have been investigated just in mixtures with fatty acids or as a part of the vegetable waxes from which they are extracted.[25] Nevertheless, fatty alcohols seem to have promising gelling properties. Recently, their gelling behavior and the microstructure of oleogels, as a function of gelator concentration, chain length, and operating conditions have been reported.[16,17,26] Within this context, the main objective of this chapter is to provide an overview of the available literature about the effect of adding fatty alcohols and vegetable waxes to edible oils. It is important to note that vegetable waxes have fatty alcohols as part of their constituents. Among the various types of fatty alcohols that exist, the characteristics of Policosanol are particularly discussed. Policosanol is a particular mixture of fatty alcohols, well known for its hypocholesterolemic properties, and less studied as a gelator of vegetable oils. A rheological and microstructural characterization of these materials is presented, as well as a selection of examples reporting direct applications of these materials in designing new foods.

6.1.2 Fatty Alcohols in Foods and Their Use as Gelators

Fatty alcohol is the term commonly used to define aliphatic hydrocarbons containing a hydroxyl group that is usually located in the terminal carbon and, less frequently, within the hydrocarbon chain.[27] According to IUPAC nomenclature, their name is based on the number of carbon atoms in the chain and contains the suffix "-ol"; the trivial name is obtained from the aliphatic hydrocarbons with the same length as the alkyl chain. The alkyl chain can be linear or branched and when branches are present, they usually occur close to the end of the chain opposite to the hydroxyl group. The number of carbon atoms within the chain ranges typically between 6 and 36 and the physical properties, such as solubility, melting point, and density, vary with the chain length. Most fatty alcohols are saturated, and a limited number are monounsaturated.[27] In the chemical industry, fatty alcohols with 6–7 atoms are used as plasticizers, whereas alcohols with higher molecular weight are used for their excellent cleaning properties to produce detergents, solvents, and cosmetic products.[28] Fatty alcohols can be obtained either through catalytic hydrogenation of fatty acids,[29] from crude oil and palm kernel, or through the use of engineered bacteria able to overproduce fatty acids from renewable biomass and to transform them into fatty alcohols.[30,31]

A source of long-chain (20–36 carbons) aliphatic alcohols is the wax fraction from sugarcane, spinach, wheat germ, beeswax, and pomace olive oil,[32,33] and a recent study also investigated the extraction of these alcohols from insect wax.[34] These alcohols have beneficial health effects such as lowering of cholesterol levels or the reduction of platelet aggregation.[32] A commercial mixture of long-chain fatty alcohols from natural waxes, known as "Policosanol", is composed mainly of octacosanol (C28), triacontanol (C30), and hexacosanol (C26) and is already used as an ingredient of dietary supplements owing to its nutraceutical properties.[33]

The potential use of fatty alcohols as gelators was investigated by Daniel and Rajasekharan[12] in 2003. These authors observed that saturated fatty alcohols with chain lengths between 20 and 30 carbons were able to gel sunflower oil, lavender oil, and diesel fuel, although their gelling efficiency was lower than that of fatty acids with the same chain length. Gandolfo *et al.*[15] investigated the properties of alcohols with 16, 18, 20, and 22 carbons, observing that these alcohols were able to gel sunflower oil at concentrations of about 2%. The oleogels obtained were harder than those obtained using the corresponding fatty acids at the same concentration. Nevertheless, a mixture of alcohols and acids with the same chain length resulted in oleogels harder than those obtained using the alcohols or the acids individually. Valoppi *et al.*[17] studied peanut oil oleogels developed using C14 and C22 alcohols. They observed that solubility in peanut oil decreased as chain length increased leading, during the gelation, to a larger number of crystals and, in turn, to a harder system as shown by the mechanical tests. Systems containing alcohols from C18 to C22 exhibited a similar firmness whereas C16 yielded a weaker gel, probably due to the lower solid content developed by this alcohol. When investigating the effects of the cooling rate, the authors observed that the fastest ramp (*i.e.*, 40 $^{\circ}C\,min^{-1}$) resulted in the formation of a large number of small crystals, tentatively resulting in the formation of a well-structured and strong crystal network. Consequently, a general reduction in firmness was observed for oleogels developed with high cooling rates.

Lupi *et al.*[16] investigated a mixture of long-chain fatty alcohols (Policosanol) ranging mainly between C26 and C30. They found that Policosanol was effective in olive oil structuration even at very low concentration (0.1% w/w), confirming that long-chain fatty alcohols seem the most suitable for use in oleogelation.

6.2 Rheological Characteristics and Modeling of Fatty Alcohol Oleogels

As already indicated, several molecules including mono-, di-, and triacylglycerols, fatty acids, lecithin, waxes, *etc.* have been reported in the literature for their capacity to structure edible oils.[4,5,14,19,35–39] In addition to these gelators, the use of fatty alcohols as gelator has also been reported by some researchers.[3,15] For instance, Gandolfo *et al.* studied the gelation of sunflower oil using fatty alcohols of varying aliphatic chain lengths.[15] The results revealed that the critical gelation concentration was about 2% (wt/wt), and the resultant oleogels were harder than the fatty acid oleogels with the same amount of gelator. The crystal network of fatty alcohol-based oleogels was usually observed as a stack of large platelets,[3] which was eventually responsible for the rheological fingerprint of the resultant system.

In addition to pure fatty alcohol-based oleogels, a mixture of fatty alcohols and fatty acids has also been investigated in the literature, mainly for producing food gels.[14,15,40] A synergistic increase in rheological properties (*i.e.*, elastic modulus) was observed for an oleogel based on a mixture of fatty acid and fatty alcohol (3 : 7 ratio of acid : alcohol).[14] The elastic modulus (G') of fatty alcohol-based oleogels has been correlated with the gelator fraction (X_O) by using a modified form of the fractal equation originally proposed by Tang and Marangoni.[41] This form of the fractal equation is given in eqn (6.1).[16,42,43]

$$G' = \lambda \cdot (X_O)^{\frac{1}{3-D}} \tag{6.1}$$

where λ represents the strength of interactions between the crystal flocs, and D is the fractal dimension of the structure. Moreover, the experimental temperature for the crystallization onset (T_{co}) as a function of gelator fraction has been fitted using a 'modified fractal model' (eqn (6.2))[16] for which an asymptotic trend at high Policosanol fractions was predicted. This equation is qualitatively similar to the model previously described by Tang and Marangoni[44] and fit the T_{co} behavior for a wider solid fat fraction interval.[16]

$$T_{co} = a\left(1 - e^{-bX_p^c}\right)^{\frac{1}{(3-D)}} \tag{6.2}$$

In eqn (6.2) a, b and c represent the fitting parameters, and D is the fractal dimension.

6.2.1 Policosanol as a Gelator: Rheological Properties

As previously introduced, Policosanol is typically obtained as a mixture of fatty alcohols.[33] Initially, it was reported for use as a dietary supplement after extracting it from different natural sources such as beeswax, wheat germ, sugar cane wax, and rice bran.[33,45] Policosanol is known to impart various positive effects on consumer health such as preventing cardiovascular diseases, decreasing low density lipoprotein in the blood stream, reducing lipid levels, and improving athletic performance.[46] Later on, its potential use as a gelator was described in the literature,[16,43] making it a suitable candidate for producing oleogels for food and organogels for pharmaceutical applications.

Thermo-rheological experiments done in olive oil showed the gelling ability of Policosanol extracted from rice bran, observing an onset of crystallization at very low concentrations (0.1%, wt/wt) and a transition between a suspension of crystal aggregates and a gelled system showing solid-like behavior between 0.3% and 0.5% (Figure 6.1a and b).[16] This concentration assured the formation of a 3D network by having a large number of crystals. The frequency sweep tests verified the formation of stronger oleogels with a Policosanol content higher than 0.5%, as highlighted by the slope change of complex modulus curves that reached an almost independent behavior as a function of the frequency of oscillation (see Figure 6.2).[16] However, in some other studies, a higher content of Policosanol (7–15%) was reported to be used for structuring the oil phase (*i.e.*, for rice bran oil and soybean oil).[26,47]

The rheological properties of Policosanol-based oleogels are found to be significantly dependent on the gelator concentration. The onset of crystallization temperature, gelation temperature, and melting temperature were observed to increase as a function of increasing Policosanol fraction.[47,48] For instance, the onset of crystallization temperature as a function of Policosanol fraction, both experimentally obtained and obtained by model

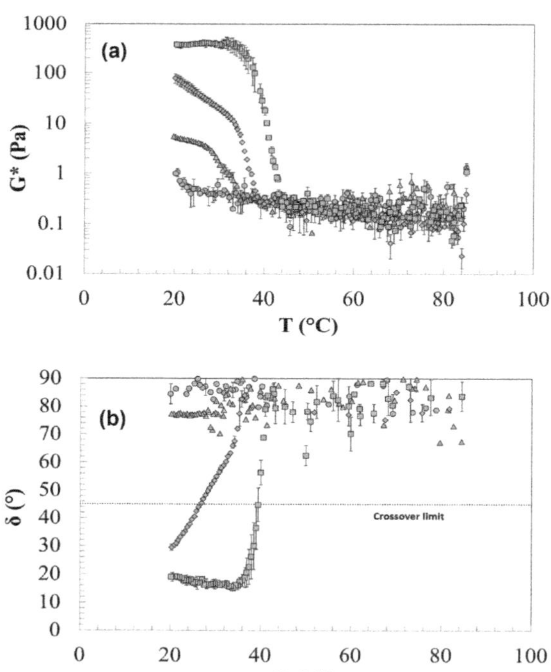

Figure 6.1 Complex modulus G^* (a) and phase angle δ (b) as a function of temperature for oleogels at different Policosanol concentrations: 0.1% (circles), 0.3% (triangles), 0.5% (diamonds), and 1% (squares). "Crossover limit" line indicates the threshold value where δ becomes 45° and the crossover between G' and G'' occurs. Bars represent the standard deviation. Reproduced from ref. 16 with permission from Elsevier, Copyright 2013.

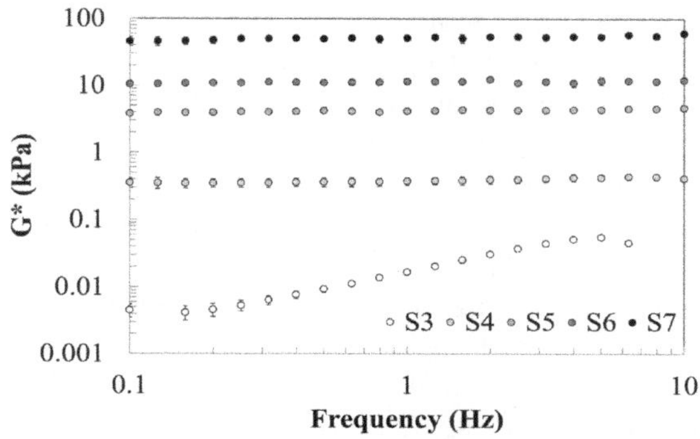

Figure 6.2 Complex modulus (G^*) as a function of frequency for oleogels at 25 °C containing different amounts of policosanol: 0.5% (S3), 1% (S4), 2% (S5), 2.5% (S6), and 3% (S7).
Reproduced from ref. 16 with permission from Elsevier, Copyright 2013.

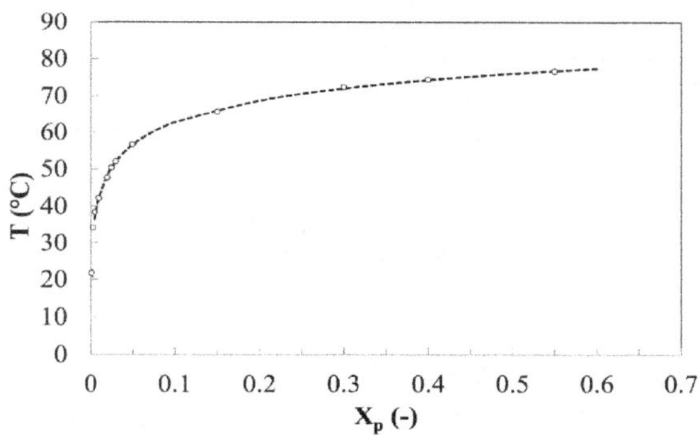

Figure 6.3 Temperature for the onset of crystallization (T_{co}) as a function of Policosanol concentrations. The dashed line represents the empirical model fitting using eqn (6.2). Bars represent the standard deviation and are smaller than the symbol size.
Reproduced from ref. 16 with permission from Elsevier, Copyright 2013.

fitting using eqn (6.2), is shown in Figure 6.3. For this system the parameters provided by eqn (6.2) were $a = 101$ °C (±8), $b = 2.9$ (±0.3), $c = 0.13$ (±0.01), and $D = 2.75$. These results showed that an increase in Policosanol concentration from 0.1% up to 0.55% resulted in an increase in the onset of crystallization temperature from 21.7 °C to 76.7 °C.[47,48] Similar results were observed for the rheological properties including the complex modulus and yield stress.[16,26] Thus, an increase in complex modulus at 1 Hz from 0.017 kPa to 51.42 kPa was observed by increasing the Policosanol concentration from 0.5% to 3% (Figure 6.2). The concentration of Policosanol also affects the solvent mobility, oil binding capacity, gelation time, and textural parameters (*i.e.*, hardness, stickiness, and firmness).[47] These results show that the final properties of Policosanol oleogels, including consistency and strength, can be finely tuned by varying the gelator concentration.

Lupi *et al.* compared the rheological characteristics of oleogels based on Policosanol with MG oleogels using different edible oils, including virgin olive oil, extra virgin olive oil, and sunflower oil.[42] The Policosanol oleogels, whose structure is attributable to van der Waals weak interactions, displayed independent rheological behavior as a function of type of oil, slower structuration rate, and higher onset of crystallization temperature as compared to the MG oleogels (see Figure 6.4). Thus, the addition of Policosanol resulted in stronger gels (G^* of 37.38 kPa at 20 °C) than the MG oleogels (G^* of 0.096 kPa at 20 °C), which might be associated with the enhanced alkyl chains in Policosanol oleogels.[42] This study showed that the rheological properties of Policosanol oleogels were significantly different from MG oleogels. The particular rheological properties of Policosanol oleogels have been used to stabilize meat sauces.[49] Thus, the oleogel containing the lower amount of Policosanol (0.1%) provided rheological characteristics (*i.e.*, moduli, viscosity, yield stress, *etc.*) similar to the benchmark product, and an increase in complex modulus (*i.e.*, gel strength) was observed by increasing the Policosanol concentration.[49] Therefore, for industrial applications and in contrast to MG oleogels, Policosanol oleogels can be used to stabilize meat sauces.

The gelling ability of Policosanol has also been reported in non-edible oils such as silicon oil.[50] The results of temperature ramp tests showed the formation of crystal Policosanol and their aggregated crystal network at very high temperatures. This is in comparison with the MG-based silicon oil system, which did not show any gelation behavior for the whole temperature interval investigated (10–80 °C; Figure 6.5).[50] Moreover, the formation of crystal aggregates in Policosanol-based oleogels resulted in higher complex modulus values (49.80 kPa at 20 °C) than the MG-based oleogel (0.002 kPa at 20 °C). This peculiar ability verifies the potential of Policosanol as a gelator for a wide range of applications involving both edible and non-edible oils.

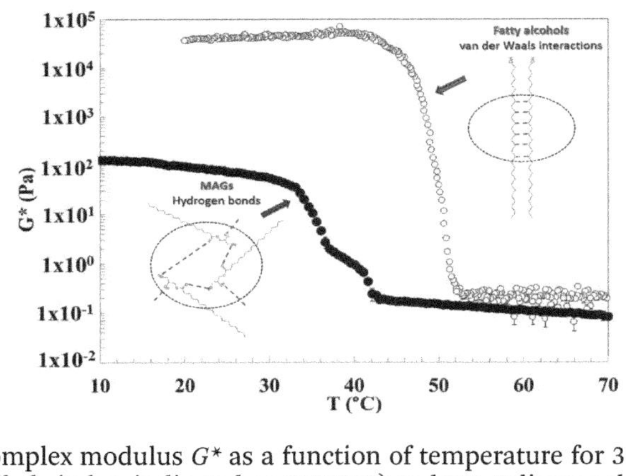

Figure 6.4 Complex modulus G^* as a function of temperature for 3% MGs oleogels (filled circles, indicated as "MAGs") and 3% Policosanol oleogels (empty circles, indicated as "fatty alcohols"). The schematics represent the interactions between the gelator molecules.
Reproduced from ref. 42 with permission from Elsevier, Copyright 2016.

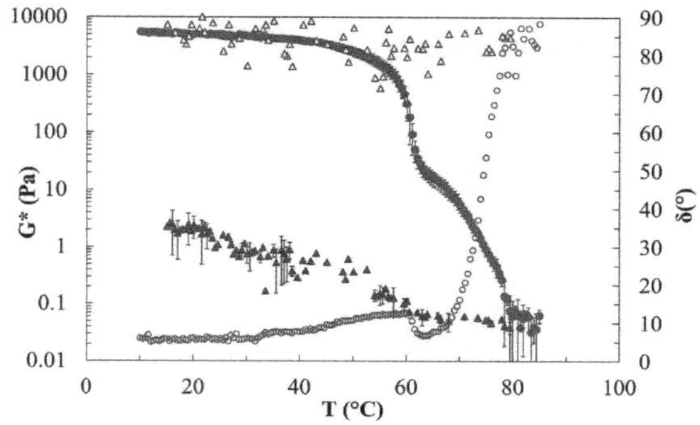

Figure 6.5 Complex modulus (G^*; filled symbols) and phase angle (δ; empty symbols) as a function of temperature for silicon oil-based oleogels with Policosanol (purple circle) and MGs (red triangle). Bars represent the standard deviation.
Reproduced from ref. 50 with permission from Elsevier, Copyright 2018.

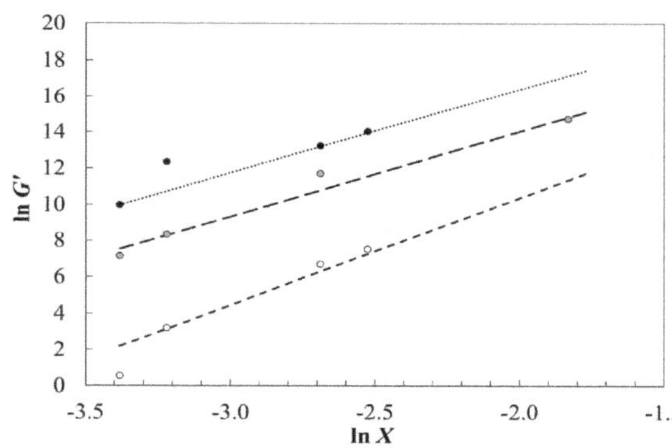

Figure 6.6 Storage modulus G' as a function of gelator fraction X for oleogels prepared with GS (white circles), Policosanol (black circles), and a mixture of Policosanol and GS (gray circles). Dashed lines represent the empirical model fitting (*i.e.*, linearized form of eqn (6.1)).
Reproduced from ref. 43 with permission from Elsevier, Copyright 2016.

In addition to the Policosanol oleogels, mixtures of glyceryl stearate (GS) and Policosanol have been used to tune the rheological properties of the resultant mixed oleogels.[43] The storage modulus as a function of gelator fraction for Policosanol and mixed GS–Policosanol oleogels is shown in Figure 6.6 (*i.e.*, experimental data and model fitting using eqn (6.1)). The results revealed the intermediate values of storage modulus as a function of gelator concentration for mixed oleogels as compared to the Policosanol or GS oleogels. In particular, the parameters of the linearized form of eqn (6.1) for the GS oleogels were $\ln(\lambda) = 22.2$ kPa (± 0.7) and $D = 2.831$ (± 0.008), for Policosanol oleogels $\ln(\lambda) = 25.7$ kPa (± 0.3) and $D = 2.785$ (± 0.006), and for the mixed GS-Policosanol oleogels (equiponderant) $\ln(\lambda) = 23.47$ kPa (± 0.02) and

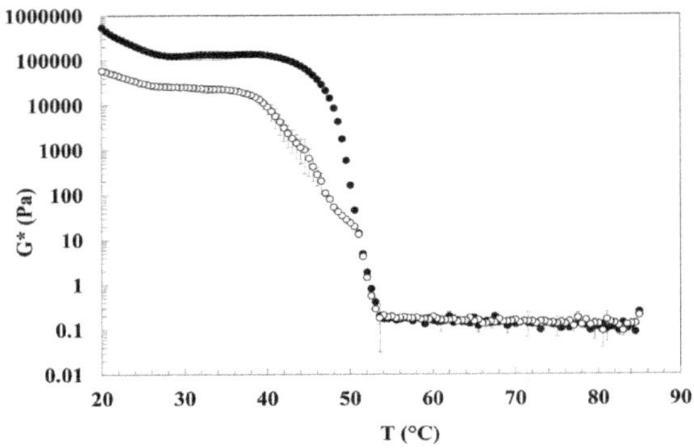

Figure 6.7 Complex modulus as a function of temperature for oleogels with 3.4% of Policosanol (empty circles) and a mixture of 3.4% Policosanol and 3.4% glyceryl stearate (GS; filled circles). Bars represent the standard deviation. Reproduced from ref. 43 with permission from Elsevier, Copyright 2016.

$D = 2.788$ (± 0.003).[43] Moreover, the characteristic temperatures (*i.e.*, onset of crystallization and gelation) of mixed oleogels were completely controlled by the Policosanol fraction whereas the gel strength (*i.e.*, complex modulus), at temperatures significantly lower than the crystallization temperature, was determined by the GS (*i.e.*, higher values of complex modulus for mixed oleogels (545.2 kPa at 20 °C) than for Policosanol oleogels (58.80 kPa at 20 °C) using the same amount of Policosanol (0.034%) (Figure 6.7).[43] This interesting mixed oleogel system has also been utilized to prepare a bigel system (*i.e.*, a mixture of oleogel and hydrogel), by varying the oleogel content.[51,52] The bigel systems with a larger content of oleogel phase showed higher modulus values and better thermal stability due to the incorporation of a stronger oleogel phase within a weaker hydrogel phase.[51]

6.3 Microstructural Characteristics of Fatty Alcohol Oleogels

The microstructural investigation of oleogels is imperative, in order to fully understand the bulk rheological fingerprint of oleogels. The microstructure of fatty alcohol-based oleogels was typically found to be a stack of platelet-like crystals.[3,15] However, the crystalline morphology of fatty alcohol was found to be modified by adding fatty acids, *i.e.*, transition from platelet-like to needle-like crystals, which also resulted in better rheological/mechanical properties of the mixed oleogels. This modification in the crystal habit of fatty alcohols was attributed to the decrease in interfacial tension, which eventually resulted in a lower critical radius required for the development of stable nuclei and, hence, more smaller needle-like crystals.[3,14]

The formation of needle- or fiber-like crystals has also been observed for Policosanol oleogels, with strong dependence on the Policosanol concentration.[26] For instance, the incorporation of 7% of Policosanol in soybean oil

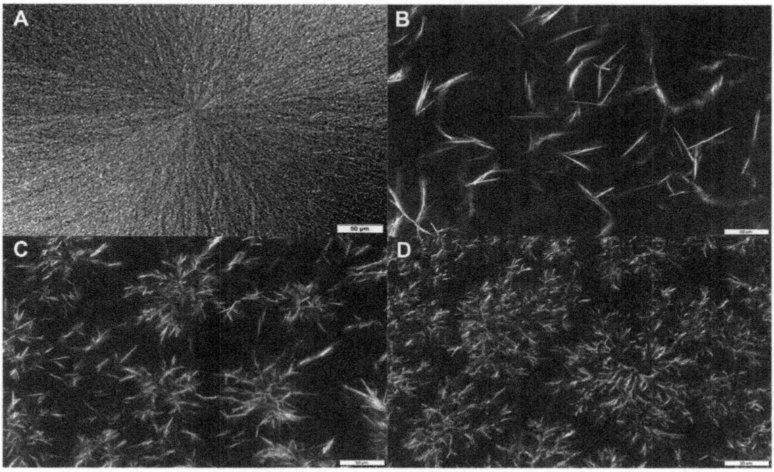

Figure 6.8 Microscopic images of (a) pure Policosanol, and soybean oil-based oleogels with 7% (b), 10% (c), and 12% (d) of Policosanol. The scale bar represents 50 μm.
Reproduced from ref. 26 with permission from Elsevier, Copyright 2018.

resulted in large non-interacting needle-like crystals (Figure 6.8b) with an area of about 148.3 μm^2, which were quite similar to the crystalline structure of pure Policosanol (Figure 6.8a). At a higher concentration of Policosanol (10%), smaller spherulitic- or dendritic-like crystals with an area of about 133.6 μm^2 were observed (Figure 6.8c). This change in microstructure as a function of gelator content was linked to the enriched nucleation behavior and increased supersaturation, which eventually resulted in smaller individual crystals and dense crystalline aggregates.[26] By further increasing the Policosanol content (12%), similar crystalline morphology (*i.e.*, dendritic-like crystals) with complex dense aggregates and even more branching was clearly evident (Figure 6.8d).[47,53]

In addition to the Policosanol content, the cooling rate used for gelation also has a significant influence on the crystalline network of the resultant oleogel.[53] By increasing the cooling rate from 2 $^{\circ}C\,min^{-1}$ to 25, the crystal area was observed to decrease from 166.4 to 60.9 μm^2. This decrease in crystal size was associated with the higher supersaturation or time-dependent undercooling, which resulted in a faster nucleation rate resulting in a large number of smaller crystals.[53] The same behavior, as a function of cooling rate, was evident for both lower and higher concentrations of Policosanol.[53] However, the size of crystals was found to be independent of the type of oil (*i.e.*, virgin olive oil, extra-virgin olive oil, and sunflower oil) for Policosanol oleogels.[42] Moreover, larger fiber-like crystals were observed for Policosanol oleogels as compared to MG oleogels, which was attributed to the difference in interactions between the gelator molecules.[42]

6.4 Application of Fatty Alcohols and Vegetable Wax Oleogels in Food Systems and as Delivery Agents for Bioactive Compounds

One of the most interesting uses of oleogels in food materials is the substitution of hard fats with an oil-based material with elastic properties, which is able to give to the final product texture, spreadability, shelf life, and a mouthfeel comparable to traditional fat-based foods. In the last 10 years nearly a thousand scientific papers have been published on this subject, but only a little over twenty in the previous ten years.[54] This highlights that the interest in research on this topic increased as a consequence of the decision of the US Food and Drug Administration (FDA) according to which *trans* fats are no longer "generally recognized as safe" (GRAS), and that partially hydrogenated oils should be banned from foods as of January 2020.[54] Therefore, the food industry has focused its research on the appropriate replacement of *trans*-fats with natural saturated fats, commonly obtained by the fractionation of vegetable oils, mainly palm oil. The recent controversy concerning the use of the saturated fraction of palm oil and its potential negative effect on the health of consumers and the environment[55,56] is, nowadays, pushing the food industry toward developing new strategies for producing healthier foods, with a low impact on the environment, but, obviously, without increasing costs too much. In this scenario, oleogel utilization is widely considered as a convincing way to respond to this problem. Fatty alcohol oleogels are mostly produced by structuring vegetable oils with waxes (both animal and plant based), containing hydrocarbons, fatty acids, and fatty alcohols as well as esters and other impurities.[57] Moreover, Carnauba Wax (CW), Rice Bran Wax (RBW), and other plant-based waxes are all food additives approved by the US FDA (United States Food and Drug Administration).[58]

Oleogels are intended to be used as semi-finished materials for producing doughs or spreadable products. In the latter case, important technological characteristics like firmness, hardness, spreadability, oxidative stability, stickiness, and sensory properties like mouth coating, fatty, waxy, grittiness, milky, rancid, sweet, salty, and so on, in comparison with conventional spreads are interesting information to be added to the rheological properties and microstructural information. Despite the large number of papers published in recent years, the literature regarding the sensory properties of oleogels produced with pure Policosanol is very scarce, whereas more detailed information is reported in the case of materials produced with sugar cane wax (SCW), beeswax (BW), RBW, and CW.[16,59,60] This lack of information on pure Policosanol oleogels used for technological applications is most likely due to its higher cost compared to that of the waxes from which Policosanol derives.

Hardness was evaluated by Buitimea-Cantúa *et al.*[61] as the main textural parameter for understanding the characteristics of CW (a natural plant source of sterols, esters, fatty alcohols, fatty acids, and resins) in soybean oil oleogels. The authors prepared oleogels with three kinds of CW, finding that hardness can discriminate among samples and that the type of wax, concentration, and storage time significantly affected the hardness and rheological properties of the resulting oleogel.[61]

Oleogels produced with fatty alcohols were used as the fat bases for preparing baked goods like cookies. Hwang *et al.*[62] showed the sensory properties of cookies prepared with vegetable oils, *i.e.*, olive oil, flaxseed oil, or soybean oil, structured with natural waxes, *i.e.*, sunflower wax (SW), RBW, BW, or candelilla wax. According to their results, soybean oil-oleogels were able to give the cookies a fracturability comparable to that of cookies prepared with commercial margarine (benchmark). Cookies prepared with SW combined with olive oil and flaxseed oil showed a hardness comparable to the benchmarks. Yilmaz and Öğütcü[63] investigated the sensory properties of cookies produced with hazelnut oil-BW and hazelnut oil-SW oleogels (5% of the gelator in vegetable oil) and compared them with cookies made with commercial bakery shortening (CBS). According to the results given by 12 panelists, cookies made with CBS showed a bite hardness significantly lower than those of cookies made with the oleogels, in agreement with the textural fracturability values. A similar result was found for the brittleness during chewing; even in this case, cookies produced with CBS were significantly less brittle than the other two samples. On the contrary, no significant differences were found in the case of the investigation of 'non-sandiness', *i.e.*, the cookie roughness during chewing, and the 'easy to chew' and 'mealiness' properties of the cookies need to be determined to completely match the consumer demands. Finally, Pandolsook and Kupongsak[64] investigated the production of fats based on oleogels of bleached RBW to replace margarine in cookie production. The authors found that both the hardness and color of cookie dough were affected by replacement of margarines with oleogels, whereas the spread factor, diameter, and width were not significantly different.

In recent years, a number of studies also defined the sensory characteristics of oleogels used in meat products. For example, Franco *et al.*[65] in their paper described the properties of meat sausages prepared by replacing 20% and 40% pork backfat with linseed oil oleogel structured with BW and γ-oryzanol and β-sitosterol. According to the authors, these new products were less acceptable than the control, suggesting deepening the research to meet the consumer requests. On the other hand, Aliasl Khiabani *et al.*[66] proposed the study of CW-based oleogels with adipic acid (AA) to produce new cakes and beef burgers. According to the results, a formulation of oleogel produced with 2% CW and 4% AA gave acceptable textural and sensory properties, in terms of odor, taste, texture, color, and overall acceptance, with the same level of quality as un-substituted products.

Moreover, as already discussed, Policosanol has also been used to structure vegetable oils, *i.e.*, a mixture of olive oil and sunflower oil, being the oil phase of spicy meat sauces potentially applicable as dressings.[49] Sauce samples were prepared with two levels of Policosanol added to the oil mixture, *i.e.*, 0.1% and 0.5% on a total basis, and the lowest amount was enough

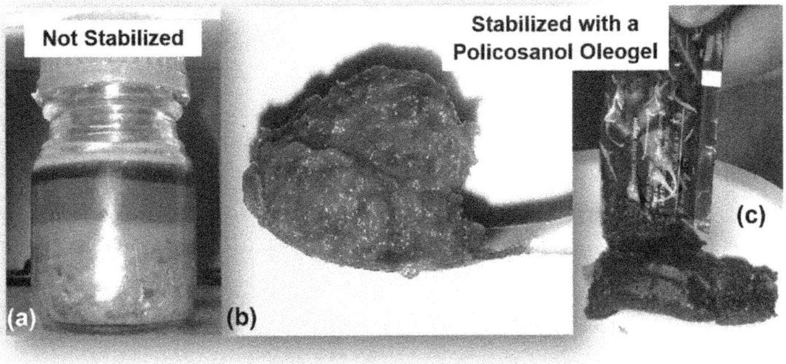

Figure 6.9 An image of spicy meat sauce produced without the addition of stabilizers (a), stabilized with 0.1% Policosanol oleogel (the solvent used is a mixture of sunflower and olive oil) (b), and after pasteurization (c). Image credit: Lupi F. R., Gabriele D., Seta L., Baldino N., and de Cindio B.

to give stability to the suspensions. Moreover, this low content of Policosanol has proved to be sufficient to stabilize these materials even after cycles of pasteurization of the samples which underwent considerable thermal changes (from 25 °C to about 95 °C). A sample of unstable sauce (after centrifugation) is shown in Figure 6.9a; the sauce stabilized with a Policosanol oleogel is shown in Figure 6.9b; the sample appeared stable even after pasteurization (Figure 6.9c).

Another interesting use of fatty alcohol oleogels was proposed by Bemer *et al.*[67] The authors suggested the addition of RBW or ethylcellulose (*i.e.*, a polymeric gelator[68]) to structure a liquid phase made of skimmed milk, soybean oil, salt, and non-fat dry milk. The main aim was to produce low-fat oleogel cream cheese to be investigated through rheological and textural characterization coupled with a microstructure analysis. The proposed results confirmed that RBW oleogels show textural properties comparable to a full-fat benchmark. On the other hand, ethylcellulose provides a lowered adhesiveness and storage modulus in comparison to the benchmark. The length of the carbon chain of fatty alcohols allows the compound to acquire different properties and, therefore, to be used in different applications. Valoppi *et al.*[17] investigated the potential ability of fatty alcohols, with chain lengths ranging from C14 to C22, to gel peanut oil in order to prevent oil separation during storage, increasing its stability. The authors found a direct relation between the product stability and the length of the fatty alcohol chain. Optimization of oleogel composition can produce materials characterized by sensory profiles similar to benchmarks, even if, in some cases, more research must be carried out to completely match the consumer demands.

Oleogels can also be used for the delivery of target molecules that can be properly tuned *via* ingestion. In fact, oleogels produced with natural waxes, rich in fatty alcohols, are food-grade, highly available at relatively low costs, and moreover, a low amount of gelator is required to structure vegetable oils with a remarkable oil-binding capacity.[69] Waxes can be hydrolyzed in the gastro-intestinal tract, but, nevertheless, they are poorly digestible owing to their resistance to lipase action.[70] Research on this specific topic is still ongoing; in fact, it is not completely understood whether the scarce digestibility of waxes can limit the use of the derived oleogels as carriers for nutraceutical applications. For example, Limpimwong *et al.*[71] studied the digestibility of RBW oleogels in rats. Animals fed with oleogels showed a different TG profile in serum, liver, and feces compared with the control group. According to the authors, the structured matrix of oleogel reduced the digestion and absorption of lipids, which were excreted in feces, indicating that oleogelation can contribute to reducing lipid digestion. An example of Policosanol used for nutraceutical purposes was given by Lupi *et al.*[48] for the controlled delivery of ferulic acid, an antioxidant agent, throughout its stabilization in policosanol–olive oil oleogel pills. The formulation of gels was optimized thanks to rheological tests. It was found that a minimum weight fraction of 0.03 of Policosanol was requested to suspend ferulic acid particles, and that their delivery is adequately controlled with a small amount of gelator added to liquid oil (1–3%). A Policosanol oleogel was also successfully used by Tian and Acevedo[53] to protect Retinyl Palmitate from photodegradation, to increase the physical UV-blocking action, and to delay or hinder free radical-mediated reactions. Retinyl Palmitate is the ester of vitamin A, an important nutrient for consumers; it is extensively used in foods, drugs, and cosmetics, but its low stability against chemical and photochemical agents requires adequate preservation.

Thus, to conclude, according to the above-described case-studies, it seems that fatty alcohol oleogels are valuable materials to replace hydrogenated and/or saturated fats in foods production and for designing new materials for the delivery of nutraceutical compounds. Policosanol, fatty alcohols, and waxes are considered to be very good gelators, rheological modifiers, and stabilizers for food applications.

6.5 Conclusions

The increasing request for food-grade fats perceived as safe by consumers has pushed researchers to develop new products with a reduced content of saturated fats. In this context, oleogels are innovative materials thanks to which a wide range of healthy foods or nutraceutical products can be introduced into the market. Oleogels are semi-solid fats at room temperature, produced by allowing the structuration of liquid oils through the self-assembly action of gelator molecules. Among the different kinds of gelators, LMWGs promote solvent structuration through short-range weak interactions. Fatty alcohols can be used as gelators specifically addressed to the food market, and, in particular, Policosanol, *i.e.*, a mixture of fatty alcohols mainly composed of octacosanol, is a very interesting compound already known as a dietary supplement. The length of the carbon chain which, together with the hydroxyl group, forms a fatty alcohol can affect the rheological and microstructural properties of the gel and, consequently, its final industrial use.

References

1. L. S. K. Dassanayake, D. R. Kodali and S. Ueno, *Curr. Opin. Colloid Interface Sci.*, 2011, **16**, 432–439.
2. S. Pakseresht and M. Mazaheri Tehrani, *Food Rev. Int.*, 2020, 1–23.
3. E. D. Co and A. G. Marangoni, *J. Am. Oil Chem. Soc.*, 2012, **89**, 749–780.
4. F. R. Lupi, D. Gabriele, B. de Cindio, M. C. Sánchez and C. Gallegos, *J. Food Eng.*, 2011, **107**, 296–303.
5. F. R. Lupi, D. Gabriele and B. de Cindio, *Food Bioprocess Technol.*, 2012, **5**, 2880–2888.
6. S. Da Pieve, S. Calligaris, A. Panozzo, G. Arrighetti and M. C. Nicoli, *Food Res. Int.*, 2011, **44**, 2978–2983.
7. F. R. Lupi, V. Mancina, N. Baldino, O. I. I. Parisi, L. Scrivano and D. Gabriele, *Food Funct.*, 2018, **9**, 3278–3290.
8. F. Valoppi, S. Calligaris, L. Barba, N. Šegatin, N. Poklar Ulrih and M. C. Nicoli, *Eur. J. Lipid Sci. Technol.*, 2017, **119**, 1500549.
9. A. Bot, R. den Adel, C. Regkos, H. Sawalha, P. Venema and E. Flöter, *Food Hydrocolloids*, 2011, **25**, 639–646.
10. A. J. Martins, M. A. Cerqueira, L. M. Pastrana, R. L. Cunha and A. A. Vicente, *J. Sci. Food Agric.*, 2019, **99**, 3318–3325.
11. A. B. Matheson, V. Koutsos, G. Dalkas, S. Euston and P. Clegg, *Langmuir*, 2017, **33**, 4537–4542.
12. J. Daniel and R. Rajasekharan, *J. Am. Oil Chem. Soc.*, 2003, **80**, 417–421.
13. S. S. Sagiri, V. K. Singh, K. Pal, I. Banerjee and P. Basak, *Mater. Sci. Eng., C*, 2015, **48**, 688–699.
14. H. M. Schaink, K. F. van Malssen, S. Morgado-Alves, D. Kalnin and E. van der Linden, *Food Res. Int.*, 2007, **40**, 1185–1193.
15. F. G. Gandolfo, A. Bot, E. Floter and E. Flöter, *J. Am. Oil Chem. Soc.*, 2004, **81**, 1–6.
16. F. R. Lupi, D. Gabriele, V. Greco, N. Baldino, L. Seta and B. de Cindio, *Food Res. Int.*, 2013, **51**, 510–517.
17. F. Valoppi, S. Calligaris and A. G. Marangoni, *Eur. J. Lipid Sci. Technol.*, 2017, **119**, 1600252.
18. K. K. Upadhyay, C. Tiwari, A. J. Khopade, H. B. Bohidar and S. K. Jain, *Drug Dev. Ind. Pharm.*, 2007, **33**, 617–625.
19. M. Pernetti, K. Vanmalssen, D. Kalnin and E. Floter, *Food Hydrocolloids*, 2007, **21**, 855–861.
20. Y. A. Shchipunov, *Colloids Surf., A*, 2001, **183**, 541–554.
21. A. Puşcaş, V. Mureşan, C. Socaciu and S. Muste, *Foods*, 2020, 9.
22. A. R. Patel, *Curr. Opin. Food Sci.*, 2017, **15**, 1–7.
23. H. Sawalha, G. Margry, R. den Adel, P. Venema, A. Bot, E. Flöter and E. van der Linden, *Eur. J. Lipid Sci. Technol.*, 2013, **115**, 295–300.
24. S. Calligaris, G. Mirolo, S. Da Pieve, G. Arrighetti and M. C. Nicoli, *Food Biophys.*, 2014, **9**, 69–75.
25. J. F. Toro-Vazquez, J. A. Morales-Rueda, E. Dibildox-Alvarado, M. Charó-Alonso, M. Alonzo-Macias, M. M. González-Chávez, M. Charo-Alonso,

M. Alonzo-Macias and M. M. Gonzalez-Chavez, *J. Am. Oil Chem. Soc.*, 2007, **84**, 989–1000.
26. Y. Tian and N. C. Acevedo, *Food Chem.*, 2018, **255**, 252–259.
27. S. M. Mudge, *Soap Deterg. Assoc.*, 2005, **132**, 1–141.
28. M. A. Sánchez, G. C. Torres, V. A. Mazzieri and C. L. Pieck, *J. Chem. Technol. Biotechnol.*, 2017, **92**, 27–42.
29. J. Dahlin, C. Holkenbrink, E. R. Marella, G. Wang, U. Liebal, C. Lieven, D. Weber, D. McCloskey, H.-L. Wang, B. E. Ebert, M. J. Herrgård, L. M. Blank and I. Borodina, *Front. Genet.*, 2019, **10**, 747.
30. Z. Jin, A. Wong, J. L. Foo, J. Ng, Y.-X. Cao, M. W. Chang and Y.-J. Yuan, *Biotechnol. Bioeng.*, 2016, **113**, 842–851.
31. S. Fillet, J. Gibert, B. Suárez, A. Lara, C. Ronchel and J. L. Adrio, *J. Ind. Microbiol. Biotechnol.*, 2015, **42**, 1463–1472.
32. A. Fernández-Arche, A. Marquez-Martín, R. de la, P. Vazquez, J. S. Perona, C. Terencio, C. Perez-Camino and V. Ruiz-Gutierrez, *J. Nutr. Biochem.*, 2009, **20**, 155–162.
33. S. Irmak, N. T. Dunford and J. Milligan, *Food Chem.*, 2006, **95**, 312–318.
34. J. Ma, L. Ma, H. Zhang, Z. Zhang, Y. Wang, K. Li and X. Chen, *PLoS One*, 2018, **13**, e0197343.
35. S. Calligaris, S. Da Pieve, G. Arrighetti, L. Barba, S. Da Pieve, G. Arrighetti and L. Barba, *Food Res. Int.*, 2010, **43**, 671–677.
36. S. da Pieve, S. Calligaris, E. Co, M. C. Nicoli and A. G. Marangoni, *Food Biophys.*, 2010, **5**, 211–217.
37. F. R. Lupi, D. Gabriele, D. Facciolo, N. Baldino, L. Seta and B. de Cindio, *Food Res. Int.*, 2012, **46**, 177–184.
38. N. K. O. Ojijo, I. Neeman, S. Eger and E. Shimoni, *J. Sci. Food Agric.*, 2004, **84**, 1585–1593.
39. J. F. Toro-Vazquez, J. Morales-Rueda, V. A. Mallia and R. G. Weiss, *Food Biophys.*, 2010, **5**, 193–202.
40. C. Blach, A. J. Gravelle, F. Peyronel, J. Weiss, S. Barbut and A. G. Marangoni, *RSC Adv.*, 2016, **6**, 81151–81163.
41. D. Tang and A. G. Marangoni, *Trends Food Sci. Technol.*, 2007, **18**, 474–483.
42. F. R. Lupi, V. Greco, N. Baldino, B. de Cindio, P. Fischer and D. Gabriele, *J. Colloid Interface Sci.*, 2016, **483**, 154–164.
43. F. R. Lupi, A. Shakeel, V. Greco, N. Baldino, V. Calabrò and D. Gabriele, *LWT – Food Sci. Technol.*, 2017, **77**, 422–429.
44. D. Tang and A. G. Marangoni, *J. Colloid Interface Sci.*, 2008, **318**, 202–209.
45. F. Francini-Pesenti, F. Brocadello, D. Beltramolli, M. Nardi and L. Caregaro, *Complement. Ther. Med.*, 2008, **16**, 61–65.
46. Z. Xu, E. Fitz, N. Riediger and M. H. Moghadasian, *Nutr. Res.*, 2007, **27**, 212–217.
47. S. Pandolsook and S. Kupongsak, *J. Food Meas. Charact.*, 2020, **14**, 2078–2086.
48. F. R. Lupi, D. Gabriele, N. Baldino, P. Mijovic, O. I. Parisi and F. Puoci, *Food Funct.*, 2013, **4**, 1512–1520.
49. F. R. Lupi, D. Gabriele, L. Seta, N. Baldino and B. de Cindio, *Eur. J. Lipid Sci. Technol.*, 2014, **116**, 1734–1744.

50. F. R. Lupi, M. P. De Santo, F. Ciuchi, N. Baldino and D. Gabriele, *Food Res. Int.*, 2018, **111**, 399–407.
51. F. R. Lupi, A. Shakeel, V. Greco, C. Oliviero Rossi, N. Baldino and D. Gabriele, *Mater. Sci. Eng., C*, 2016, **69**, 358–365.
52. F. R. Lupi, M. P. De Santo, F. Ciuchi, N. Baldino and D. Gabriele, *Rheol. Acta*, 2017, **56**, 753–763.
53. Y. Tian and N. C. Acevedo, *RSC Adv.*, 2020, **10**, 2526–2535.
54. H.-S. Hwang, *Biocatal. Agric. Biotechnol.*, 2020, **26**, 101657.
55. F. Marangoni, C. Galli, A. Ghiselli, G. Lercker, C. La Vecchia, C. Maffeis, C. Agostoni, D. Ballardini, O. Brignoli, P. Faggiano, R. Giacco, C. Macca, P. Magni, G. Marelli, W. Marrocco, V. L. Miniello, G. F. Mureddu, N. Pellegrini, R. Stella, E. Troiano, E. Verduci, R. Volpe and A. Poli, *Int. J. Food Sci. Nutr.*, 2017, **68**, 643–655.
56. K. Obidzinski, R. Andriani, H. Komarudin and A. Andrianto, *Ecol. Soc.*, 2012, **17**, 25.
57. A. R. Patel, M. Babaahmadi, A. Lesaffer and K. Dewettinck, *J. Agric. Food Chem.*, 2015, **63**, 4862–4869.
58. A. I. Blake, E. D. Co and A. G. Marangoni, *J. Am. Oil Chem. Soc.*, 2014, **91**, 885–903.
59. W. F. Tinto, T. O. Elufioye and J. Roach, Waxes, in *Pharmacognosy: Fundamentals, Applications and Strategy*, ed. S. Badal and R. B. T.-P. Delgoda, Academic Press, Boston, 2007, pp. 443–455.
60. K. J. Murphy, D. A. Saint and P. R. C. Howe, *J. Am. Coll. Nutr.*, 2008, **27**, 476–484.
61. G. V Buitimea-Cantúa, S. O. Serna-Saldívar, E. Pérez-Carrillo, T. Jordânia Silva, D. Barrera-Arellano and N. E. Buitimea-Cantúa, *LWT*, 2021, **136**, 110267.
62. H.-S. Hwang, M. Singh and S. Lee, *J. Food Sci.*, 2016, **81**, C1045–C1054.
63. E. Yılmaz and M. Öğütcü, *Food Funct.*, 2015, **6**, 1194–1204.
64. S. Pandolsook and S. Kupongsak, *J. Food Eng.*, 2017, **214**, 182–192.
65. D. Franco, A. J. Martins, M. López-Pedrouso, M. A. Cerqueira, L. Purriños, L. M. Pastrana, A. A. Vicente, C. Zapata and J. M. Lorenzo, *J. Sci. Food Agric.*, 2020, **100**, 218–224.
66. A. Aliasl khiabani, M. Tabibiazar, L. Roufegarinejad, H. Hamishehkar and A. Alizadeh, *Food Chem.*, 2020, **333**, 127446.
67. H. L. Bemer, M. Limbaugh, E. D. Cramer, W. J. Harper and F. Maleky, *Food Res. Int.*, 2016, **85**, 67–75.
68. A. K. Zetzl, A. J. Gravelle, M. Kurylowicz, J. Dutcher, S. Barbut and A. G. Marangoni, *Food Struct.*, 2014, **2**, 27–40.
69. P. K. Okuro, T. P. Santos and R. L. Cunha, *Trends Food Sci. Technol.*, 2021, **111**, 55–67.
70. J. L. Hargrove, P. Greenspan and D. K. Hartle, *Exp. Biol. Med.*, 2004, **229**, 215–226.
71. W. Limpimwong, T. Kumrungsee, N. Kato, N. Yanaka and M. Thongngam, *J. Funct. Foods*, 2017, **39**, 250–256.

CHAPTER 7

On the Applicability of the Oryzanol–Sitosterol Structuring System

M. SCHARFE* AND E. FLÖTER

Department of Food Process Engineering, Technical University Berlin, Berlin, Germany
*Email: maria.scharfe@tu-berlin.de

7.1 Introduction

Structured lipids are a crucial element in achieving specific textures in many food products such as margarine. Moreover, they can stabilize surfaces, generate macroscopic hardness, and affect disintegration characteristics and melting behavior.[1,2] The fatty acid (FA) composition of triacylglycerols (TAGs) defines their physical state. Plant oils generally contain a mixture of saturated FAs (SAFAs), and mono- and poly-unsaturated fatty acids (MUFAs, PUFAs). Their composition and portion determine a fat's state since SAFAs tend to self-assemble into crystalline nanoplatelets (CNPs) at ambient temperature. CNPs stack and aggregate to form a three-dimensional network, immobilizing liquid TAGs.

Although proclaimed differently for many years, a high intake of SAFAs could not be directly correlated to an increased risk of coronary heart diseases (CHDs). However, a diet rich in MUFAs and PUFAs reduces the risk of CHD incidents significantly.[3,4] Thus, increasing their share in food products is a possible approach to support dietary improvement.

Food Chemistry, Function and Analysis No. 31
Development of *Trans*-free Lipid Systems and their Use in Food Products
Edited by Jorge F. Toro-Vazquez

Published by the Royal Society of Chemistry, www.rsc.org

Since the beginning of the 1980s, palm oil has advanced as the primary source of TAGs to create functional fat phases with an annual volume of 73.3 MMt in 2018 (38.6% of total edible oil production).[5] Palm oil comprises roughly 50% SAFAs (primarily palmitic acid) and 50% MUFAs and PUFAs (primarily oleic acid and linoleic acid). In addition, high and low melting palm oil fractions and their combinations deliver desired product properties such as hardness or melting temperature. Therefore, considering the low production price, high oil yield, availability, and unique FAs composition, palm oil has a distinctive position in the global market. Nevertheless, palm oil is at least debatable from an environmental perspective since it is grown in tropical regions where large, diverse rainforest areas have been cut down since the 1980s.[6]

Consequently, the consumer perception of palm oil in western countries is often controversial. Growing ecological awareness and partly flawed information is conducive to sustaining this critical viewpoint in industrialized countries.[7,8] Thus, fat alternatives that provide structure to food or cosmetics have to be considered.

Hydrogenated plant oils are another cost-effective source of semi-solid fats. However, deliberate partial oil hydrogenation results in high *trans* fatty acid levels (TFA). TFAs increase low-density lipoprotein and simultaneously lower high-density lipoprotein.[9] Therefore, they boost the risk of developing CHD by 21%.[10] Hence, public health authorities worldwide suggested limiting TFA contents in food products and recommended that intake should not exceed 1% of the total nutritional energy.[11,12] Nowadays, partial hydrogenation is much less common practice, and the primary intake of TFAs relates to convenience products, fried foods, and baked goods. The EU Commission Regulation 2019/649 asserts that the non-natural TFA content in food products should not exceed 2 g per 100 g fat.[13] Convenience products, baked goods, or fried foods can overstep this threshold value.[14] The TFA content in these products can be reduced to a minimum by varying standard recipes and developing suitable quality guidelines.[15] However, TFA intake predominately depends on the individual's consumption behavior, and an accurate number is generally hard to estimate.[15–18]

In combination with an increasing demand for healthy, organic, sustainable, and regional food products, all of this led to the momentum in finding alternatives to current structuring routes. One route might be gelling liquid edible oils rich in MUFAs and PUFAs with a suitable structuring agent called an oleogelator. An oleogelator should be dissolvable in a non-polar liquid such as edible oil but have limited solubility to create a solid structure from now on called an oleogel or organogel. The gels are distinguished from the traditional structuring route since they are not based on crystalline TAGs but complex structures such as fibers, ribbons, and tubules. The gelling is possible utilizing I) crystalline particles, II) self-assembled networks, III) polymer networks, and IV) emulsions (examples are given in Figure 7.1).

More than 3000 publications concerning oleo-/organogelation indicate that many gelators have been identified in the past 20 years. The list comprises numerous studies addressing all structuring routes mentioned above. However, the sum of product launches is still insubstantial, considering the global scientific effort (more than 8000 patents).

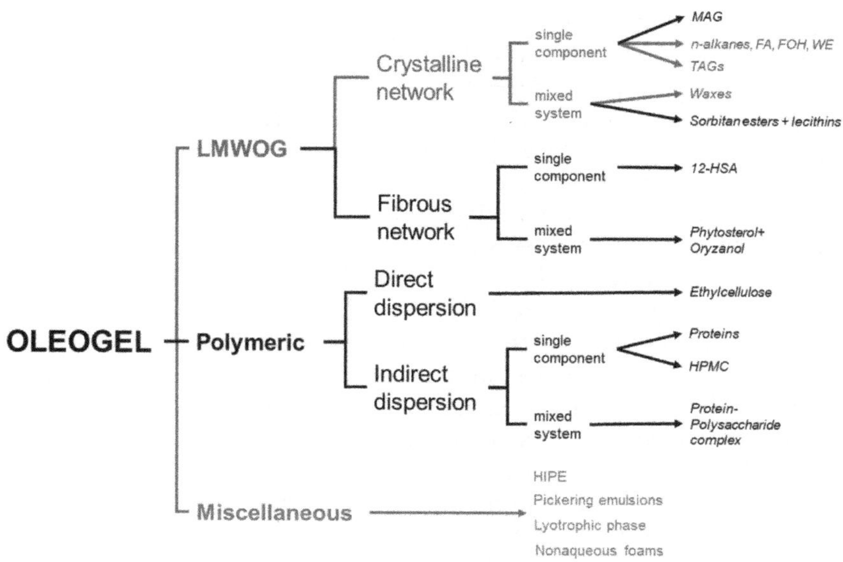

Figure 7.1 Classification of oleogelator classes and examples.

Besides, potential alternatives have to satisfy particular needs tailored to meet the individual food product's requirements. Moreover, according to Co and Marangoni[19] and Bot and Flöter[20,21] reasonable candidates for oil structuring should preferentially

- be of food grade
- provide a functional structure to edible oil
- be affordable
- not interfere with other ingredients in a product
- be versatile, *e.g.*, offer the possibility to design specific product properties such as melting temperature
- ensure a similar taste and mouthfeel.

Meeting all these criteria is inconceivable. The list also fails to include consumer acceptance, which remains a crucial parameter for successful product applications of oleogels. Therefore, the system has to offer similar sensory sensations or other exceptional benefits. Unfortunately, while oleogel systems disintegrate when consumed, lipids based on crystalline TAGs melt. Thus, the sensation perceived by the consumer is different. Consequently, the last two properties on the list above are the most critical and challenging to achieve. This illustrates the obstacles which have to be overcome to replace conventional SAFA-based lipid phases successfully. It is challenging to satisfy both functional and economic demands.

A recent study applied enzymatic glycerolysis to cottonseed oil to obtain solid fats with similar textural and thermal properties to palm oil.[22] The process converts the oil into a mixture of up to 34% monoglycerides (MAGs) and 49% diglycerides (DAGs). While the FA composition of the oil is not changed, MAGs and DAGs show emulsifying properties and higher melting points than TAGs. Consequently, a crystalline network forms, immobilizing the remaining TAGs. The European Food Safety Authority (EFSA) re-evaluated the food additives MAGs and DAGs as safe with no relevant adverse toxicologic or carcinogenic effects. Hence, there is no need to establish a numerical acceptable daily intake (ADI). However, the panel noted that only 0.8–3.5% of the daily fat intake are mono- and diglycerides.[23] Possibly, a much higher exposure could have different effects on human digestion and absorption of nutrients in the gastrointestinal tract. Other studies report that a high intake of emulsifiers such as polysorbate-20 modifies the delicate exchange of nutrients between the gut microbiome and the host. Consequently, they can change the gut lining thickness and promote the development of metabolic disorders.[24–26] However, no such data is available on MAGs and DAGs.

Another oil structuring system rests on combining a plant sterol (*e.g.*, β-sitosterol) and a plant sterol ester (γ-oryzanol). The approach is exceptional since the structure of the equimolar mixture formed in a suitable solvent cannot be predicted from an individual component's behavior. While each component alone does not provide structure to edible oils, the blend forms a transparent to slightly hazy gel. The molecules self-assemble into tubules with a 9.5 ± 0.9 nm diameter and a wall thickness of 2.3 ± 0.8 nm.[27,28] Subsequently, the tubules link to form a three-dimensional network. The combination structures oils at concentrations as low as 3% w/w, and both sterols offer several health benefits.

Interestingly, the system was initially discovered by accident while searching for new formulations to design sterol-rich margarine formulations.[29] A solid gel was found the next day after mixing the liquid waste of an oryzanol-rich batch with a plant sterol mix formulation.

This chapter aims to provide a comprehensive yet critical review of this system regarding its potential for food applications. To this end, it will present the properties of the individual components, emphasizing their availability, production, and price. Moreover, the system will be rated regarding its health benefits, digestion, and legal situation. Next, the physical chemistry of the binary system will be addressed briefly, followed by an introduction to the ternary system. Here, oleogel formation and dissolution and essential modifications and methods of characterization will be discussed. Interactions of the oleogel with other components such as minor oil components or water will be highlighted to assess the gel's functionality within food matrices. Therefore, the role of the lipid phase in the products will be considered as well as its physical state during processing. Lastly, studies concerning product applications will be reviewed.

7.2 β-sitosterol and γ-oryzanol

7.2.1 γ-oryzanol

Oryzanol is a blend of ferulate esters of triterpene alcohols and phytosterols.[30] However, cycloartenyl ferulate, 24-methylenecycloartanyl ferulate, and campesteryl ferulate comprise 80% of the blend.[31] The consumption of γ-oryzanol relates to several positive effects, such as lowering the plasma and serum cholesterol, reducing cholesterol absorption, and increasing muscle mass.[32–35]

As the name implies, rice, or mainly its by-products, is the crude material used to obtain γ-oryzanol. The bran, which is removed during the polishing of paddy rice, comprises 16–32% oil, 15–30% proteins, up to 28% fiber, and numerous vitamins and minerals.[36,37] Rice bran has a high nutritional value but is primarily used as animal feed due to its high fiber content and rapid deterioration.[38,39] However, in some countries such as India, crude rice bran oil (RBO) is extracted from the bran with an annual production volume of 1.2 MMt (2014).[39] Crude RBO has a high level of bioactive compounds such as phytosterols, tocopherols, and tocotrienols.[37] It contains 1.5–2.6% oryzanol and up to 8% waxes (typically 2–5%).[40–42]

Currently, oryzanol is recovered from wet and dry soapstocks from crude RBO refining.[41] The soapstock mainly contains water (65–70%, w/w), soap (20–22%, w/w), and 7–7.5% w/w unsaponified matter with up to 42% sterols and 20% oryzanol.[43] Numerous processes based on solid–liquid extraction, crystallization and precipitation, and supercritical fluid extraction (SCFE) are patented for oryzanol production.

Commonly, a combination of the first two is applied to wet or dry soapstocks. Initially, the cake is acidified, and oryzanol is extracted using a suitable solvent mixture such as hexane and isopropanol, diethyl, or methanol.[44] Consequently, oryzanol with purities between 40–98% is recovered from the organic phase by crystallization.[41,45] The yield can be optimized using a combination of continuous chromatography of the organic solution followed by precipitation.[46] Interestingly, the saponification of crude rice bran seems to halve oryzanol yield due to hydrolyzation of the ester bond under harsh alkali conditions. Hence, direct extraction from crude RBO is favorable but not economically feasible.[44] Although these methods are relatively easy to realize considering the equipment used, they are time- and solvent-consuming. Moreover, the yield or purity are often not fully explained or determined in publications and can vary significantly with soapstock composition.

An alternative method based on SCFE with carbon dioxide extracts oryzanol directly from crude RBO with high purities and yield (Table 7.1).[44,47] In contrast to organic solvents, CO_2 is cheap, fireproof, and safe for food use. Also, the extract crystallizes directly once the supercritical fluid evaporates. Subsequently, oryzanol is obtained by suitable solid–liquid extraction. On the downside, the equipment for SCFE is relatively expensive.

Table 7.1 indicates hypothetical production volumes of oryzanol extracted from rice bran oil by SCFE or solvent and precipitation. The calculations are based on an annual rice production of 755 MMt (2019,) presuming that the milling process removes 8% bran (~60.4 MMt) from the paddy rice.[5,36,48]

Since the demand for rice bran oil has grown over the past ten years and continues to grow, the exploitation of its valuable by-products is likely to grow comparably. Nevertheless, actual production volumes of oryzanol from

Table 7.1 Hypothetical oryzanol production: rice production 755 MMt,[5] 60.4 MMt rice bran assuming rice bran content of 8%,[48] amount oleogel: 4% (w/w = total sterol concentration on oil of which 60% w/w are oryzanol, price per t oleogel based on an estimated sunflower oil price of 890 $ per ton (Nov 2020).

Extraction technique	Yield [mg per g rice bran]	Production [MMt]	Price [1000 $ per ton][a]	Amount oleogel [MMt]	Price per ton oleogel [$ per ton]
Solvent	1.68 ± 0.02^{b}	0.105	15–102	4.3	1230–3330
SCFE	5.39 ± 0.20^{c}	0.337	219–548	13.8	6130–13 590

[a]Information obtained from ref. 50 and 51.
[b]Hexane : isopropanol (1 : 1), 60 °C, 60 min, information obtained from ref. 44.
[c]CO_2, 0.689 bar, 50 °C, 25 min, information obtained from ref. 44.

rice bran or RBO are lacking, and extraction yields and estimated production prices of oryzanol are quite variable (Table 7.1). Recovery from soapstocks seems cost-effective, but yields decrease significantly due to decomposition reactions. Nevertheless, the recovery of oryzanol and other products during RBO refining is reasonable and improves the valorization of the operation. Moreover, RBO production is far from reaching its full potential (1.2 MMt/ 9.5–12.5 MMt in 2017).[49]

7.2.2 β-sitosterol and Other Phytosterols

Phytosterols such as stigmasterol and campesterol are natural components in unrefined plant oils, present in diverse compositions and varying concentrations. Commonly the most significant share of the plant sterol fraction is β-sitosterol with 50–80%.[37] Like oryzanol, its consumption is connected to positive effects on health. Additionally, both sitosterol and oryzanol are used as dietary supplements. As the market for health-boosting supplements is steadily growing, sales are presumably increasing likewise, raising production volumes while simultaneously causing declining production costs.

Phytosterols are extracted from side-product streams of commercial oil refining. Precisely, they are recovered from condensates obtained during deodorization. While the distillate constitutes only 0.3–0.5% of the total processed oil volume, it can contain 8–20% phytosterols, depending on the type of oil used.[37]

Sterols can also be separated from tall oil, a by-product of the wood pulp manufacturing process. The tall oil pitch contains approximately 5–15% phytosterols after distillation, of which 40–65% are β-sitosterol.[52] In 2018, 1.46 MMt of chemical wood pulp was produced.[53] On average, 30–50 kg of tall oil is recovered per ton of wood pulp.[54] Consequently, there is a potential to produce at least 219 630 t of mixed phytosterols with a high content of β-sitosterol from tall oil (based on 30 kg tall oil per t wood pulp, 5% sterol content). Unfortunately, the production of tall oil is still limited to about 1.2 MMt (2017),[55] facilitating an annual production of 60 000 tons of sterols. However, no data on the actual number of mixed sterols recovered from tall oil is available, nor numbers about global production volumes and prices.

7.2.3 Legislation

Oleogelators need approval from the respective state authorities since they are deemed direct food additives. Within the European Union (EU), regulation (EC) No 1333/2008 (16.12.2008) is applied to approve new food additives, which are subsequently assigned an E number. For this, the additive needs to fulfill specific criteria:

- no risk to human health (based on scientific evidence)
- technologically required
- not misleading the consumer

The Codex Alimentarius Commission (international numbering system for food additives, INS) provides a collection of recognized standards for good food practice, safety, and transparency. It was founded by the FAO and later joined by the WHO and adapted and extended the E number system. Moreover, the FAO and WHO release the Joint FAO–WHO Expert Committee (JECFA) Report on food additives regularly. Besides assessing new additives, the reports constantly re-evaluate existing ingredients.

However, approvals are typically granted for specific food applications and comprise an acceptable daily intake (ADI). The determinations are adapted from existing nutritional studies, including long-term feeding studies on animals or humans. Therefore, the authorization of a new food additive is a lengthy procedure.

Moreover, the EU established in 1997 the Novel Food legislation (according to EC Nr. 1829/2003) for food products comprising:

- a new or modified molecular structure
- microorganisms, fungi, algae, or extracts of the aforementioned
- novel plants, animals, or their extracts
- a novel process that causes a change in food structure, composition, or metabolism.

Novel food applications have to be submitted to the EU commission, which can then consult the EFSA. The proposal includes details about a food's composition, production processes, and scientific evidence confirming that the product does not pose any danger to human health. The novel food is then added to a union list once approval is granted. The procedure implies that approval and admission procedures can be tortuous if the candidate is substantially different from an existing product or ingredient. Moreover, in the EU, every health claim linked to the consumption of food products or ingredients needs to be approved separately according to Regulation EC 1924/2006.

In the USA, the FDA approves and regulates food additives unless listed as "generally recognized as safe" (GRAS). According to the Federal Food, Drug, and Cosmetic Act (FFDCA), experts must issue GRAS approval, which is only

Table 7.2 Overview of the legislation situation of phytosterols and their esters for use in food products in the US and Europe.

USA-GRAS approved[a] Products	ADI	Europe-Novel food[b] Products	ADI
Meat, poultry, fish, dried beans, pea, nuts and seeds, grain products, fruit and vegetable products, oils, dressings, sugars, sweets, beverages	2–3 g	Dressings, mayonnaise, milk-type products, spreadable fats	3 g

[a]Information available in ref. 57.
[b]Open access regulation documents provided by EU, ref. 58.

granted for the intended use. Food additives are then listed by their Chemical Abstracts Number (CAS) and FDA regulation number (according to the US Code of Federal Regulations).

Table 7.2 shows the current legislation of phytosterols and their esters in the EU and USA. However, approval and ADI can be significantly different in other countries or unions but usually follow scientific output and other countries' legislation. The utilization of sterols and sterol esters is restricted for specific food products. Consequently, the oleogelators must be re-assessed by the FDA and EFSA (and Novel Food) again when utilized for structuring edible oils. Additional feeding or intervention studies might have to be performed depending on their amount in the final product.

Nevertheless, most feeding studies conduct trials with exposures much higher than the estimated intended consumption. For example, in studies investigating the toxicological potential of plant waxes, the no-observed-adverse-effect levels (NOALs) were 10–50 times higher than the recommendations given by the EFSA (22 mg per kg bodyweight per day). Furthermore, the suggested limitations are typically based on the metabolism of the respective ingredient, amongst other things. Hence, the following section will provide necessary information about the digestion of phytosterols and their esters.

7.2.4 Digestion

According to the WHO, 41 million people die each year of non-communicable diseases, accounting for 71% of all deaths. Of these, 17.9 million can be associated with cardiovascular diseases (CD).[59] Common risk factors such as nutrition and physical activity play a crucial role in developing CD.[60] Replacing SAFAs with PUFAs (ω-3, EPA, DHA) reduces serum cholesterol and the LDL/HDL ratio but does not cause an overall increase of cholesterol or blood pressure and thus a higher risk of developing CHD.[3] One could argue that their positive effect diminishes if they are included in a complex food matrix. However, PUFA adsorption and, thus, blood cholesterol reduction was found to be comparably effective in patés and sausages.[61] During digestion, food structures have to be broken down to give lipases access to TAGs. The enzymes break down TAGs into FAs, MAGs, and DAGs in the small intestine. Subsequently, MUFAs and PUFAs can be absorbed by intestinal epithelial cells from micelles of aggregated lipid metabolism products.

Hence, oleogels have to decompose similarly to release TAGs and provide valuable MUFAs and PUFAs. They have been used for the controlled delivery of pharmaceuticals since the release of lipophilic substances such as ibuprofen can be retarded by incorporating it into a hydrophobic network.[62–64] However, the release and bioavailability of TAGs from oleogels are much more complicated throughout the disintegration of a food matrix. Here, physiochemical properties such as interfacial phenomena play a significant role.[65,66] In general, lipase activity is impeded due to the immobilization of TAGs in an oleogel network. Thus, the release of FAs is hampered, which was observed for different structuring systems.[64,65] In line with that, lipid digestibility decreased in rats subjected to a high-fat diet based on wax oleogels.[67] In contrast, sponge cake bread with wax oleogels leads to an increase *in vitro* starch digestion, which might be associated with an increase in short-range crystallized starch structures.[68] Nevertheless, the variation in digestibility of macronutrients is likely to be particular for each type of oleogelator, indicating the issue's complexity. Within the scope of this work, the release from different food matrices cannot be evaluated, but some excellent reviews address this topic.[69–72]

However, regarding legislation, the question of greater significance is the metabolism of the network-providing molecules. The system discussed in this chapter is a mixture of different molecules. Luckily, their digestion is likely comparable due to structural similarities. Therefore only essential information about their digestion will be provided in the following.

Novel food ingredients have to undergo clinical studies before they are deregulated for food use unless their absorption is believed to be similar to approved substances (*e.g.*, waxes of different origins). Plant sterols and their esters are approved for food use (see Legislation). Hence, information about the absorption and digestion of the oleogelators addressed in this chapter can be obtained from the official legislation reports (WHO, FAO, EFSA, FDA).

Since the chemical structure of plant sterols (PS) like β-sitosterol is similar to cholesterol, they potentially influence its absorption, metabolism, and excretion in the intestine (by displacement).[56] Indeed, in the presence of phytosterols, cholesterol absorption is suppressed.[73] However, the absorption rate of PS mixtures from the gastrointestinal tract is only about 5%, depending on the mixture, and thus much lower than cholesterol (55–60%).[56,73–75] Absorbed phytosterols/stanols are excreted unchanged or converted to bile acids in the liver and subsequently excreted.[56] Oil in water emulsions (10/90) containing a mixture of phytosterol/phytosterol esters showed slower lipolysis, possibly due to restriction of TAGs in the gel structure (substrate limitation)[65] or adsorption of lipase on the hydrophobic surface of the elements of scaffolding (enzyme restriction).[76] Additionally, oxidation of PS should be avoided since the modification eliminates their LDL-cholesterol-lowering effect.[77]

The absorption of phytosterol esters such as γ-oryzanol in the intestinal tract is poor, similar to other plant sterols.[78,79] After passing the intestinal mucosa, sterol esters are distributed into various tissues and metabolized to *i.a.* ferulic acid and phytosterols.[80] The ferulic acid moiety can then inhibit an enzyme involved in cholesterol production in the liver and, therefore, lower cholesterol levels.[81] However, sterol esters are not exclusively hydrolyzed after absorption but can also be broken up in the gastrointestinal tract resulting in a similar cholesterol-lowering effect to phytosterols.

Thus the official panels concluded that phytosterols and their esters could be utilized safely in food products and provide an additional cholesterol-lowering effect in hypercholesterolemic patients.[74,82] Moreover, no negative impact on the human fecal bacterial profile was found during an interventional study where patients consumed phytosterol margarine (8.6 g sterols per day) for four weeks.[83,84] However, food products comprising increased levels of phytosterols/-stanols may lower the absorption of β-carotene (by 20%) and α-tocopherol (levels were still within the normal range).[85,86]

7.3 Physical Chemistry of the Binary Mixture

The properties of the three binary sub-systems of the sterol/sterol ester oleogel – β-sitosterol/oil, γ-oryzanol/oil, and β-sitosterol/γ-oryzanol – provide insights that are important to understand these oleogels. To this end, the sub-systems shall be briefly discussed in this section. At first, it is worthwhile to mention that both oleogelators in their pure form show a complicated crystallization behavior. For example, it seems impossible to reproduce the crystalline state of γ-oryzanol arising after long-term storage. Studies in our lab confirmed that the initial melting point given in the literature (136 to 142 °C) could only be reconfirmed after lengthy stabilization procedures. Furthermore, the DSC data indicate multiple solid phases and a rather complex crystallization behavior that has – to our knowledge – not yet been elucidated entirely.

Similarly, aged samples of β-sitosterol reveal intricate crystallization patterns. However, the literature value (approx. 142 °C) can be replicated somewhat more easily. For both components, DSC thermograms on cooling and heating in the presence of a small portion of liquid oil (5%, w/w) still reveal several crystal structures but are better reproducible when identical protocols are followed. Due to the complexity identified, discussing the binary systems of the sterol components and oil in detail is beyond the scope of this chapter.

Even though it is challenging to establish the solubilities of an oleogelator in oil appropriately, the information is vital for forming the equimolar structure. If the solubilities of the individual gelator in the continuous phase differ significantly, they need to be considered for the dosage. However, the solubility data of the pure component can only indicate the state of equilibrium between a dissolved oleogelator and oleogelator bound in the equimolar self-assembled structure. This is related to the fact that the solubility curves, after all, describe the relative stability of the solid-state and dissolved state. The binary system composed of β-sitosterol and γ-oryzanol has been discussed in detail by Sawalha *et al.*[87] A plausible binary phase diagram was derived from DSC and corresponding X-ray data (Figure 7.2).

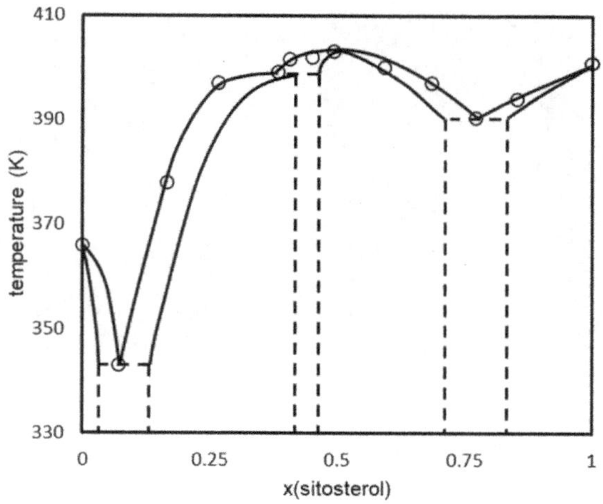

Figure 7.2 Phase diagram of the binary β-sitosterol and γ-oryzanol system. Dissolution temperatures from DSC (circles) as a function of mole fraction. Solid–solid two-phase regions: Rectangular zones with a dashed line. Adapted from ref. 87 with permission from John Wiley & Sons, Copyright © 2008 American Oil Chemists' Society (AOCS).

As Figure 7.2 reveals, the crystallization behavior of the mixtures of β-sitosterol and γ-oryzanol is somewhat complicated. Two single-phase areas with a limited mutual solid solubility have been identified at mole fractions close to the pure components, respectively. In the intermediate composition range, the authors postulated that two mixed solid phases could be found. One predominantly contains γ-oryzanol (12.8 to 42% molar sitosterol), and the other (46 to 71% molar sitosterol) is in the balanced composition range. It needs to be mentioned that the maximum dissolution temperature found corresponds to the equimolar mixture of the two oleogelators, which is thus preferably used for oleogelation. The phase diagram indicates that equimolar mixtures are exceptionally stable since they result in the highest dissolution temperatures (Figure 7.2). The analysis of Sawalha *et al.* showed that the solid structure is different from the one identified in oleogels.[87] In line with that, Wendt *et al.* demonstrated that the binary phase diagram is even more complicated by claiming that several low-temperature solid–solid transitions had been observed during detailed DSC analysis.[88]

Assuming that tubular structures resulting from the equimolar self-assembly of β-sitosterol and γ-oryzanol are the basis for oleogelation, it is crucial to understand the other phases that are potentially competing with this structure. In the paragraphs above, the liquid and solid states have been discussed. However, for β-sitosterol, another solid phase must be considered since it tends to crystallize into a compound crystal with water to form hemi- and mono-hydrates. This has been studied in the context of gel formation by Sawalha *et al.*[89] Bonsdorff-Nikander *et al.* studied the physical changes of β-sitosterol crystals in oily suspension during heating.[90] In both publications, it is stated that hydrates are energetically favorable at low temperatures.

Figure 7.3 Light microscope images of β-sitosterol crystals. Left: β-sitosterol-hydrate formed in an oleogel exposed to water, scale bar 500 micron. Right: β-sitosterol crystal anhydrous, scale bar 100 micron.

In an oily environment, the transition of the β-sitosterol hydrate to a crystalline anhydrate seems to be completed at temperatures as high as 90 °C.[90] In the presence of γ-oryzanol a different phase transition occurs in the oleogel ternary system when exposed to water. In this case the β-sitosterol hydrates disappear to the benefit of tubules emerging at a temperature of 70 °C.

The transition temperatures mentioned above and the conversions of the β-sitosterol-hydrate to crystalline β-sitosterol at 90 °C and to the self-assembled equimolar β-sitosterol/γ-oryzanol structure at 70 °C, confirm that the tubular structure is energetically favorable in the ternary system. Figure 7.3 shows growing β-sitosterol-hydrate crystals (left) following exposure of the gel to water and anhydrous β-sitosterol crystals (right) to illustrate the different structures.

However, a detailed mechanism of the effect of water on gel formation and gel properties has not been determined yet. Nevertheless, it was shown that a reduction in the water activity below 0.9 is sufficient to shift the equilibrium between the β-sitosterol-hydrate and the β-sitosterol/γ-oryzanol tubules to the latter even at low temperatures.[89]

7.4 The Ternary Mixture

7.4.1 Oleogel Structure, Formation and Dissolution

Ignoring that neither seed oils nor β-sitosterol or γ-oryzanol are truly pure components, one can, in a first approximation, consider the oleogel as a ternary system. Before discussing the properties relevant to product development, such as gel hardness and oil binding capacity, the following section provides the fundamentals of this unique structure. After the sterol/sterol-ester system was discovered, other phytosterols were screened for their ability to form oleogels in combination with oryzanol.[91] It was found that oryzanol self-assembles into the same hollow tube structure when combined with ergosterol, stigmasterol, cholesterol, and cholestanol and consequently forms oleogels with comparable properties.[92] From this list of sterols, cholesterol is the only one available in

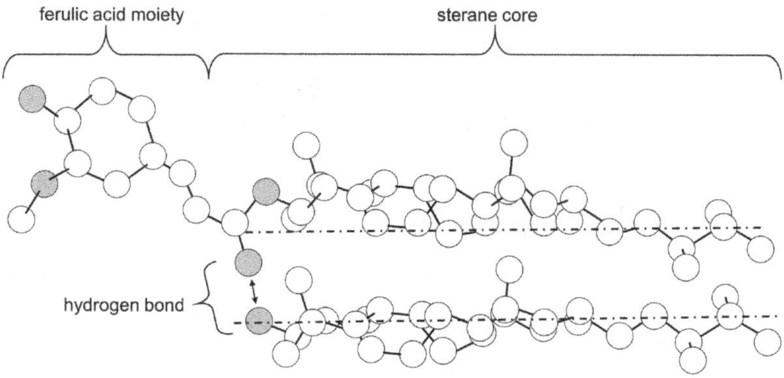

Figure 7.4 Sketch of the molecular arrangement of β-sitosterol and γ-oryzanol in a fibrillar structure.
Reproduced from ref. 102, https://doi.org/10.1002/ejlt.202000213, under the terms of the CC BY 4.0 license, https://creativecommons.org/licenses/by/4.0/.

large quantities. However, cholesterol will unlikely evolve as a candidate for food product development due to its distinct negative perception.

The β-sitosterol/γ-oryzanol system is different from most other oleogel structuring systems since the nature of the networks' building blocks was not understood when the gel-functionality was discovered. This does not imply that other systems are trivial or well recognized. But for most of the systems mentioned in Figure 7.1, the nature of the building blocks is reasonably understood. However, the system at hand has now been studied quite elaborately. The earlier work was mainly based on X-ray diffraction and resulted in a structural hypothesis that has since been confirmed and refined rather than challenged.[27,93,94] It still holds that sterols and γ-oryzanol stack in an equimolar tubular structure. The intermolecular hydrogen bonds prevent fully parallel stacking of the sterane cores, resulting in a tilted molecular arrangement that generates a spiral-like structure.[27,93,95] Figure 7.4 illustrates the essential molecular position and the exposure of the ferulic acid moiety.

Over the past decade, a substantial amount of data on the tubule structure was generated based on, for example, small-angle X-ray scattering (SAXS),[93] small-angle neutron scattering (SANS),[27] atomic force microscopy (AFM),[96] scanning electron microscopy (SEM),[97] and molecular dynamics (MD) simulations.[28] The overall consensus on the structure formed converges towards a tubule with an outer diameter of approximately 10 nm. The core is supposedly 'hollow' and has a diameter of approximately 5 nm. The high-density wall is built of the stacked sterol cores and has a thickness of approximately 1.5 nm. The outer shell is less dense since it consists of ferulic acid moieties sticking out of the basic structure and measures between 0.2 and 1.2 nm.[27] A more detailed discussion on the fundamental data concerning the tubular structure and aggregation can be found elsewhere.[98] The information discussed in this chapter will be limited to the aspects relevant to food applications.

In Figure 7.5, DSC dissolution temperatures of the 'ternary' system β-sitosterol, γ-oryzanol, and sunflower oil (SF) are depicted. The data represent mixtures of the two oleogelators in combination with 0, 10, 60, 84, and 92% (w/w) SF from top to bottom. It has to be noted here that in line with Figure 7.2 all compositions are considered in mole fractions. This implies that the mole fraction of sunflower oil (based on 891.5 g mol^{-1} for SF) varies from 4.9 to 7.0% from left to right in Figure 7.5 for 10% (w/w) SF. This variation originates from differences in molecular weights. That results in the ranges from 42.1 to 50.4% (60%, w/w), 70.9 to 78.0% (84%, w/w), and 84.3 to 88.6% (92% w/w). The rectangular areas indicate the 2-phase regions found in the binary system shown in Figure 7.3. The dashed curves are included to guide the eye. Considering that the different curves cover the binary system β-sitosterol/γ-oryzanol and the combination of these components with sunflower oil, the evolution of the different dissolution temperatures is remarkable. Throughout the different oil levels, the maximum dissolution temperatures remain practically unchanged at equimolar oleogelator composition. That is surprising considering the structural data given by Sawalha *et al.*[87,89] The authors reported that SAXS patterns change significantly on increasing oil inclusion for the equimolar mixture. At oil levels of 27% molar or above, the diffraction pattern indicates the system's typical tubular structures. Besides illustrating that the gel's disintegration

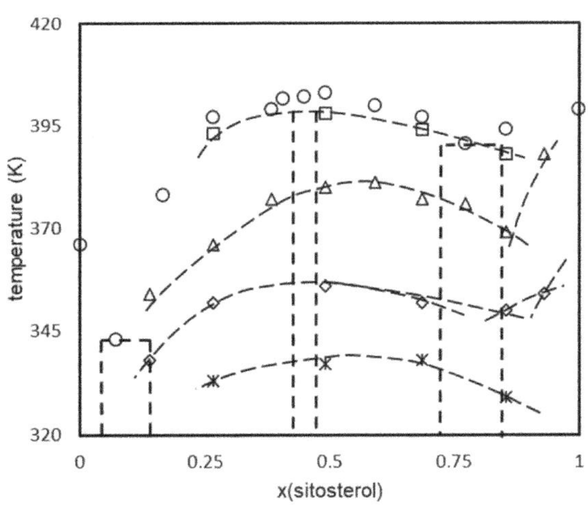

Figure 7.5 Dissolution temperatures for 'ternary' mixtures of sitosterol, γ-oryzanol, and sunflower oil (SF) over sitosterol mole fraction within the sterol mixture. 'Binary' subsystem β-sitosterol, γ-oryzanol. Rectangular zones with a dashed line: Solid–solid two-phase regions; circles: dissolution temperature. Dissolution temperatures in SF: squares: from 4.9 to 7.0% molar (left to right, 10% w/w) SF; triangles: from 41.1 to 50.4% molar (60% w/w) SF); diamonds: from 70.9 to 78.0% molar (84% w/w) SF; stars: from 84.3 to 88.6% molar (92% w/w) SF.

Adapted from ref. 87 with permission from John Wiley & Sons, Copyright © 2008 American Oil Chemists' Society (AOCS).

temperature can be varied deliberately by overall oleogelator composition, the data confirm that the system's behavior is far from simple, considering the whole composition range. Furthermore, all the data on sterol/sterol ester oleogels suggest that the structural nature of the tubules is reasonably robust concerning the variation of the system's composition, indicated by the flat curves of dissolution temperatures.

In contrast to the consensus on the tubules' existence and nature, the mechanism of gel formation is not fully understood yet. This is understandable since the measurement procedure might influence the results when applied during the transition process (*e.g.*, applying shear *vs.* quiescent conditions). Data by den Adel *et al.*[99] and Rogers *et al.*[94] do not allow the kinetics of the tubule formation and gelation process to be concluded unambiguously. However, it was argued that the two processes occur at the same time. At first, the nucleation and growth of tubules have to be initiated by generating supersaturation. Once tubules exist, their limited interconnectivity is critical to establish a space-filling network. Both AFM and cryo-SEM data demonstrate that the tubules are flexible and tend to arrange in bundles over longer stretches (see also Figure 7.8D).[28,89,96] In contrast to the retrogradation in starch gels, which is related to local crystalline regions, no structural changes related to bundling and actual gel formation have been identified yet.[100] However, the insights generated by MD simulations support the hypothesis that the driving force for bundling and consequently network formation are weak interactions between ferulic acid moieties of different tubules.[95] Depending on the specific system, the mixture of sterol and sterol ester gels edible oils at oleogelator concentrations of 3 to 4% w/w. Varying the solvent type and minor polar components affect the spatial arrangement of tubules and, consequently, macroscopic properties such as hardness.[96,97] MD simulations confirmed that these effects are most pronounced for components able to engage in hydrogen bonds.[101]

In addition, early data on the system suggested that the application of shear during, *e.g.*, rheological measurements, influences structure formation greatly.[91] However, gel formation kinetics are considered non-critical for food application because higher oleogelator levels are anticipated in product applications.

Regarding the gel–sol transition, it seems appropriate to revisit the arguments brought forward by Bot *et al.*,[98] who stated that the loss of firmness is exclusively related to the dissolution of the tubules. They noticed that the diffraction pattern of the tubules vanished simultaneously with the structure break-down when both are observed at the same temperature. However, recent work on the effects of minor components on gel structure and dissolution suggested that inter-tubular interactions play a significant role in the state of the gel.[96,101,102] This was confirmed using MD simulations on the gels in the presence of water or glycerol.[95,101] Moreover, the disintegration pattern (peak height and fronting) determined using DSC thermograms change as a function of minor polar oil components.[96,103] The effects are not primarily related to modifications of the solubility of the oleogelators since the final temperature of the gel–sol transition appears to be invariant with different levels of polar components. In summary, this suggests that the loss of structure on increasing temperature is a combination of reduced tubule–tubule interactions and continuous dissolution of the sterol and sterol ester.[96] It needs to be mentioned that both processes co-occur – depending on the scanning rate – above a specific temperature, indicated by the shape of the DSC thermograms.

It is essential to realize that β-sitosterol/γ-oryzanol-based oleogels are strongly related to the presence of hydrogen bonds when considering applications. This renders them prone to variations induced by minor components such as water. Additionally, water at water-activity values (a_W) above 0.9 disturbs the basic gel structure due to the formation of β-sitosterol hydrates, making the system less suited for emulsion type products. Details on product applications will be discussed in Section 7.5.

7.4.2 Methods of Characterization

The characterization of oleogels either intends to extend knowledge or defines relevant properties for product applications. Although both are somewhat connected, no substantial application knowledge in food products exists, meaning it is often unclear whether a specific characteristic is relevant for a particular application. Nevertheless, a distinct set of typical methods has emerged, following increasing publications on oleogel systems.

Among these are the critical gelling concentration (CGC) at a specific temperature, specifying the mass of an oleogelator needed to yield a non-leaking soft solid. Secondly, the oil binding capacity (OBC) defines the network's ability to hold liquid oil when exposed to external forces. Moreover, regularly the macroscopic hardness, transition temperatures, and rheological behavior are determined, and attempts have been made to visualize the network structure without artifacts. These characteristics should elucidate the underlying interactions and structure–function relations to create a comprehensive understanding of oleogel systems. Unfortunately, the experimental strategies primarily lack structural unity and are often not debated in a holistic frame to provide valuable information about a structuring system.

7.4.2.1 CGC – Solubility and Gel–Sol Transition

The CGC is specified as the minimum concentration of a gelator to produce a non-flowing state. The definition is subjective, considering the typical experimental procedure in many publications. After dissolving the oleogelator at a sufficiently high temperature, the solution is left to gel in an uncontrolled manner at ambient temperature. The process is executed at varying oleogelator concentrations (often 0.5% steps). The CGC is postulated to be the lowest concentration at which the sample is visually not flowing when turned upside-down. This concentration is regularly used to characterize the oleogel system further. However, the CGC appears to have limited scientific relevance (visual determination) and is impractical for applications (no information about consistency and sensitivity).

A relevant approach should be coherent and include parameters to describe an oleogel system more comprehensively. Among these could be the minimum concentration needed to reach equality of storage (G') and loss modulus (G'') at a specific temperature and the solubility of the gelator over a broader concentration and temperature range. The latter can be interpreted to determine the quantity of oleogelator dissolved in the liquid at any temperature (assuming equilibrium conditions). The amount of solid material at a given temperature can hence be determined using simple mass balances. The use of this approach would allow us to establish the relation of the fraction of dispersed phase and temperature, commonly used in fat technology. Moreover, supersaturation – a critical parameter for process design – could be calculated.

Differential scanning calorimetry (DSC) is a widespread analysis to establish the solubility of solid material in a liquid by measuring its dissolution temperature (gel–sol transition). However, reliable results highly depend on the measurement protocol, particularly scan rate (heating rate) and sample stabilization. Nevertheless, the solubility of the oleogelators and their gels can be studied using gel–sol transition thermograms. This information can be a prerequisite for interpreting, for example, the role of non-triglyceride species in an oil on interactions in the sterol/sterol ester network.[96] Yet, the interpretations become increasingly complicated if one or more components crystallize (*e.g.*, sitosterol hydrated crystals) instead of forming structuring elements. The determinations in these multi-component systems are challenging since specific crystallization and melting events must be correctly assigned to the respective structural elements.

Unfortunately, DSC is not easily applicable for weak exothermal phase transitions, such as the gelation of edible oil using sterols and sterol esters. Still, if DSC is not applicable, rheological methods like the dynamic mechanic thermal analysis (DMTA) can provide insight into sol–gel transitions. Here, a temperature sweep within the linear viscoelastic region (LVE) can be performed to compute the crossover of storage (G') and loss modulus (G'').[104] Before that, an amplitude sweep must be performed on both the liquid (oil) and the solid (oleogel) to determine the LVE's limits. By performing DMTA analysis with solutions containing varying oleogelator concentrations, the CGC at the desired temperature can be determined by extrapolating the crossover points for the different concentrations tested.

7.4.2.2 Sol–Gel Transition

The same methods mentioned at the end of the previous section, DSC and DMTA, might be applied to determine the sol–gel transition temperatures at defined oleogelator concentrations. However, the structuring kinetics of supersaturated liquid oily solutions depend on the structuring system (crystals or binary self-assembly) and differ considerably. For food applications, knowing the solidification process can be vital to create a desired food structure. In spreads, ice cream, and confectionery products, the lipid phase must undergo a phase transition from liquid to semi-solid during the manufacturing process.[2] In other processes, the fat (or oleogel) state is of minor importance. Among these are products in which a semi-solid lipid phase is necessary from the beginning of the process, like in comminuted meat products.[105,106]

The characterization of sol–gel transitions might be simple for a pure oleogel system. However, during processing, the solidification kinetics are influenced by external factors such as product composition, mechanical forces, and temperature profiles. It is thus impossible to reproduce all possible product scenarios. Nevertheless, knowledge of oleogel transition kinetics (gel–sol and sol–gel) is crucial for developing efficient application strategies.

Rheology and DSC are direct characterization methods that are commonly used to determine sol–gel transitions. However, non-exothermal gelation cannot be captured with DSC, as explained in the previous section. Sitosterol/oryzanol oleogels do not produce evaluable thermograms at low oleogelator concentrations. Even at 16% (w/w) total oleogelator concentration, only weak signals can be obtained, and repeatability is often insufficient.[93,96] Moreover, in DSC, transition solely occurs under quiescent conditions. Hence, sol–gel transition temperatures occasionally appear unrealistically low. Usually, rheometers can operate at a broad range of cooling rates and shear forces simultaneously and offer the possibility to apply shear *via* oscillation and rotation. Hence, specific procedures from manufacturing processes may be adapted by variation of these parameters. Therefore, rheology is the most reliable and versatile way to characterize sol–gel transition in oleogels.

7.4.2.3 Firmness and Deformation

Macroscopic properties such as the hardness could, at first glance, be the crucial parameter for product development. A penetration test primarily assesses the hardness of oleogels avoiding displacements by the sample container. Typically, a measuring probe (often cylindrical) penetrates the gel while monitoring the force. These tests provide a quick impression of the gel's stiffness or determine its cohesiveness and brittleness. However, the tests comprise numerous settings such as penetration depth (compression *vs.* destruction) and speed, probe, and sample geometry, which considerably impact the values determined. Moreover, gel hardness is always a function of the production procedure, *e.g.*, cooling rate or shear forces. Hence, comparison of data sets is only feasible for a particular measurement and manufacturing protocol.

Moreover, most of the data presented in the literature focus on the absolute values obtained with, for example, different solvents or the addition of structure-enhancers. While understanding modifications of gel hardness by solvent and non-triglyceride species is undoubtedly important, a comprehensive dose–response relation is often lacking. That would indirectly allow the oleogelator's solubility to be determined by extrapolation. Moreover, the amount needed to generate a particular structure is crucial concerning the food additive legislation (see the section on 'Legislation') and costs of the oleogelator.

An incremental increase of oleogelator concentration does not always result in linear growth of gel hardness.[91,106–108] While it certainly gives more insight into the relative potency of different structuring systems, it significantly depends on the type of oleogelator and the concentration range of interest (Figure 7.6, left). Figure 7.6 shows the dose–response of penetration hardness for ethylcellulose (EC), sunflower wax (SFX), and sitosterol/oryzanol oleogels. EC yields the firmest and SFX the softest structure at high concentrations (>15%, w/w, Figure 7.6, left). However, these concentrations are hardly relevant for food applications. Hence, Figure 7.6 (right) zooms into the dose–response between 5 and 15% (w/w). While the increase appears linear for SFX and sitosterol/oryzanol gels in this concentration range, EC hardness seems to surge drastically above a critical amount of oleogelator. Extrapolation of linear regressions showed that the amount of dissolved oleogelator was 3.7, 4.7, and 9.5% (w/w) for SFX, sitosterol/oryzanol, and EC, respectively. Hence, the minimum amount for structure development is lowest for SFX and highest for EC. Moreover, the slopes provide insight into the dose–response relation, which describes the generation of firmness per increase of oleogelator mass percent ($\Delta F/\Delta\%$ w/w). Between 5–15% w/w, the response is constant for sitosterol/oryzanol ($\approx$12.7) and SFX ($\approx$5.7) but varies significantly for EC ($\approx$23 between 10 and 15% (w/w) oleogelator). Although these values do not offer the precision of solubility studies by DSC, they provide a valuable reference point for product developers.

In addition to standard compression and penetration tests, the force needed to induce flow between the sample and a suitable sample container

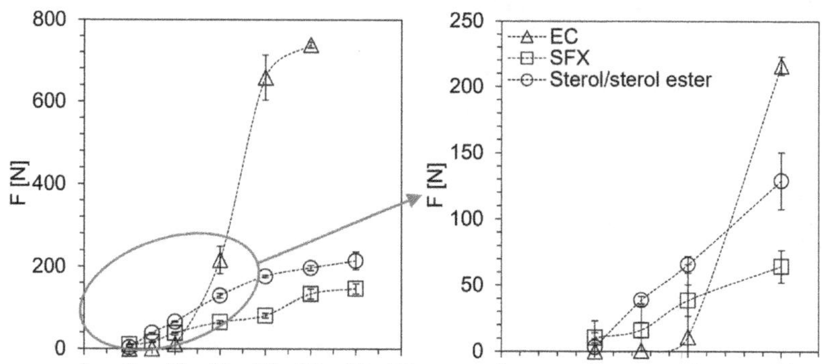

Figure 7.6 Dose-dependent structure increase (penetration hardness) of ethylcellulose (EC), sunflower wax (SFX), and sitosterol/oryzanol oleogels with canola oil. Left: full concentration range; right: concentration range relevant for application in food products. Lines are to guide the eye.

is of interest for some food applications.[109–111] Here, a cylindrical probe penetrates a cylindrical container. The annulus between the inner and outer cylinder needs to be sufficiently narrow to induce a flow upon penetration. The test is executed with a constant penetration velocity and enables the flow behavior of complex fluids and soft solids to be determined.[111] Similar to penetration tests, the results are affected by the geometry of the experimental setup. Moreover, the surface roughness of the container and the probe might cause slippage and thus result in experimental artifacts.

Besides penetration and back extrusion tests, gel deformation characteristics can be assessed using large deformation rheology such as amplitude sweeps. Moreover, the results provide more detailed information about network properties, allowing the connection points of the structural elements and their breakdown characteristics to be evaluated. The deflection can be operated at two pre-settings: controlled shear deformation (CSD) and controlled shear stress (CSS). Figure 7.7 shows a typical stress sweep response of sitosterol/oryzanol oleogels. The storage modulus (G') is invariable within the linear viscoelastic region (LVE) and can be associated with the gel's stiffness and the number of junction points in the network. The curve declines continuously once brittle fracturing occurs in the sample (for gels $\gamma_{max} \approx 0.5\%$),[104] indicating a gradual breakdown of the structure.

In contrast, the loss modulus (G'') increases above γ_{max} until reaching a distinct maximum and then declines less sharply than G'. The extent of the maximum ($\Delta G''$) is related to the amount of deformation energy (internal friction) that is lost due to the breakup of individual bonds and the formation of freely moving network fragments in micro-cracks of the network. These fragments are no longer incorporated in the network structure, although the surrounding material is still firm. Consequently, a larger $\Delta G''$ relates to weak network bonds, resulting in a more significant formation of fragments. As the strain increases further, the viscous behavior begins to dominate, and the sample starts to flow at $G' = G''$ (flow point).

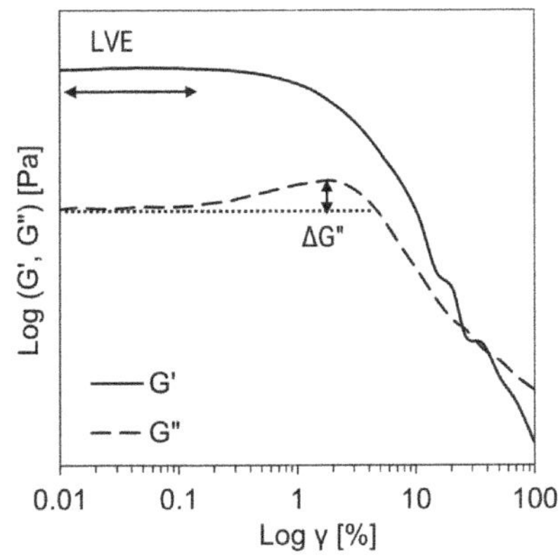

Figure 7.7 Typical response of a 10% (w/w) oleogel (1 : 1 molar ratio sitosterol : oryzanol) to a strain sweep at 20 °C and $\omega = 10$ rad s^{-1}.

In general, a gel characterized by $G' > G''$ in the LVE and a distinct maximum of G'' exhibits a steady, three-dimensional network under quiescent conditions. The breakup of the structure takes place with some delay due to the fragment formation. Macro cracks finally cause the rupture of the entire sample, and the viscous behavior prevails. A gel's behavior under increasing strain is crucial for successful product development since most food production processes involve shearing. Moreover, natural minor oil components and deterioration products significantly modify the gel's stress response and macroscopic network properties (see the section 'Influence of Other Components').

7.4.2.4 Oil Binding Capacity

Oil binding capacity (OBC) is considered an essential property of oleogels. Before discussing the different methods used to evaluate OBC, it is worthwhile to review the underlying concept.

To this end, the functionality of the structured lipid phase in a product has to be considered. For example, in water-continuous emulsions (*e.g.*, ice cream), the structure needs to resist destabilization mechanisms such as emulsion break-up and coalescence. These would otherwise result in phase separation and droplet size increase, influencing the emulsion's macroscopic properties. In fat-continuous products like spreads or chocolate, the oleogel has to immobilize the liquid oil, similar to fat crystal networks. In other applications, such as dough for baked goods, the gel is dispersed and intensively sheared during mixing.

Consequently, oleogels have to be steady against diverse physical and physico-chemical forces caused by interphases and food matrix elements. The light of processing conditions and product formulation, one has to ask which contributions could be relevant for oil exudation. Hereafter, a few simple answers will be given and explained:

1. mechanical forces such as compression
2. inertial force (viscosity) and cohesion of the oil when the network is, *e.g.*, compressed
3. adhesion forces between the surface of the scaffolding elements and the oil
4. surface tension and the formation of a concave liquid surface concerning the gels environment.

Compression (1) does not seem to be the dominant factor for OBC since compression tests and the subsequent storage never resulted in visible oil loss in sitosterol/oryzanol oleogels in our lab. However, compression forces do not exclusively impact the scaffolding elements so that the oil is exuded from the volume reduced three-dimensional network. Instead, the network breaks down before releasing any oil. Hence it is fair to assume that other forces (2–4) prevail when compressing the gel.

However, in a gel with a very high concentration of oleogelator (30% w/w EC), oil leakage was noticeable during the test. Here, the tremendous force ($\approx$1 kN) needed to compress the network resulted in the oil's exudation. Once the sample was discharged, the oil was absorbed back into the network.

The second hypothesis implies that the flow through the porous network changes with the continuous phase's compositional properties, particularly viscosity and cohesion forces. This has been studied superficially in our lab, and a change in viscosity did not relate to the changes in the gel's behavior.

Unfortunately, it remains tough to unravel the phenomena listed under points 3 and 4 indicated above. In either case, the strength and type of the interactions between the scaffolding elements and the oil are essential. Considering the gel's outer surface (4), the contact angle between the scaffolding surface and the oil with regard to the outer atmosphere is vital analogous to capillary effects. Preliminary experiments addressing the impact of the surrounding atmosphere on interfacial tension included submersion of a gel in air (the standard practice), water and water containing emulsifiers, reducing the interfacial tension. The results showed no significant effect concerning the gels' integrity.

It is even more challenging to examine adhesion forces between the continuous phase and the gels' surface (3) since they depend on the molecular interactions of the oil constituents with the scaffolding elements. Section 7.4.3 (Influence of Solvent Composition), *i.e.* aims to unravel some interactions between polar oil components and parts of the tubular structure formed by sitosterol and oryzanol. Even though the mechanisms discussed occur on a microscopic scale, they could modify a gel's macroscopic properties. Hence, the oil perfusion rate through the gel – a method discussed further on in this section – might offer a meaningful assessment.

The standard procedure to evaluate OBC is centrifugation. The method is typically operating at high centrifugal forces (>3000*g*) for significant periods (>10 min). Subsequently, the supernatant is discarded, and the remaining sample volume or mass is related to the starting material to define the OBC. Here, the OBC relates to the density difference between the network and the continuous phase. However, the network stiffness and resistance to compression and the forces described in (3) and (4) also significantly impact the results of this method.

Interestingly, experiments in our labs showed that disproportionate centrifugation did not result in substantial oil exudation in commercially available liquid margarine (2% high melting fat, 20% dispersed water phase). This is surprising since this product reportedly (consumer complaints) suffers from free oil. In another study, palm oil shortening was centrifuged for 4 hours at 4000 rpm at 25, 40, and 70 °C, and unexpectedly, the great variety in temperature did not compare to the minor changes in OBC (1.1 to 10.6%).[112] Hence, centrifugation seems ineligible to produce consistent results for different structuring systems. Moreover, the centrifugal force does not relate to actual processing conditions, nor does it represent deterioration phenomena during storage sufficiently.

Table 7.3 Method parameters for the determination of oil loss according to ref. 114 and 115. N/A – not available.

%CLX (wt/wt)	Filter Ø [mm]	Ø sample [mm]	Herzberg [s per 100 ml]	mass [g]	T_{gel} [°C]	t_{gel} [h]	T_{meas} [°C]	t_{meas} [h]	Mech. force
1	125	80	1420	100	23–25	24	40	24	No
6	125	10	150	N/A	5, 25	24 + 24	25	168	Yes

Omonov *et al.* used a different approach and developed a method based on the weight increase caused by oil absorption into a filter paper holding structured fat with a specified mass and surface area.[113] Systematic, yet different, results were obtained for canola oil comprising 10% of fully hydrogenated canola or fully hydrogenated sunflower oil. Furthermore, different microstructures induced by varying crystallization temperature were clearly manifested in the OBC data. In two studies, versions of this method were also applied to determine oil loss in candelilla wax (CLX) oleogels.[114,115] Table 7.3 illustrates the differences in sample preparation and measurement protocol. Surprisingly, the oil loss was 20% for both procedures, which does not imply that the same oil mass was released, independent of the method. If anything, the data illustrates that different method conditions (*e.g.*, sample size, filter type, and size) and sample manufacturing processes (*e.g.*, % w/w oleogelators, gelling time, and temperature) randomly result in the same oil loss. This renders the method insignificant if no standard procedure is developed and applied by all scientists. Moreover, a standard method would also allow the OBC of different structuring systems to be compared.

To this end, our lab evaluated the OBC using the centrifugation and filter paper method for three oleogel systems (SFX, EC, and sitosterol/oryzanol). A significant drawback of the latter procedure is oil adhering to the scale while weighing the soaked filter papers. Moreover, any filter paper test bears the risk of oleogelator dissolving into the oil reservoir due to long test times. Hence, the method was modified by introducing scaled stripes of filter paper into the top of the gel to monitor the oil 'climbing'. The speed at which the oil is rising vertically (perfusion rate) can be calculated. In contrast to previously reported methods, the oil has to overcome the gravitational force.

Moreover, the oil is not depleted from the gel network but replenished by submerging the bottom surface of the sample into a container filled with oil.

The centrifugation method was modified by placing the gel in a centrifuge container with a fine mesh at an intermediate height of the tube. During centrifugation, the oil is forced through the mesh and accumulates in the space beneath the mesh. Subsequently, the remaining gel and the mesh are removed, and the vial, including the exudated oil, is weighed.

Figure 7.8 (A–C) depicts the results obtained by centrifugation (A), the 'standard' filter paper method (B), and the modified filter paper method (oil climbing filter paper out of the gel) (C). All gels contained 10% w/w of either EC, sitosterol/oryzanol (1 : 1 molar), or SFX. The detailed procedures for all methods can be found elsewhere.[116] However, it should be mentioned that

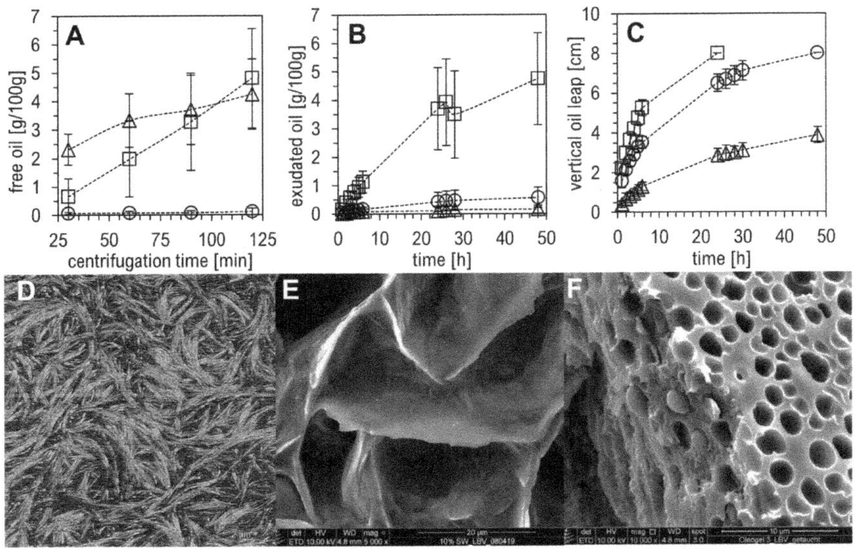

Figure 7.8 (A)–(C) Comparison of different methods to determine OBC for oleogels with 10% (w/w) structurant in canola oil oleogels. Symbols: Squares EC, triangles sitosterol/oryzanol, circles SFX. A: centrifuge method, B: standard filter paper method, (C) modified filter paper method. (D)–(F) visualization of oleogel microstructure. (D) AFM image sitosterol/oryzanol (scale bar 2 μm). (E) cryo-SEM image SFX (scale bar 50 μm). (F) cryo-SEM image EC (scale bar 10 μm). Graphics E and F kindly provided by Birgit Hetzer (MRI Karlsruhe).

each data point in Figure 7.8 represents 12 repetitions. The superiority of the proposed filter paper procedure is striking regarding the data's systematic behavior, reproducibility, and discriminative power. Nevertheless, the duration of this method is – at this early stage of development – impractical.

Interestingly, the centrifugation-based method found SFX to best hold the oil, the standard filter test indicates a good holding capacity for SFX and sitosterol/oryzanol, while the filter paper perfusion test results in distinct results indicating that sitosterol/oryzanol was superior over the other structurants. This is likely related to the differences in applied force and method principles, as discussed at the beginning of this chapter. However, it is beyond the scope of this paper to elaborate on these findings.

7.4.2.5 Microstructure

The elucidation and identification of the building blocks in different oleogel systems are intricate, considering the diversity of their scaffolding elements. Hence, issuing reliable statements about the interactions between the network elements through imaging technology is even more challenging.

Insufficient data often denies drawing reasonable conclusions regarding structural hypotheses. However, some systems have been studied quite intensively with decent consistency of the data obtained. These are gels based on the binary system of β-sitosterol and γ-oryzanol[27,28,89,92,93,96,117] and ethylcellulose.[118–121] X-ray diffraction and neutron scattering were first applied to β-sitosterol/γ-oryzanol gels and revealed a 1 : 1 molar self-assembled tubular structure.[27,93] Subsequently, other methods such as atomic force microscopy,[28,96] electron microscopy,[89,117] and molecular modeling[28] consistently support this hypothesis.

Electron microscopes with a sufficient resolution allow for very detailed '3D imaging'. However, the sample preparation is fairly intense, rendering these methods prone to artifacts. On the other hand, atomic force microscopy (AFM) can be utilized to visualize a gel's surface directly without sample preparation. On the downside, the reproduction of the gel's surface does not provide information about the scaffolding element's spatial arrangement. Furthermore, using the AFM phase or height profiles, the tube size determined by neutron scattering was confirmed.[27,96]

Additionally, the size of bundles can be measured. Figure 7.8D shows an oleogel with 16% w/w sitosterol/oryzanol in canola oil. Tubule bundles fill the whole micrograph, and they seem to align parallel but branch, twist, and turn as they connect with other bundles. Unfortunately, the length of a single bundle cannot be determined with any known imaging technology. Nevertheless, the bundles' thickness on the gel surface can be derived from the height profiles and varies considerably within one sample (30–600 nm).[96] Interestingly, minor oil components modify the bundle thickness and their spatial arrangement. The reasons for this will be discussed in more detail in the following section.

Similar to scanning electron microscopy, AFM does not provide accurate information about the spatial distribution of a network's structural elements. Consequently, relying on a single method to make meaningful statements about the network arrangement is not advisable. Moreover, consistent interpretation and association of thermo-physical data obtained by, *e.g.*, DSC and rheology, and structural hypothesis based on visualization are far from trivial.

7.4.3 Influence of Solvent Composition

The influence of solvent composition on β-sitosterol/γ-oryzanol oleogel properties has been studied for more than a decade. Although crucial parameters have been identified, it is still not entirely certain to what extent they contribute to the changes observed. However, as explained in the section *Methods of characterization*, knowledge about, for example, oleogel transition and oleogelator solubility are vital for research and product development. Unfortunately, they seem, to some extent, to be related to solvent composition. Figure 7.9 shows the parameters that influence oleogel properties such as gel formation and dissolution, network structure, and hardness. The illustration indicates that the parameters are mutually dependent, making it challenging to design and execute meaningful experiments that consider their effects independently.

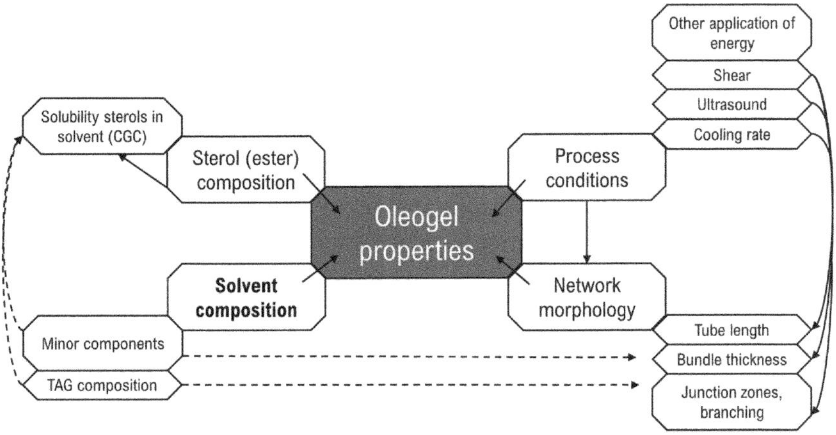

Figure 7.9 Factors influencing sterol/sterol-ester oleogel properties.

In a first approach, Sawalha *et al.* utilized solvents with varying polarities (decane < limonene < sunflower oil < castor oil < eugenol) and chemical compositions to produce sitosterol/oryzanol oleogel emulsions (10% water-in-oil).[89] Reportedly, emulsion firmness increased with decreasing solvent polarity expressed as the dielectric constant (ε). However, the emulsion firmness was much lower than that of a pure oleogel, which was addressed to the crystallization of sitosterol into a hydrated crystal due to increased water solubility in polar solvents. This resulted in a partial loss of the tubular structure. Bot *et al.* first recognized the recrystallization by relating X-ray diffraction patterns of oleogel emulsions to crystalline signatures of sitosterol monohydrates.[122] However, the oleogelator concentration and a_W were crucial to avoid recrystallization and the associated oleogel collapse (see also Section 7.3).[89,123] To this end, Sawalha *et al.* added varying amounts of sodium chloride to the aqueous phase. They noticed that a_W has to be reduced to 0.9 or lower, which is valuable information for product development.[89] Another study conducted by the group addressed the dependency of solvent type (the same solvents as mentioned earlier) and oleogel properties such as gel formation.[97] They discovered that the process of molecular self-assembly is enthalpy driven and promoted in solvents with a lower polarity. Hence, tubes form at higher temperatures and lower oleogelator concentrations. The study comprised chemically diverse solvents with a wide range of permittivities (2.0–10.4), which likely modify the solubility of oleogelators, among other things. Indeed, solvent polarity impacts oryzanol solubility substantially.[124] Notably, a seven-fold increase of solvent permittivity (1.8 to 13.3) resulted in 7.4 times more oryzanol dissolved (hexane and hexanol). However, significant changes are not expected within the range of edible oils (permittivity 2.9–3.2). Moreover, solvent viscosity presumably affects oleogel formation.

S. Calligaris *et al.* studied the impact of oil unsaturation (flaxseed > sunflower > extra virgin olive oil) and viscosity on firmness and the gel–sol transition of sitosterol/oryzanol oleogels.[125] They reported a decline in gel hardness with increasing solvent viscosity caused by lower molecular mobility of the oleogelators. Hence, the probability of coinciding during molecular self-assembly (tubule formation) is claimed to decrease. However, this was not valid for extra virgin olive oil and flaxseed oil since they exhibit almost identical hardness, even though their viscosities vary. Interestingly the data showed a maximum in gel hardness when plotted over solvent permittivity.

The results obtained in these early studies illustrate a certain inconsistency and implicate that macroscopic changes in gels result from modifications on a microscopic and even molecular scale. To this end, the next paragraphs will discuss interactions within the unique tubular structure formed by sterols and sterol esters.

In Section 7.4, the molecular arrangement of the tubular structure was discussed. The ferulic acid moiety of oryzanol protrudes out of the tube into the surrounding liquid and enhances its stability by forming non-covalent interactions within the helix (intra-tube).[95] That implies that a higher oryzanol solubility in polar solvents (*e.g.*, hexanol) is likely related to stronger interactions of the ferulic acid moiety with the respective solvent.[124] The same interaction types, particularly van der Waals interactions and π–π contacts, are responsible for tube–tube (inter-tube) interactions.[95] Consequently, one can assume a shift from intra- to inter-tube interactions when individual fibrils stick together. Nevertheless, interaction points are not static, but their number is limited by the amount (n) of oleogelators in the gel.

In addition, inter-tube interactions may shift if polar molecules are introduced to the system, resulting in a modified network arrangement on a microscopic scale, impacting macroscopic properties such as gel strength.[96,101] For example, glycerol ($\varepsilon \approx 46.5$) supposedly forms hydrogen bonds with the hydroxyl group of sitosterol and the methoxy, phenol, and carbonyl groups of oryzanol.[101] These interactions improve tubule stability compared to a pure triglyceride environment. However, this does not imply an increase in oleogel network strength since any rise in tube stability caused by glycerol is neutralized by an increase in oleogelator solubility, resulting in fewer tubules. Furthermore, the effects of glycerol are reportedly concentration-dependent, whereas only high substitution levels were studied (15–50% w/w). Therefore, polar molecules might enhance network stability at low concentrations while a stabilizing effect inverts at higher levels. In addition, the strength and type of interaction are obviously related to the additive's functional groups.

This highlights the complexity of the issue since there seem to be numerous elements contributing to the changes in gel formation and properties as a function of solvent composition. However, regarding product applications comprising edible oils, significant effects of oleogelator solubility are not expected. Nevertheless, polar molecules dissolved in the oil might trigger a shift in network interactions, modifying the gel's micro- and macroscopic properties. Hence, it appears essential to consider changes due to the bulk solvent's contribution (fatty acid profile of TAGs) and polar compounds separately.

Consequently, using a standardized oil comprising only TAG molecules is necessary to extricate the different contributions and gain a deeper understanding of oleogel properties as a function of solvent characteristics.

In recent studies, this was addressed more comprehensively.[96,103,126] To this end, oils with high levels of polar components (formed during thermal treatment) and pure triglyceride oils (purified by column chromatography) were utilized in sitosterol/oryzanol oleogels. Gel–sol transition temperatures and enthalpies determined with DSC did not change substantially with changing levels of polar components. Hence, it seems justified to presume that the number of tubes and the network's binding energy are reasonably constant. Interestingly, several studies reported that the network break-up starts at low temperatures, indicated by a long fronting in DSC thermograms.[92–94,99] Considering the different interactions discussed above, two distinct mechanisms were identified during gel–sol transition: the break-up of inter-tubular connections in the lower temperature region (fronting) and the dissolution of tubes (intra-tube, peak). The disentanglement of peak areas and fronting revealed a continuous reduction of the peak with increasing minor components. However, this is balanced by increased transition energies related to the peak fronting, resulting in unchanged total areas. Furthermore, MD simulations revealed that the hydrogen bonds responsible for molecular stacking contribute little to the tube's binding energy but that interactions formed by the ferulic acid moieties are critical to the strength of this unique structure.[95] Still, it is uncertain whether the fronting and peak areas in DSC thermograms can be assigned to the distinct interactions with certainty or if the break-up of bonds overlap during the gel's dissolution.

However, the peak fronting increase likely leads to a shift in network interactions similar to those reported for glycerol.[95] Hence, in purified oils (pure TAG oil), intra-tubular relations prevail, while in oils with a high concentration of polar components, inter-tubular interactions dominate. In line with that, AFM imaging indicated generally thicker bundles in oils comprising polar components.[96] In contrast, oleogels from purified oils showed networks less ordered than the reference (untreated oil), and tube bundles were thinner. Moreover, the alignment of bundles on the gel surface is considerably different with or without polar minor components. In their absence, bundles arrange more randomly, appear shorter, and do not exhibit much branching. In contrast, the alignment and branching in a gel with polar components seem similar to those found in untreated canola oil. Although the illustration of the surface revealed changes in network arrangement as a function of oil quality, AFM cannot deliver factual information about the spatial distribution of the scaffolding elements, and hence interpretations remain vague.

In contrast, modifications of network interactions do not result in an equal change in penetration hardness. Here, a maximum was observed for briefly heated oils (few polar components).[96] This appears counterintuitive in the first place. However, considering that polar components increase interactions and lead to enhanced bundling of tubes, the effective network surface is consequently reduced. Consequently, there are larger areas of free oil, resulting in reduced network strength. This effect seems to be dose-dependent: At low minor component concentrations, the scaffolding element's surface is occupied sparely, and the inter-tube network is strengthened. At high concentrations, the surface is saturated, resulting in thick strands with a smaller scaffolding–oil interface. Similarly, in the absence of polar minor components, a fractured network with fewer junction zones is formed, resulting in a lower gel hardness. This hypothesis is supported by the increase in storage modulus with minor components.[127] Hence minor components interacting with the ferulic acid moiety promote network junction zones (see Section 7.4.2.3).

It is conceivable that the type of polar components (*e.g.*, chemical interaction points) impacts the mechanisms listed above. It would be interesting to understand how naturally occurring minor oil components modify a gel's properties concerning product development. To this end, edible oils with different iodine values (canola < sunflower < flaxseed oil) were freed from their minor components.[126] Selected polar species (oleic acid, tocopherol acetate, and a monoglyceride mix) were added to purified oils at different concentrations.[103] Their impact on oleogel formation, dissolution, network hardness, deformation behavior, and microstructure was studied. In general, the FA composition of oils appears only to have a negligible impact when compared to changes observed in samples with polar components present. Briefly, in the absence of minor components, the impact of oil viscosity dominates. A lower viscosity (=higher iodine value) results in higher sol–gel transition temperatures and softer gels. For example, sol–gel transition temperatures declined about 1.5 K with increasing iodine value. In comparison, they decreased 7–8 K and 5–8 K in samples containing 5% w/w oleic acid or tocopherol acetate, although the viscosity was similar to the reference. This is likely related to sterols or sterol esters interacting with the polar molecules, which hamper the self-assembly.

Also, distinct mechanisms were observed depending on the type of minor component: The addition of the monoglyceride mix significantly increased solvent viscosity, and at a concentration ≥2.5% (w/w), a composite structure was formed, hindering the development of the sitosterol/oryzanol network. Unfortunately, no synergistic effect was detected, in contrast to ethylcellulose and monoglyceride oleogels.[128]

A dose-dependent response similar to oils containing deterioration products occurred in samples containing tocopheryl acetate and oleic acid. The gel hardness was highest at low concentrations and declined as the concentration increased. Interestingly, the storage modulus in the LVE increased substantially with tocopheryl acetate. At the same time, it was relatively constant for all concentrations of oleic acid. This implies that the behavior is *i.a.* related to the functional groups of the additive. Oleic acid and tocopherol acetate both participate in hydrogen bonds at their carbonyl group (–C=O). However, oleic acid is also a hydrogen-bond donor (–OH). Hence, the interaction between oleic acid molecules might outweigh the tendency to interact with oryzanol's ferulic acid moiety at high concentrations. However, this hypothesis is speculative and cannot be proven at this point.

Nevertheless, solvent composition substantially influences oleogel formation, network properties, and appearance. No significant effects on oleogelator solubility can be expected in the range of solvent polarity of edible oils. Solvent viscosity affects molecular self-assembly and gel hardness but seems to have a marginal impact compared to the effects of polar minor components. These change inter-fibrillar interactions and network appearance depending on their type and concentration, resulting in modified macroscopic gel properties such as hardness. Moreover, they retard network formation due to solute–solute contacts. Hence, it is crucial to pay close attention to the composition of the continuous phase to understand and predict these effects. Therefore, a comprehensive characterization of oils, such as fatty acid composition, minor components in the oil or originating from other ingredients, and state of deterioration, is required. Only then can comparability of different studies on oleogelation be achieved, allowing for substantiated product development.

7.5 Food Applications

Edible oleogels offer the possibility to improve the nutritional value (fatty acid profile) of food products. They can be used as solid fat replacers since they immobilize liquid plant oils rich in MUFAs and PUFAs in a 3D network. Additionally, they potentially resolve issues occurring in existing products, *e.g.*, recrystallization of oil migration in chocolate. Finally, oleogels could simplify production, and new products might emerge, such as heat-resistant chocolate (HRC).[129]

A considerable number of research articles have dealt with food applications of sitosterol/oryzanol oleogels (Table 7.4). However, the fabrication of a product with comparable characteristics in the lab does not guarantee that the technologies will be appropriate on an industrial scale. Therefore, characterization methods discussed in the previous section have to be relevant and discriminative to enable a suitable choice of production routes. Only then can

Table 7.4 Selection of oleogel food applications.

Food category	Fat function	Product	References
Confectionary products	Snap, appearance, melting sensation	Oil migration in Confectionary fillings	130
Meat products	Snap, texture	Sausages (Frankfurter-, Breakfast-, Bologna-type), Hamburger	131–134
Dairy products	Richness, creaminess, 'scoopability', stabilization	Ice cream	123, 135, 136
	Richness	Yogurt	137

future innovations be pursued. Hence, it is advantageous not to focus on a single composition but to include various compositions and process parameters. Nevertheless, product application limitations might arise due to the availability of sufficient quantities of the oleogelators (see Section 7.2.1 and 7.2.2) and the consumer acceptance. Moreover, the tubular network structure's destabilization mechanisms, such as sitosterol monohydrate formation and shear application during processing, have to be considered.

This section will discuss product applications of sitosterol/oryzanol oleogels stressing the fat's functionality in the original product. However, the successful fabrication of a model product in a lab environment is far from being proof of the principle of industrial manufacturability. Therefore, a detailed understanding of the structured lipid phase's functionality during the manufacturing process is of utmost importance. Moreover, it is essential to understand at which process stage the rheological properties of the oleogel matter, if there are kinetic restrictions during phase transition and the gel is incorporated into the product matrix. Moreover, it will be discussed whether the utilization of an oleogel is reasonable or if liquid oils in their native state are sufficient for a particular application.

7.5.1 Confectionery Products

The fat determines the final product's appearance, snap and melting sensation, and a range of process conditions in chocolate and confectionery products. For example, during conching at 60–65 °C, the fat needs to be liquid to enable the coating of cocoa particles. Hence the viscosity is lower, and the mobility of ingredients in the mass increases. This promotes the discharge of undesired flavors and prevents particle aggregation. Subsequently, during crystallization, the temperature control must be precise to prevent the crystallization of undesired polymorphs.

Hence, cocoa butter substitutes in chocolate-like applications have to meet a series of particular requirements, especially during processing. Therefore, it comes as no surprise that oleogels have not replaced cocoa butter yet. However, sitosterol/oryzanol, EC, and wax oleogels have been utilized in filled confectionaries successfully,[130,138,139] heat resistant chocolate,[105,121,129,140,141] and fat-reduced chocolate.[142]

Among the key factors promoting fat bloom in chocolate products are likely erratic storage conditions such as temperature fluctuations.[143] In confectionaries, oil migration into the outer chocolate layer accelerates recrystallization further. This is likely due to the dissolution of cocoa butter TAGs in the liquid fraction of the filling. Consequently, they can migrate to the product's surface through diffusion and capillary forces, where they recrystallize. Hence, immobilization of the liquid oil *via* gelation might hamper the migration process. This hypothesis was recently revised by gelling the filling or the chocolate layer or forming a distinct oleogel barrier between the filling and the chocolate shell.[130] To this end, γ-oryzanol/β-sitosterol oleogels (10 or 25% w/w) have been used in nougat pralines, and oil migration was observed for 24 weeks at 10, 18, and 28 °C.[130] The relative liquid oil level was monitored by (DSC). The gel inclusion level was 2.5% in the filling, and the chocolate cover and the barrier layer were pure oleogel and made up 14% of the total sample.

No oil migration was detectable at 10 °C, while it was highest at 28 °C. The gelled chocolate and the three-layer system had higher or similar oil migration at 18 °C and 10% oleogelator compared to the reference. Interestingly, oil migration at 28 °C was lowest when the gel was utilized in the chocolate layer (25% w/w oleogelators). Gelling of the filling successfully inhibited the migration even at low oleogelator concentrations. Moreover, the additional oil introduced by the gel's presence did not promote a higher oil migration rate.

Hence, saturated fatty acids in filled confectionaries can be reduced by gelling the oleaginous filling with γ-oryzanol/β-sitosterol. Oil migration could be suppressed even at elevated temperatures allowing the storage of pralines in tropical climes.

7.5.2 Meat Products

The fat globule size and the fat to protein proportion are vital factors defining sensory sensation in comminuted meat products. Therefore, the myosin protein needs to form a network to obtain a stable batter, supporting the large fat globules (<100 μm).[144]

Beef fat is semi-solid during the mincing process due to the low temperatures of approximately 8–12 °C. Nevertheless, the intensive mechanical treatment in the cutter comminutes the globules. This results in the broad fat globule size distribution in cooked sausages responsible for their snap and texture.[106] Studies substituting meat fat in sausages with liquid edible oils showed a substantial increase in firmness and chewiness.[144–146] The liquid oil exclusively forms small droplets (<20 μm) with a narrow size distribution during mincing. Hence, the surface area for protein adhesion is substantially bigger, enabling enhanced protein adhesion at the interface. Consequently, a more homogeneous and stiffer protein network filled with fat globules is formed.[106] The sensory properties of these products are unlikely to meet consumer expectations.

However, pre-gelled oleogels can be added to the batter, and, similar to animal fat, excessive droplet breakup is avoided during mincing. Nevertheless, before cooking, the fat globule size distribution must match the original to obtain sausages comprising a higher nutritional value but comparable sensory properties.

Panagiotopoulou *et al.* substituted 50% of animal fat in frankfurter sausages (20% total fat) with sitosterol/oryzanol oleogels.[131] These authors reported no differences in textural changes or significant increases in fat oxidation levels. Although the gels might melt during the cooking process, protein adhesion and denaturation likely prevent structure loss.

In another study, gelled linseed oil was used to replace pork fat (20 or 40% w/w) in fermented sausages.[133] Although the dry-curing process is lengthy

(49 days), and linseed oil usually contains more than 70% PUFAs, the sausages did not show any detectable off-flavors caused by oil oxidation. However, depending on the level of substitution, the sausages showed significant changes in sensory properties. The same group partially exchanged 25 and 75% of the solid fat in hamburger patties with linseed oil oleogels (sitosterol/oryzanol).[134] Regarding hardness, cohesiveness, and chewiness, no substantial differences were observed. However, a preference test showed less acceptance among panelists compared to the reference.

7.5.3 Margarine and Spreads for Direct Consumption

The solid fat in margarine stabilizes liquid oil by immobilization within a 3D network. Moreover, this network stabilizes water droplets and thus contributes to emulsion stabilization and ensures spreadability. The emulsion has to be stable throughout the complete product lifecycle. The fat should also provide structure at ambient temperature but dissolve at body temperature to avoid a waxy mouthfeel. The dissolution overlaps with water droplet destabilization and consequently phase inversion, which is crucial for flavor release.

Therefore, it is challenging to replace hard stocks in margarine since most oleogels do not dissolve but disintegrate in the mouth. However, oleogels based on a non-crystalline network such as EC or sterol/sterol-ester offer the possibility to simplify the margarine production process. In conventional margarine production, scraped surface heat exchanger units (SSHE) apply intense shear and rapid cooling rates, simultaneously breaking up droplets and forming the crystalline fat network. However, in oleogel margarine, liquid emulsions could be packed directly after emulsification if the droplet size of the aqueous phase (3–10 μm) can be maintained throughout the network development and storage. Hence the scraped surface heat exchanger units (SSHE) become obsolete.

Bot *et al.* successfully utilized β-sitosterol and γ-oryzanol (16 or 32% w/w) oleogels in Pickering-type emulsions without the addition of emulsifiers by mixing a hot oil phase into a preheated water phase (10, 30, 60%) under high shear conditions.[147] Subsequently, the mixture rapidly solidifies during cooling because of high oleogelator concentrations, which prevent water droplet coalescence and phase inversion. Nevertheless, sitosterol monohydrate crystals form at the water–oil interface, causing a partial loss of the tubular structure formed by the sterol/sterol-ester system (see the section on β-sitosterol and other phytosterols). Interestingly, the reorganization reportedly leads to substantially firmer emulsions, and phase inversion could not be observed. Hence, the rod-like hydrate crystals seem to contribute to the emulsion stabilization as well.[89] Nevertheless, they are less efficient in preventing coalescence as light microscopy revealed significantly coarser water droplets after one week of storage.[147] However, recipes can be adapted to avoid structure loss by utilizing high oleogelator concentrations (>16% w/w) and reduced water activity (<0.9).[89]

Similarly, the sterol/sterol-ester system could be used in shortening for shortcrust products. In contrast to bread and cake-like doughs, individual starch grains must be coated by the fat to prevent the formation of an extended gluten structure network. That and a sufficient amount of solid structure ensure the typical stiffness and snap in the final product (*e.g.*, cookies).[148] Hence, good mixing properties of the fat are desirable during dough production. To this end, shortenings without water, a soft texture at ambient temperature, and a defined ratio of solid fat to liquid oil are used. A high water content would result in excessive gelatinization of the starch and loss of the rigid structure.

In contrast, in pastries and cakes, air bubbles and water must be incorporated and stabilized in the dough during the mixing process. Hence, common cake shortenings regularly contain emulsifiers such as mono- and diglycerides, which increase the amount of air incorporated, provide a more tender structure to the baked product, and retain moisture. During baking, the water evaporates, gelatinizes the starch, and the dough rises. Our lab experiments showed that muffins with acceptable sensory properties and a finely porous, fluffy structure could be produced using β-sitosterol and γ-oryzanol oleogels. Comparable results were also obtained using sunflower or beeswax wax oleogels. However, the doughs showed some de-oiling, and after baking, slightly harder muffins with a lower volume and larger pores were obtained (compared to typical baking margarine). For comparison, liquid canola oil was also utilized and resulted in muffins with a high final volume (higher than the reference), large pores, good moisture retainment, and soft texture.[149] Hence, the utilization of oleogels for cake dough is questionable since liquid oil produces satisfactory products, which were even preferred over oleogel muffins in sensory evaluation.

7.5.4 Dairy Products

Ice cream is a multiphase system comprising a partly solidified aerated oil-in-water emulsion. The mixture is based on a concentrated sugar solution, proteins, and hydrocolloids (in the frozen state). It contains ice crystals and delicately dispersed semi-solid fat droplets supporting the air/water interface's stabilization by forming agglomerates. The addition of emulsifiers such as mono- and diglycerides aids the formation of this complicated network.

Before freezing the premix, a stable emulsion must develop where proteins stabilize fat droplets to avoid creaming. Therefore, the mixture is homogenized at elevated temperatures (75 °C). Subsequently, during ripening (4 °C, 4 h), the fat solidifies and forms flocs that aggregate and possibly form a weak network. Subsequently, air bubbles (10–150 μm) are introduced by whipping the emulsion. The volume increase is expressed as overrun, relating the product's excess volume to the initial emulsion volume – a more significant air incorporation results in softer but less creamy ice creams. Commonly, low-temperature scraped surface heat exchangers (−5 °C) initiate water crystallization while simultaneously introducing air into the whipped emulsion. Small ice crystals and air bubbles and uniform distribution are preferable since they increase creaminess and decrease the melting rate.

Milk fat was substituted by sunflower oil or sunflower oil oleogels containing 8 or 12% w/w sitosterol/oryzanol in ice creams with 4 or 8% total fat content.[136] Ice creams manufactured with the 12% w/w oleogelator showed the highest overrun of all samples, while the lowest was obtained when straight sunflower oil was utilized (42.2 and 27.5%, respectively). Supposedly, air bubbles are not sufficiently stabilized by proteins since they rather adsorb at the oil–water interface. In contrast, more emulsifiers could be available in oleogel ice cream premixes since the oleogel droplets solidify rather quickly upon cooling at higher sterol concentrations (>10% w/w). Consequently, more emulsifiers are available to stabilize the air interface. Nevertheless, sitosterol monohydrate formation in the premix must be prevented by reducing the water activity in the continuous phase below 0.9.[89] Interestingly, milk proteins were considered to successfully suppress monohydrate formation in model systems by adsorbing at the oleogel–water interface and forming a protective shell. The utilization of low esterified pectin aids this process by immobilizing the water in the continuous phase.[123] Once ice crystals are formed, the water activity is low, and loss of the tubular structure is not expected. Interestingly, the utilization of sunflower oil and oleogel retarded the beginning of ice cream melting. Still, as the melting continued, the rates accelerated and were much higher than in milk fat ice creams.[136] In conventional ice creams, ice crystals are entrapped within the fat network enabling a controlled release. Unfortunately, it is uncertain whether sitosterol/oryzanol droplets aggregate and form a network stabilizing the air bubbles. Nevertheless, the results indicate that sitosterol/oryzanol oleogels produce stable emulsions and ice creams. Nevertheless, more data is needed regarding individual stabilization mechanisms, storage stability, and organoleptic properties.

In contrast to ice creams, yoghurt is considered a gel in which a protein network entraps the aqueous phase. The texture of yoghurt, such as its richness, is linked to its fat content and fat globule distribution. Hence, the milk is homogenized at elevated temperatures (~70–75 °C) to obtain fat droplets with diameters around 2 μm before the inoculation with lactic acid bacteria.[150] The droplets should remain stable throughout the fermentation process. At a pH of 4.6, caseins coagulate and form a three-dimensional network restraining the fat droplets since they interact with hydrophobic protein domains.[151]

Abumweis *et al.* reported a substantial reduction in LDL caused by the addition of phytosterols to yoghurt.[152] This renders milk fat substitution in yoghurt with sitosterol/oryzanol oleogels a promising method to develop a health-boosting food product. To this end, sterol/sterol ester solutions (varying ratios, 20 and 35% w/w in sunflower oil) were emulsified into an aqueous phase containing skimmed milk and polysorbate 20.[137] Subsequently, the emulsions were fermented to obtain yoghurts with 2 or 4%

lipid content. All formulations showed characteristic gelation and acidification kinetics during the production process. Unfortunately, ungelled samples containing high oryzanol ratios (100 and 80%) showed rheological properties (G' values) similar to the reference. This is likely related to a crystalline network visible in optical micrographs, which supports the continuous phase.[137] Yogurts comprising an equimolar ratio (60 : 40) of sterol ester : phytosterols had the lowest G'. This might result from a partial loss of the tubular structure due to the crystallization into hydrated sitosterol crystals. No supportive crystal network was formed in these samples since the individual phytosterol concentration and supersaturation are lower in the equimolar mixture.

Unfortunately, the actual state of fat droplets in the yoghurt (gelled or a shell of hydrated crystals) remains unclear since the authors neither provide information about that nor perform sensory evaluations and storage tests. Nevertheless, gelling of the oil phase does not seem crucial to obtain yoghurts with rheological properties similar to the milk-fat based reference. Recent studies conducted in our labs showed similar results. Stable yoghurt formulations (10% w/w fat content, 14 days storage at 5 °C) with satisfying rheological characteristics were produced employing liquid edible oils (unpublished data). However, the macroscopic structure could be manipulated by structuring the lipid phase.

In summary, the utilization of gelled oil phases is, in most cases, promising. The stabilization of interfaces and distribution within the food matrix is essential in fat-continuous products. Contrarily, fat often functions as a stabilizer and procurer in water-continuous products. Indeed, the state of the fat and its functionality during and after processing have to be studied cautiously. Here, the sterol/sterol ester system is susceptible since the presence of water might cause the gel to fall apart. Nevertheless, this problem can be addressed by developing suitable formulations. Moreover, products containing specific amounts of phytosterols and their esters offer health benefits proven by scientific evidence, which can be claimed and used to boost commercialization.

7.6 Conclusion

This chapter represents an attempt to evaluate the information available on oleogels based on a combination of β-sitosterol and γ-oryzanol concerning their applicability in food products. It is well established that the best structuring results are obtained when the two oleogelators are present equimolarly since the two oleogelators self-assemble alternately into 'hollow' tubules of approximately 10 nm diameter. These tubules are the building blocks of the three-dimensional scaffolding based on some areas of tubule bundling. Both the self-assembly into tubules and their bundling strongly depend on the formation of hydrogen bonds. Gel structures can be achieved at oleogelator dosing levels as low as 3% (w/w). The hardness of such gels and their disintegration temperature can be manipulated *via* typical dose–response relations. The capacity of the gels to immobilize oil is competitive to other structuring systems. Variations of the system's characteristics appear to be based on interference with hydrogen bonds, for example, the presence of different minor components. The presence of water with an a_W-value above 0.9 has a detrimental effect on the tubular structure because β-sitosterol hydrate crystals form alternatively. This feature of the structuring system certainly restricts its application in emulsion-based products. Additional features of the gel, thermoreversibility, transparency and non-reversible structure break down on shear could be either beneficial or prohibitive depending on the specific application.

Both oleogelators are natural components and can be extracted as side streams from sizable primary product streams. β-Sitosterol is a native component in many edible oils but can also be obtained from tall oil originating from wood pulp production. γ-Oryzanol is present in rice bran and can hence be isolated from rice bran oil. Both components currently have relatively high supply costs driven by cumbersome isolation processes. However, these could be subject to future changes once the interest in the components warrants a more targeted and large volume-oriented operation.

The application of oleogels in food products is subject to food legislation. Both oleogelators considered here have GRAS and Novel Food status. Ignoring the products listed, a simple back-of-the-envelope calculation based on the ADI and an inclusion level of 6% oleogelator in the oleogel reveals that an advised consumption per day should not exceed 50 g.

In contrast to the fact that no food products with β-sitosterol/γ-oryzanol oleogels are yet in the food market, successful prototypes have been described in numerous publications. That holds for other structuring systems as well. The fact that β-sitosterol and γ-oryzanol structure oil in their native state and carry intrinsic health benefits yields this system beneficial over many other oleogelators. Even though this illustrates the hypothetical potential of oleogels, they are by default more expensive than palm oil due to the bulk substitution of palm oil by more costly seed oils, which certainly hampers product introductions.

In summary, the β-sitosterol/γ-oryzanol oil-structuring system remains peculiar. Some of its characteristics, such as sensitivity to water and shear, and price, currently limit applications substantially. On the other hand, some characteristics, transparency, health effects, and naturalness offer clear benefits so that it remains exciting when products will be introduced.

References

1. W. Kloek, PhD thesis, Wageningen University, 1998.
2. A. Bot and E. Flöter, in *Edible Oil Processing*, ed. G. Calliauw, R. J. Hamilton and W. Hamm, Wiley-Blackwell, Chichester, West Sussex, 2013, pp. 223–250.
3. J. A. Nettleton, I. A. Brouwer, J. M. Geleijnse and G. Hornstra, *Ann. Nutr. Metab.*, 2017, **70**, 26–33.

4. R. P. Mensink, P. L. Zock, A. D. M. Kester and M. B. Katan, *Am. J. Clin. Nutr.*, 2003, **77**, 1146–1155.
5. Food and Agriculture Organization of the United Nations (FAO), *STAT Crop Production Worldwide*, http://www.fao.org/faostat/en/#data/QC, (accessed February 2021).
6. H. K. Gibbs, A. S. Ruesch, F. Achard, M. K. Clayton, P. Holmgren, N. Ramankutty and J. A. Foley, *Proc. Natl. Acad. Sci. U. S. A.*, 2010, **107**, 16732–16737.
7. *The World of Organic Agriculture: Statistics and Emerging Trends 2019*, 2019, ed. H. Willer and J. Lernoud, pp. 220–255.
8. G. Thompson, *HortTechnology*, 2000, 663–674.
9. D. Mozaffarian and R. Clarke, *Eur. J. Clin. Nutr.*, 2009, **63**, 22–33.
10. R. J. de Souza, A. Mente, A. Maroleanu, A. I. Cozma, V. Ha, T. Kishibe, E. Uleryk, P. Budylowski, H. Schünemann, J. Beyene and S. S. Anand, *BMJ*, 2015, **351**, 3978–3994.
11. World Health Organization, *Guidelines: Saturated Fatty Acid and Trans-fatty Acid Intake for Adults and Children: (Draft Issued for Public Consultation in May 2018)*, Geneva, 2018.
12. Bundesinstitut für Risikobewertung, Höhe der derzeitigen trans-Fettsäureaufnahme in Deutschland ist gesundheitlich unbedenklich, 2013, https://mobil.bfr.bund.de/cm/343/hoehe-der-derzeitigen-trans-fettsaeureaufnahme-in-deutschland-ist-gesundheitlich-unbedenklich.pdf (accessed November 2021).
13. European Commission, *Scientific and Technical Assistance on Trans Fatty Acids*, 2019.
14. G. Marakis, C. Fotakis, E. Tsigarida, S. Mila, L. Palilis, S. Skoulika, G. Petropoulos, A. Papaioannou and C. Proestos, *J. Consum. Prot. Food Saf.*, 2020, **15**, 373–381.
15. World Health Organization, *Replace Trans. Fat.*, 2018, https://cdn.who.int/media/docs/default-source/nutritionlibrary/replace-transfat/1-replace-framework-updated-june-2019-ke.pdf?sfvrsn=47e47367_2 (accessed November 2021).
16. R. Micha, S. Khatibzadeh, P. Shi, S. Fahimi, S. Lim, K. G. Andrews, R. E. Engell, J. Powles, M. Ezzati and D. Mozaffarian, *BMJ*, 2014, **348**, 2272–2284.
17. V. Chavasit, J. Photi, P. Dunkum, T. Krassanairawiwong, M. Ditmetharoj, S. Preecha and F. Martinez, *J. Clin. Hypertens.*, 2020, **22**, 1347–1354.
18. S. Kamel and H. Al Otaibi, *Curr. Res. Nutr. Food Sci.*, 2018, **6**, 748–756.
19. E. D. Co and A. G. Marangoni, *J. Am. Oil Chem. Soc.*, 2012, **89**, 749–780.
20. A. Bot and E. Flöter, in *Edible Oleogels: Structure and Health Implications*, ed. A. G. Marangoni and N. Garti, AOCS Press, Urbana, 1st edn, 2011, vol. 1, ch. 3, pp. 49–79.
21. A. Bot and E. Flöter, in *Edible Oleogels*, Elsevier, 1st edn, 2018, vol. 1, ch. 2, pp. 31–63.
22. R. A. Nicholson and A. G. Marangoni, *Nat. Food*, 2020, DOI: 10.1038/s43016-020-00160-1.

23. M. Younes, P. Aggett, F. Aguilar, R. Crebelli, B. Dusemund, M. Filipič, M. J. Frutos, P. Galtier, D. Gott, U. Gundert-Remy, G. G. Kuhnle, J.-C. Leblanc, I. T. Lillegaard, P. Moldeus, A. Mortensen, A. Oskarsson, I. Stankovic, I. Waalkens-Berendsen, R. A. Woutersen, M. Wright, P. Boon, D. Chrysafidis, R. Gürtler, P. Mosesso, P. Tobback, A. M. Rincon, Z. Horvath and C. Lambré, *EFSA J.*, 2017, **15**, e05045.
24. P. D. Cani and A. Everard, *Trends Endocrinol. Metab.*, 2015, **26**, 273–274.
25. M. M. W. Aponso, G. de Silva and A. Abeysundara, *J. Pharmacogn. Phytochem.*, 2017, **6**, 485–487.
26. Z. Celebi Sözener, L. Cevhertas, K. Nadeau, M. Akdis and C. A. Akdis, *J. Allergy Clin. Immunol.*, 2020, **145**, 1517–1528.
27. A. Bot, E. P. Gilbert, W. G. Bouwman, H. Sawalha, R. den Adel, V. M. Garamus, P. Venema, E. van der Linden and E. Flöter, *Faraday Discuss.*, 2012, **158**, 223–239.
28. A. B. Matheson, V. Koutsos, G. Dalkas, S. Euston and P. Clegg, *Langmuir*, 2017, **33**, 4537–4542.
29. H. Ritter and R. Volkmer, *Eur. Pat.*, 918465A1, 1997.
30. E. J. Rogers, S. M. Rice, R. J. Nicolosi, D. R. Carpenter, C. A. McClelland and L. J. Romanczyk, *J. Am. Oil Chem. Soc.*, 1993, **70**, 301–307.
31. Z. Xu and J. S. Godber, *J. Am. Oil Chem. Soc.*, 2001, **78**, 645–655.
32. S. Eslami, N. M. Esa, S. M. Marandi, G. Ghasemi and S. Eslami, *Ind. J. Med. Res.*, 2014, **139**, 857–863.
33. A. C. Fry, E. Bonner, D. L. Lewis, R. L. Johnson, M. H. Stone and W. J. Kraemer, *Int. J. Sport Nutr.*, 1997, 7, 318–329.
34. A. L. Gerhardt and N. B. Gallo, *J. Nutr.*, 1998, **128**, 865–869.
35. G. Yoshino, T. Kazumi, M. Amano, M. Tateiwa, T. Yamasaki, S. Takashima, M. Iwai, H. Hatanaka and S. Baba, *Curr. Ther. Res., Clin. Exp.*, 1989, **45**, 543–552.
36. R. M. Saunders, *Food Rev. Int.*, 1985, **1**, 465–495.
37. F. Shahidi and A. E. Bailey, *Bailey's Industrial Oil and Fat Products*, Wiley, New York, Chichester, 2005.
38. A. G. Gopala Krishna, *Inform*, 2013, **24**, 260–264.
39. G. Ahmad Nayik, I. Majid, A. Gull and K. Muzaffar, *J. Rice Res.*, 2015, **3**, 151.
40. G. S. Seetharamaiah and J. V. Prabhakar, *J. Food Sci. Technol.*, 1986, **23**, 270–273.
41. A. V. Narayan, R. S. Barhate and K. S. M. S. Raghavarao, *J. Am. Oil Chem. Soc.*, 2006, **83**, 663–670.
42. N. A. Venkatadri, B. Rajendrakumar Suresh, I. Tyakal Nanjundiah, T. Purnima Kaul, R. Karumanchi Sreesaila Mallikarjuna Srinivasa, R. Appu Rao Gopal and P. Vishweshwariah, *WO Pat.*, 2004/055040A1, 2002.
43. T. Akiya, *Agric. Biol. Chem.*, 1962, **26**, 180–186.
44. Z. Xu and J. S. Godber, *J. Am. Oil Chem. Soc.*, 2000, **77**, 547–551.
45. M. Patel and S. N. Naik, *J. Sci. Ind. Res.*, 2004, **63**, 569–580.
46. M. Saska and G. J. Rossiter, *J. Am. Oil Chem. Soc.*, 1998, **75**, 1421–1427.
47. N. T. Dunford and J. W. King, *J. Food Sci.*, 2000, **65**, 1395–1399.

48. Food and Agriculture Organization of the United Nations (FAO), http://www.fao.org/3/x5043e/x5043e0a.htm, (accessed February 2021).
49. Food and Agriculture Organization of the United Nations (FAO), http://www.fao.org/faostat/en/#data/FBS, (accessed February 2021).
50. K. Triratanasirichai, M. Singh and A. K. Anal in *Food Processing By-products and their Utilization*, ed. A. K. Anal, John Wiley & Sons, Ltd, Chichester, UK, 2017, pp. 277–293.
51. N. T. Dunford and J. W. King, *J. Food Sci.*, 2000, **65**, 1395–1399.
52. Food and Agriculture Organization of the United Nations (FAO) http://www.fao.org/fileadmin/templates/agns/pdf/jecfa/cta/69/Phytosterols.pdf, (accessed February 2021).
53. Food and Agriculture Organization of the United Nations (FAO), http://www.fao.org/faostat/en/#data/FO, (accessed February 2021).
54. P. Bajpai, *Biermann's Handbook of Pulp and Paper: Volume 1: Raw Material and Pulp Making*, Elsevier, Amsterdam, 2018.
55. L. J. Konwar, J.-P. Mikkola, N. Bordoloi, R. Saikia, R. S. Chutia and R. Kataki in *Waste Biorefinery: Potential and Perspectives*, ed. A. Pandey, D.-J. Lee, S. Mohan, S. K. Khanal and T. Bhaskar, Elsevier, Amsterdam, 2018, pp. 85–103.
56. EFSA, *EFSA J.*, 2012, **10**, 2659–2697.
57. U.S. Food and Drug Administration, https://www.cfsanappsexternal.fda.gov/scripts/fdcc/index.cfm?set=FoodSubstanFoo, (accessed February 2021).
58. European Commission, Commission Implementing Regulation (EU) 2017/2470 of 20 December 2017 Establishing the Union List of Novel Foods in Accordance with Regulation (EU) 2015/2283 of the European Parliament and of the Council on Novel Foods: C/2017/8878, 2017.
59. World Health Organization, *Global Action Plan for the Prevention and Control of Noncommunicable Diseases: 2013–2020*, World Health Organization, Geneva, Switzerland, 2013.
60. GBD 2015, Risk Factors Collaborators, *Lancet*, 2016, **388**, 1659–1724.
61. G. Delgado-Pando, P. Celada, F. J. Sánchez-Muniz, F. Jiménez-Colmenero and B. Olmedilla-Alonso, *Eur. J. Nutr.*, 2014, **53**, 83–93.
62. C. M. O'Sullivan, S. Barbut and A. G. Marangoni, *Trends Food Sci. Technol.*, 2016, **57**, 59–73.
63. S. Murdan, *Expert Opin. Drug Delivery*, 2005, **2**, 489–505.
64. K. Iwanaga, T. Sumizawa, M. Miyazaki and M. Kakemi, *Int. J. Pharm.*, 2010, **388**, 123–128.
65. N. Duffy, H. C. G. Blonk, C. M. Beindorff, M. Cazade, A. Bot and G. S. Duchateau, *J. Am. Oil Chem. Soc.*, 2009, **86**, 733–741.
66. A. Speranza, M. G. Corradini, T. G. Hartman, D. Ribnicky, A. Oren and M. A. Rogers, *J. Agric. Food Chem.*, 2013, **61**, 6505–6515.
67. W. Limpimwong, T. Kumrungsee, N. Kato, N. Yanaka and M. Thongngam, *J. Funct. Foods*, 2017, **39**, 250–256.
68. J. Alvarez-Ramirez, E. J. Vernon-Carter, Y. Carrera-Tarela, A. Garcia and C. Roldan-Cruz, *LWT*, 2020, **130**, 109701.

69. J. E. Norton, Y. Gonzalez Espinosa, R. L. Watson, F. Spyropoulos and I. T. Norton, *Food Funct.*, 2015, **6**, 663–678.
70. D. J. McClements, E. A. Decker, Y. Park and J. Weiss, *Food Biophys.*, 2008, **3**, 219–228.
71. D. J. McClements, E. A. Decker and Y. Park, *Crit. Rev. Food Sci. Nutr.*, 2009, **49**, 48–67.
72. A. J. Martins, A. A. Vicente, L. M. Pastrana and M. A. Cerqueira, *Food Sci. Human Wellness*, 2020, **9**, 31–39.
73. EFSA – European Food Safety Authority, *EFSA J.*, 2011, **9**, 2005.
74. D. Turck, J. Castenmiller, S. de Henauw, K. I. Hirsch-Ernst, J. Kearney, A. Maciuk, I. Mangelsdorf, H. J. McArdle, A. Naska, C. Pelaez, K. Pentieva, A. Siani, F. Thies, S. Tsabouri, M. Vinceti, F. Cubadda, T. Frenzel, M. Heinonen, R. Marchelli, M. Neuhäuser-Berthold, M. Poulsen, J. R. Schlatter, H. van Loveren, W. Gelbmann and H. K. Knutsen, *EFSA J.*, 2020, **18**, 6135.
75. European Union, https://ec.europa.eu/food/sites/food/files/safety/docs/sci-com_scf_out56_en.pdf, (accessed February 2021).
76. M. E. Lowe, *J. Lipid Res.*, 2002, **43**, 2007–2016.
77. Y. T. Liang, W. T. Wong, L. Guan, X. Y. Tian, K. Y. Ma, Y. Huang and Z.-Y. Chen, *Atherosclerosis*, 2011, **219**, 124–133.
78. K. B. Wheeler and K. A. Garleb, *Int. J. Sport Nutr.*, 1991, **1**, 170–177.
79. W. Ling and P. Jones, *Life Sci.*, 1995, **57**, 195–206.
80. E. Kobayashi, J. Ito, S. Kato, K. Sawada, M. Matsuki, H. Hashimoto, T. Miyazawa and K. Nakagawa, *Food Funct.*, 2016, **7**, 4816–4822.
81. C. Perez-Ternero, M. Alvarez de Sotomayor and M. D. Herrera, *J. Funct. Foods*, 2017, **32**, 58–71.
82. T. A. Miettinen, T. E. Strandberg and H. Gylling, *Arterioscler., Thromb., Vasc. Biol.*, 2000, **20**, 1340–1346.
83. R. Ayesh, J. Weststrate, P. Drewitt and P. Hepburn, *Food Chem. Toxicol.*, 1999, **37**, 1127–1138.
84. J. Weststrate, R. Ayesh, C. Bauer-Plank and P. Drewitt, *Food Chem. Toxicol.*, 1999, **37**, 1063–1071.
85. European Union, https://ec.europa.eu/food/sites/food/files/safety/docs/sci-com_scf_out143_en.pdf, (accessed February 2021).
86. R. P. Mensink, S. Ebbing, M. Lindhout, J. Plat and M. M. van Heugten, *Atherosclerosis*, 2002, **160**, 205–213.
87. H. Sawalha, P. Venema, A. Bot, E. Flöter, R. d. Adel and E. van der Linden, *J. Am. Oil Chem. Soc.*, 2015, **92**, 1651–1659.
88. A. Wendt, presented in part at the 14th EFL Congress, Ghent, October, 2016.
89. H. Sawalha, R. den Adel, P. Venema, A. Bot, E. Flöter and E. van der Linden, *J. Agric. Food Chem.*, 2012, **60**, 3462–3470.
90. A. von Bonsdorff-Nikander, S. Lievonen, L. Christiansen, M. Karjalainen, J. Rantanen and J. Yliruusi, *AAPS Pharm. Sci. Technol.*, 2005, **6**, E413–E420.
91. A. Bot and W. G. M. Agterof, *J. Am. Oil Chem. Soc.*, 2006, **83**, 513–521.

92. A. Bot, R. den Adel, E. C. Roijers and C. Regkos, *Food Biophys.*, 2009, **4**, 266–272.
93. A. Bot, R. den Adel and E. C. Roijers, *J. Am. Oil Chem. Soc.*, 2008, **85**, 1127–1134.
94. M. A. Rogers, A. Bot, R. S. H. Lam, T. Pedersen and T. May, *J. Phys. Chem. A*, 2010, **114**, 8278–8285.
95. G. Dalkas, A. B. Matheson, H. Vass, A. Gromov, G. O. Lloyd, V. Koutsos, P. S. Clegg and S. R. Euston, *Langmuir*, 2018, **34**, 8629–8638.
96. M. Scharfe, Y. Ahmane, J. Seilert, J. Keim and E. Flöter, *Eur. J. Lipid Sci. Technol.*, 2019, **121**, 1800487.
97. H. Sawalha, G. Margry, R. den Adel, P. Venema, A. Bot, E. Flöter and E. van der Linden, *Eur. J. Lipid Sci. Technol.*, 2013, **115**, 295–300.
98. A. G. Marangoni, *Edible Oleogels: Structure and Health Implications*, Elsevier, Cambridge, 2018.
99. R. den Adel, P. C. M. Heussen and A. Bot, *J. Phys.: Conf. Ser.*, 2010, **247**, 012025.
100. K. Frost, D. Kaminski, G. Kirwan, E. Lascaris and R. Shanks, *Carbohydr. Polym.*, 2009, **78**, 543–548.
101. A. Matheson, G. Dalkas, R. Mears, S. R. Euston and P. S. Clegg, *Soft Matter*, 2018, **14**, 2044–2051.
102. M. Scharfe and E. Flöter, *Eur. J. Lipid Sci. Technol.*, 2020, **122**, 2000213.
103. M. Scharfe, D. Prange and E. Flöter, *J. Am. Oil Chem. Soc.*, 2021, 1–21, DOI: 10.1002/aocs.12556.
104. T. G. Mezger, *Das Rheologie Handbuch: Für Anwender von Rotations- und Oszillations-Rheometern*, Vincentz Network, Hannover, 2016.
105. T. A. Stortz, A. K. Zetzl, S. Barbut, A. Cattaruzza and A. G. Marangoni, *Lipid Technol.*, 2012, **24**, 151–154.
106. A. K. Zetzl, A. G. Marangoni and S. Barbut, *Food Funct.*, 2012, **3**, 327–337.
107. D. Tang and A. G. Marangoni, *J. Am. Oil Chem. Soc.*, 2006, **83**, 377–388.
108. A. J. Gravelle, M. Davidovich-Pinhas, A. K. Zetzl, S. Barbut and A. G. Marangoni, *Carbohydr. Polym.*, 2016, **135**, 169–179.
109. G. H. Brusewitz and H. Yu, *J. Food Eng.*, 1996, **27**, 259–265.
110. S.-Y. Lee, I. Luna-Guzmán, S. Chang, D. Barrett and J.-X. Guinard, *Food Qual. Prefer.*, 1999, **10**, 447–455.
111. J. F. Steffe, *Rheological Methods in Food Process Engineering*, Freeman Press, East Lansing, 1996.
112. S. Kanagaratnam, M. Enamul Hoque, M. Mat Sahri and A. Spowage, *J. Food Eng.*, 2013, **118**, 90–99.
113. T. S. Omonov, L. Bouzidi and S. S. Narine, *Chem. Phys. Lipids*, 2010, **163**, 728–740.
114. A. I. Blake, E. D. Co and A. G. Marangoni, *J. Am. Oil Chem. Soc.*, 2014, **91**, 885–903.
115. T. L. T. da Silva, D. B. Arellano and S. Martini, *J. Am. Oil Chem. Soc.*, 2019, **96**, 681–691.
116. S. Theierl, Master Thesis, Technical University, Berlin, 2021.

117. H. Sawalha, P. Venema, A. Bot, E. Flöter, Y. Lan and E. van der Linden, *Food Biophys.*, 2020, DOI: 10.1007/s11483-020-09654-8.
118. V. Giacintucci, C. D. Di Mattia, G. Sacchetti, F. Flamminii, A. J. Gravelle, B. Baylis, J. R. Dutcher, A. G. Marangoni and P. Pittia, *Food Hydrocolloids*, 2018, **84**, 508–514.
119. T. Laredo, S. Barbut and A. G. Marangoni, *Soft Matter*, 2011, **7**, 2734–2743.
120. A. K. Zetzl, A. J. Gravelle, M. Kurylowicz, J. Dutcher, S. Barbut and A. G. Marangoni, *Food Struct.*, 2014, **2**, 27.
121. T. A. Stortz, D. C. de Moura, T. Laredo and A. G. Marangoni, *RSC Adv.*, 2014, **4**, 55048–55061.
122. A. Bot, Y. S. Veldhuizen, R. den Adel and E. C. Roijers, *Food Hydrocolloids*, 2009, **23**, 1184–1189.
123. K. Lubosch, Master thesis, Technical University, Berlin, 2015.
124. M. A. Cuevas, R. E. Shinzat, M. C. Costa, C. Rodrigues and Meirelles, *Int. Congress Food Eng.*, 2011, 1–3.
125. S. Calligaris, G. Mirolo, S. Da Pieve, G. Arrighetti and M. C. Nicoli, *Food Biophys.*, 2014, **9**, 69–75.
126. M. Scharfe, D. Prange and E. Flöter, *J. Am. Oil Chem. Soc.*, 2021, 1–14, DOI: 10.1002/aocs.12555.
127. M. Scharfe, On the importance of minor components and oil properties for sterol/sterol ester strength, presented in part at the 16th Congress, Sevilla, October 2019.
128. M. Davidovich-Pinhas, S. Barbut and A. G. Marangoni, *Carbohydr. Polym.*, 2015, **127**, 355–362.
129. T. A. Stortz and A. G. Marangoni, *Trends Food Sci. Technol.*, 2011, **22**, 201–214.
130. A. Wendt, K. Abraham, C. Wernecke, J. Pfeiffer and E. Flöter, *J. Am. Oil Chem. Soc.*, 2017, **94**, 1131–1140.
131. E. Panagiotopoulou, T. Moschakis and E. Katsanidis, *LWT – Food Sci. Technol.*, 2016, **73**, 351–356.
132. T. Moschakis, E. Panagiotopoulou and E. Katsanidis, *LWT – Food Sci. Technol.*, 2016, **73**, 153–161.
133. D. Franco, A. J. Martins, M. López-Pedrouso, M. A. Cerqueira, L. Purriños, L. M. Pastrana, A. A. Vicente, C. Zapata and J. M. Lorenzo, *J. Sci. Food Agric.*, 2020, **100**, 218–224.
134. A. J. Martins, J. M. Lorenzo, D. Franco, A. A. Vicente, R. L. Cunha, L. M. Pastrana, J. Quiñones and M. A. Cerqueira, *Eur. J. Lipid Sci. Technol.*, 2019, **121**, 1900111.
135. T. D. Kammegne, Bachelor thesis, Technical University Berlin, 2015.
136. M. E. Moriano and C. Alamprese, *LWT – Food Sci. Technol.*, 2017, **86**, 371–376.
137. T. Moschakis, I. Dergiade, A. Lazaridou, C. G. Biliaderis and E. Katsanidis, *J. Funct. Foods*, 2017, **33**, 386–395.
138. N. E. Hughes, A. G. Marangoni, A. J. Wright, M. A. Rogers and J. W. Rush, *Trends Food Sci. Technol.*, 2009, **20**, 470–480.

139. C. D. Doan, A. R. Patel, I. Tavernier, N. de Clercq, K. van Raemdonck, D. van de Walle, C. Delbaere and K. Dewettinck, *Eur. J. Lipid Sci. Technol.*, 2016, **118**, 1903–1914.
140. T. A. Stortz and A. G. Marangoni, *Food Res. Int.*, 2013, **51**, 797–803.
141. T. A. Stortz, T. Laredo and A. G. Marangoni, *Food Biophys.*, 2015, **10**, 253–263.
142. T.-A. Do, J. R. Mitchell, B. Wolf and J. Vieira, *React. Funct. Polym.*, 2010, **70**, 856–862.
143. S. T. Beckett, *The Science of Chocolate*, Royal Society of Chemistry, Cambridge, 2018.
144. M. K. Youssef and S. Barbut, *Meat Sci.*, 2009, **82**, 228–233.
145. J. Park, K. S. Rhee, J. T. Keeton and K. C. Rhee, *J. Food Sci.*, 1989, **54**, 500–504.
146. E. J. Marquez, E. M. Ahmed, R. L. West and D. D. Johnson, *J. Food Sci.*, 1989, **54**, 867–870.
147. A. Bot, R. den Adel, C. Regkos, H. Sawalha, P. Venema and E. Flöter, *Food Hydrocolloids*, 2011, **25**, 639–646.
148. I. P. Freeman and S. M. Melnikov, in *Ullmann's Encyclopedia of Industrial Chemistry*, ed. G. Belussi, *et al.*, Wiley, Chichester, 7th edn, 2010, vol. 1, ch. 3, pp. 1–24.
149. Z. Ma, Bachelor thesis, Technical University Berlin, 2020.
150. G. Corrieu and C. Béal, in *Encyclopedia of Food and Health*, ed. B. Caballero, F. Toldrá and P. M. Finglas, Elsevier, Burlington, 2nd edn, 2015, vol. 2, ch. 3, pp. 617–624.
151. R. C. Chandan, in *Yogurt in Health and Disease Prevention*, ed. N. P. Shah, Elsevier, Burlington, 1st edn, 2017, vol. 1, ch. 2, pp. 31–47.
152. S. S. Abumweis, R. Barake and P. J. H. Jones, *Food Nutr. Res.*, 2008, **52**, DOI: 10.3402/fnr.v52i0.1811.

CHAPTER 8

Structuring Vegetable Oils with Monoglycerides and Monoglyceride–Lecithin or Monoglyceride–Ethylcellulose Mixtures

JORGE F. TORO-VAZQUEZ,*[a] MAYRA AGUILAR-ZÁRATE,[a] ARACELI LÓPEZ-MARTINEZ[b] AND MIRIAM CHARÓ-ALONSO[a]

[a] Facultad de Ciencias Químicas, Laboratorio de Fisicoquímica de Alimentos, Universidad Autónoma de San Luis Potosí, Av. Dr. Manuel Nava 6, Zona Universitaria, San Luis Potosí SLP 78290, México; [b] Coordinación Académica Altiplano Oeste, Universidad Autónoma de San Luis Potosí, Salinas de Hidalgo, Carretera Salinas-Santo Domingo 200, San Luis Potosí SLP 78600, México

*Email: toro@uaslp.mx

8.1 Introduction

We consume vegetable oils directly as dressings or frying oils, or indirectly as part of the formulation of different types of food systems, a role also played by several animal fats like milk fat and lard. Triacylglycerides (TAGS) are the main component of animal fats and vegetable oils, consequently, their physical and chemical properties determine the functional properties of the fats and oils. The phase changes (*i.e.*, polymorphism and solid content as a

Food Chemistry, Function and Analysis No. 31
Development of *Trans*-free Lipid Systems and their Use in Food Products
Edited by Jorge F. Toro-Vazquez

Published by the Royal Society of Chemistry, www.rsc.org

function of temperature), the microstructure developed by crystallized TAGS (*i.e.*, rheological properties), and their mechanical properties determine the functional properties of fats and oils associated with texture and oil binding capacity. Within this context, the functional properties required by consumers in diverse food products like shortenings, margarines, vegetable creams, and spreads like peanut butter, are mainly associated with the crystallization of TAGS containing saturated and/or *trans* fatty acids. Saturated fatty acids are native to animal fats and vegetable oils, although they might be produced through full hydrogenation of vegetable oils. In contrast, *trans* fatty acids are naturally present in vegetable oils and fats just in minor quantities and are mainly produced during the partial hydrogenation of vegetable oil.[1]

If sufficient concentrations of saturated TAGS and/or TAGS containing *trans* fatty acids are present in fats and oils, a decrease in temperature below their solubility limit establishes supercooling conditions that result in TAGS nucleation and subsequent crystal growth. The crystallization of saturated TAGS and/or TAGS containing *trans* fatty acids develops a three-dimensional crystal organization with an appropriate microstructure to physically trap a significant amount of liquid phase (*i.e.*, vegetable oils). This crystallized system provides structure and viscoelastic properties to vegetable oils, providing texture to lipid-based food systems (*i.e.*, margarines, whipped cream, non-dairy coffee creamers) and to more complex food systems like processed meat (*i.e.*, comminuted meat products), confectionary and bakery products (*i.e.*, refrigerated dough). Unfortunately, there is a well-established link between the consumption of saturated fatty acids, particularly *trans* fatty acid consumption, and a higher risk of coronary heart diseases.[2,3] Nowadays enough evidence strongly links the consumption of *trans* fats with an increase in the risk of cardiovascular disease (CVD), even above the risk associated with saturated fat consumption.[2] Through meta-analysis, Hu *et al.*[3] established that replacing 2% of energy from *trans* fats with energy from mono or polyunsaturated fatty acids, decreases the risk of CVD by about 50%. Additional studies also showed that there is a strong relationship between the intake of diets rich in *trans* fatty acids and the development of a metabolic syndrome. This clinical condition involves increased blood pressure, excess body fat, abnormal cholesterol and/or TAGS levels, and the development of type II diabetes.[3–5] In 2019 the World Health Organization (WHO) considered heart disease and stroke, obesity, and the development of a metabolic syndrome among the top 10 public health problems worldwide.[6] These circumstances derived in a sustained increase in consumer awareness of the deleterious health effects associated with the consumption of foods high in saturated and *trans* fatty acids. Within this context, several health organizations worldwide issued recommendations for the food industry that include the use of appropriate labeling to guide consumer choices, limit the use of saturated fatty acids in food formulations,[7] and reduce or even eliminate the presence of *trans* fatty acids in processed foods.[8] In many countries these recommendations occurred along with the involvement of government agencies demanding effective action from the industry that resulted in producing foods low in saturated and *trans* fatty acids and, eventually, *trans*-free food products. Presently, several countries have issued regulations that limit *trans* fats and even ban their use in food products.[9] Evidently, the challenge for the food industry is to develop *trans*-free food products low in saturated fatty acids, but still with the functional and nutritional properties demanded by the current consumers, particularly from lipid-based foods. Several nutritional benefits are associated with the consumption of vegetable and marine oils high in mono- and poly-unsaturated fatty acids. Unfortunately, using current food processing conditions, these edible oils do not provide food systems with the appropriate textures associated with TAGS crystallization. Thus, structuring vegetable oils high in mono- and poly-unsaturated TAGS is essential to achieve the functional properties associated with the microstructure provided by the solid phase developed through the crystallization of saturated and *trans* fatty acids.

8.1.1 Vegetable Oil Structuring Through Molecular Self-assembly

Research dealing with edible oil structuring and its application to food systems has received considerable interest during the last 10–15 years, particularly through the development of oleogels (*i.e.*, edible organogels).[10–12] An oleogel is a class of gel composed of an edible oil phase within a three-dimensional, cross-linked network developed by the spontaneous self-assembly of low molecular weight molecules (*i.e.*, gelator molecules). This three-dimensional microstructure is stabilized through non-covalent short-range attractive forces (*i.e.*, London dispersion forces, van der Waals forces, and hydrogen bonds), that hold the micro domains of the vegetable oil in a non-flowing state. In an oleogel the molecular self-assembly occurs in a vegetable oil gelator solution, developing supramolecular structures formed by micelles, self-assembled fibrillar networks, tubules, or microplatelets that physically trap the edible oil mainly through surface tension. The self-assembly of the gelator molecules occurs without an external induction force other than the supercooling and/or supersaturation of the gelator in the oil. To trap a significant amount of edible oil the microstructure developed by saturated and *trans* fats requires a mass of crystals usually exceeding 30%. In contrast, the development of an oleogel requires very low concentration of the crystallizing/gelator molecule. This is because the supramolecular structure developed has a large solid to liquid interfacial area, resulting in a phase change of the vegetable oil at low gelator concentrations. As an example, we require just 0.5% rice bran wax or 2% candelilla wax or carnauba wax to gel olive oil.[13] Rice bran wax is essentially constituted by long-chain aliphatic esters (<92%),[14] candelilla wax mainly by *n*-alkanes (49–50%) mixed with high molecular weight alkyl esters (6% to 16%) and other minor components,[14–16] and carnauba wax by a complex mixture of long-chain alkyl esters (62% to 85%), *p*-hydroxycinnamic aliphatic di-esters (20–23%), and mono-hydric alcohols (10–12%) as the main components.[17] Within this context, it is important to note that the minimum gelator concentration required to develop an oleogel depends on several factors. This is because the molecular self-assembly and therefore whether a gel will be formed depend, in a pure system (*i.e.*, a pure molecule in solution), on the molecular structure of the gelator molecule, and in a complex system (*i.e.*, mixture of molecules like in vegetable wax), in the molecular interactions between the gelator molecules present. Additionally, the minimum gelator concentration depends on the balance between gelator–gelator and gelator–oil interactions that determine whether the gelator will be soluble (*i.e.*, no gelation), not soluble (*i.e.*, the gelator precipitates) in the oil, or able to develop a supramolecular structure that physically traps the oil forming a self-supporting structure.[18] The potential of molecular self-assembly in vegetable oil technology, resides in the replacement in food products of the solid fats associated with saturated and *trans*-fat crystallization with structured unsaturated and poly-unsaturated vegetable oils. This will improve the nutritional profile of the food system, tentatively providing similar or better functional properties than the original food.

From the above, evidently structuring vegetable oils through molecular self-assembly of edible gelators is nowadays an important research area not only from a fundamental perspective, but also from a practical/technological point of view. Although there are several types of low molecular weight molecules that self-assemble in vegetable oils and develop self-standing structures, just a few have edible characteristics. The edible gelators most studied using vegetable oils as the liquid phase include phytosterols,[19–21] lecithin,[22,23] monoglycerides,[24,25] and also *n*-alkanes, long-chain esters, fatty acids, and alcohols in the pure state[26–29] or as native complex mixtures like those present in candelilla, carnauba, rice bran, and sunflower waxes.[13,15–17,30–33] (*R*)-12-hydroxystearic acid (12-HSA) is a good gelator that has also been extensively studied in vegetable oils.[28,34–36] However, the FDA has approved 12-HSA only as an indirect additive used in materials with incidental food contact like in defoaming agents employed in the manufacture of paper and paperboard or as part of lubricants.[37] This situation limits its use as a gelator in the development of edible oleogels. Independent of the type of gelator the molecules self-assemble *via* weak non-specific (*i.e.*, dipole–dipole, dipole-induced, and instantaneous dipole-induced forces) and specific (*i.e.*, hydrogen bonds) intermolecular non-covalent interactions.[38,39]

The intermolecular forces that predominate in establishing the gelator–gelator and gelator–vegetable oil interactions and, subsequently, the topology of the three-dimensional network, the oleogel thermo-mechanical properties and stability to oil phase separation, depend mainly on the gelator's chemical structure and the type of functional groups present in the molecule (*i.e.*, apolar, polar, those prone to forming hydrogen bonds). Monoglycerides (MGs) and lecithin (LC) are two groups of molecules widely present in biological systems that share some chemical characteristics. This is because MGs and LC are constituted by glycerol (1,2,3-propanetriol) esterified with one or two fatty acids, respectively. However, while MGs have two free –OH groups in the phospholipids present in LC, *i.e.*, phosphatidylcholine (1,2-diacyl-*sn*-glycerol-3-phosphocholine), one primary hydroxyl group of the glycerol is additionally esterified to a phosphate group and to an amine (*i.e.*, phosphocholine). Thus, given their amphiphilic character the MGs and LC are industrially used in foods and cosmetics manufacture mainly as surfactant agents (*i.e.*, emulsifiers). Under particular time–temperature conditions and concentration, MGs and LC can develop oleogels either independently[23,25,40–42] or through molecular interactions between them, with other molecules including water,[24,43–45] high molecular weight polymers (*i.e.*, ethylcellulose) or with components present in vegetable waxes (*i.e.*, alkyl esters).[46–48] In all these cases the oleogels obtained showed useful functional properties. Within this framework, in this chapter we present information regarding the molecular self-assembly of MGs in vegetable oils, as well as the physical properties and stability of MG oleogels. It is important to point out that MG oleogels are metastable systems, and the polymorphic changes occurring during storage result in a deleterious effect of the MG oleogels rheological properties and, eventually, oil phase separation.[41,43,46] This series of phenomena restricts the commercial use of MG oleogels in food systems. In this chapter we discuss information about the effect of another low molecular amphipathic gelator (*i.e.*, LC) and a high molecular weight polymer (*i.e.*, ethylcellulose) in limiting the MG polymorphic changes, thus extending the MG oleogel's stability. Usually, oleogels developed by mixtures of gelators have better rheological properties with longer stability for oil phase transition than oleogels developed independently by each gelator.[31,43,46,47]

8.2 Monoglycerides as Gelator Molecules

MGs and their organic acid derivatives (*e.g.*, acetic, citric, diacetyl tartaric, and lactic acids) are the most common polar lipids used as emulsifier in food products.[49] Here we will discuss information essentially of MGs, the amphiphilic lipids constituted by a fatty acid molecule esterified to one of the primary hydroxyl groups of glycerol (*i.e.*, α-MG or *sn*-1 MG). Thus, MGs are generally described as molecules with a polar "head" with no charged functional groups and a hydrophobic "tail" formed by the alkyl chain of the fatty acid. The polar "head" is formed by the secondary and primary –OH groups, both with hydrogen bonding capabilities. The –OH group in the *sn*-2 position provides chirality to MG molecules, thus existing as D and L optically active molecules. Because of their amphipathic character MGs reduce the surface tension of water and most organic solvents, including vegetable and mineral oils, and therefore are considered as nonionic surfactants.

The esterified fatty acids forming the "tail" of commercial MGs commonly include lauric, linoleic, myristic, oleic, palmitic, and stearic, depending on the source of TAGS used in the MG production. The industrial process for MG production consists of TAGS interesterification with glycerol at temperatures above 200 °C in the presence of a catalyst [*i.e.*, NaOH or $Ca(OH)_2$]. For this process, the source of TAGS is animal fats or vegetable oils (edible refined canola, sunflower, cottonseed, coconut, palm oil, fully or partially hydrogenated palm oil, or soybean oil). The resulting product is a mixture of MGs (35–60%), diacylglyceride (35–50%), TAGS (1.0–20%), and free fatty acids (1–10%), that after some purification to eliminate the catalyst and TAGS, the resulting MG mixture has acceptable emulsifying properties. This MG mixture is Generally Recognized As Safe (GRAS) as described in the Code of Federal Regulation 21CFR184.1505,[50] and thus might be used in food products without any further purification. However, if higher MG purity is required (*i.e.*, higher than 90%) this can be achieved by molecular distillation of the MG/diacylglyceride/TAGS mixture.[51,52] The monoglyceride composition and purity of commercially available distilled MGs depend essentially on the source of TAGS, the processing conditions of the interesterification reaction, and the distilling conditions used.[53,54]

The application of MGs in food systems is mostly associated with their capacity to adsorb at interfaces favoring the formation of emulsions and foams and also to prevent starch retrogradation extending the shelf life of baking products through the development of MG–amylose complexes.[55,56] However, an important property of MGs in pure and commercial forms is their capacity of structuring polar systems (*i.e.*, aqueous phases),[57,58] organic solvents (*i.e.*, vegetable and mineral oil),[24,25,40] and mixtures between organic and polar phases (*i.e.*, vegetable oil–water),[24,59,60] through the formation of several liquid crystalline structures (*i.e.*, mesophases) developed by the MGs.

8.2.1 Self-assembly of Monoglycerides in Vegetable Oils

The molecular self-assembly of MGs has been studied mainly in melts of pure MGs[61,62] and in aqueous MG solutions.[58,63,64] Fewer studies have addressed the self-assembly of MGs in organic solvents, vegetable, or mineral oils.[25,43] The results of these studies indicate that the structure developed by MGs in solution depends, mainly, on the solvent polarity, MG concentration in the solvent (*i.e.*, water or oil), the length of the esterified fatty acid, MG composition (*i.e.*, pure, mixture of MGs as in commercial products), the presence of a co-surfactant (*i.e.*, phospholipids), temperature, and storage time.[24,25,43,62–66] The MGs represent an excellent example where the molecular architecture in interaction with the solvent properties, determines the development of distinctive self-assembly structures with particular functional properties. Within this context, the physical properties of the mesophase structures developed by the MGs in water or vegetable oils, are derived from the hydrophobic character of the alkyl chains and the polar –OH groups strong affinity by water and aversion to apolar environments (*i.e.*, vegetable oil). Thus, at low concentrations and a temperature above the

MG solubility limit in water or polar solvents, the MG molecules are in an isotropic phase. As the MG concentration increases, a decrease in temperature results in the development of micelles. The MG micelles are spherical molecular aggregates with a hydrophobic core structured by the fatty acid's alkyl chains, and a polar surface formed by the secondary and primary –OH groups.[57,67] Thus, the MG micelles are used to trap within their structure lipid-soluble molecules with biological activity, nutrients, flavors, or aromas and "solubilize" them in aqueous systems for their use in food, cosmetics, and pharmaceutical products.[57] Above the critical micelle concentration in water, the excess MGs are fully incorporated into the micelle structures until a closer packing of the polar head groups leads to a structural rearrangement into flat bilayers (*i.e.*, lamellas) with water incorporated between the polar heads.[68,69] In contrast, on cooling from the isotropic phase MG vegetable oil solutions, the MGs develop micelles with inverse organization from those micelles formed in water. At MG concentrations well above the critical micelle concentration in a hydrophobic environment (*e.g.*, 8–10% distilled MG in vegetable oil), upon cooling from the isotropic phase the MGs directly develop inverse lamellar bilayer structures. In these inverse lamellas the hydrophilic head groups are more compressed than in the lamellar structure developed in an aqueous environment (Figures 8.1 and 8.2).[25,70] The packed structural organization of the glycerol heads in the inverse lamellar structure (Lα) is responsible for (1) the characteristic single wide-angle X-ray reflections of the MG bilayer at 4.17 Å and 4.11 Å observed during the development of MG oleogels (Figure 8.2),[41] and (2) providing the

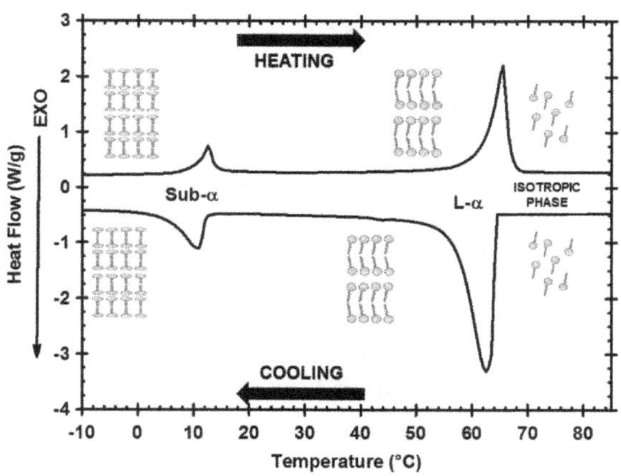

Figure 8.1 Cooling and heating thermograms of an 8% solution in safflower oil high in triolein of a distilled commercial monoglyceride constituted by a mixture of 1-mono-palmitoyl-glycerol (54.0%) and 1-mono-stearoyl-glycerol (37.66%). The corresponding exotherms and endotherms associated with the reversible Lα⇆sub-α phase transitions occurring as a function of temperature are shown, along with artistic representations of the molecular organization in monoglycerides at each stage.
Reproduced from ref. 25 with permission from Elsevier, Copyright 2014.

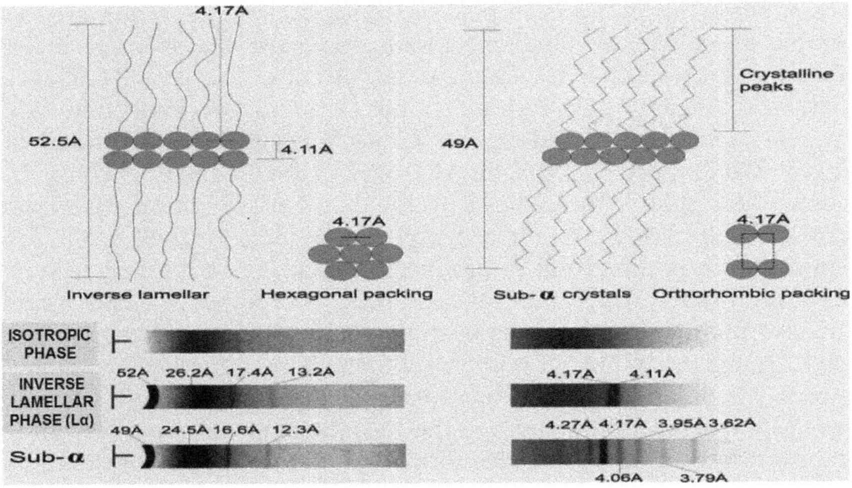

Figure 8.2 Wide-angle (WAX, right side) and small-angle (SAX, left side) X-ray diffraction patterns for a 10% monoglyceride (93% of 1-mono-stearoyl-glycerol, 7% of 1-mono-palmitoyl-glycerol) oleogel in hazelnut oil. The inverse lamellar phase (Lα) was obtained at 50 °C and the crystalline sub-α phase at 26 °C. The SAX diffraction pattern describes the lamellar ordering in the Lα and sub-α phases, and the WAX diffraction pattern corresponds to the periodic spacing between neighboring glycerol heads inside the inverse lamellar bilayer. In the sub-α crystalline phase the WAX diffraction pattern indicates the orthorhombic chain packing. Reproduced from ref. 42 with permission from Elsevier, Copyright 2018.

structural orthorhombic packing of the MG alkyl chains (sub-α phase) during the continuous cooling of monoglyceride–oil solutions (Figures 8.1 and 8.2).[25,40,41] The Lα and sub-α phase, in turn, form the microstructure responsible for providing the characteristic thermo-mechanical properties observed by MG oleogels.[25] At a temperature below the formation of the sub-α phase, the MG microstructure is stabilized by a balance between the van der Waals attraction force established along the alkyl chains and the short-range repulsive forces present throughout the inverse bilayer.

The formation and melting of the Lα and sub-α phases occurring during cooling–heating cycles of pure and commercial MGs in the neat state or in vegetable oil solution, are closely followed by differential scanning calorimetry (DSC; Figure 8.1). Within this context, it is important to note that in pure MGs with esterified fatty acids longer than 16 carbons, we observe an additional transition occurring at lower temperatures than the Lα→sub-α transition, *i.e.*, the sub-α2 phase.[62] When this occurs, upon applying cooling–heating cycles the order of the reversible phase transitions is Lα⇄sub-α1⇄sub-α2 (Figure 8.3). The Lα to sub-α1 transition involves a change in the hydrocarbon backbone from a randomly oriented mesophase structure to a well-ordered crystalline state (Figures 8.1 and 8.2). However, the molecular changes involved in the sub-α1⇄sub-α2 transition are not

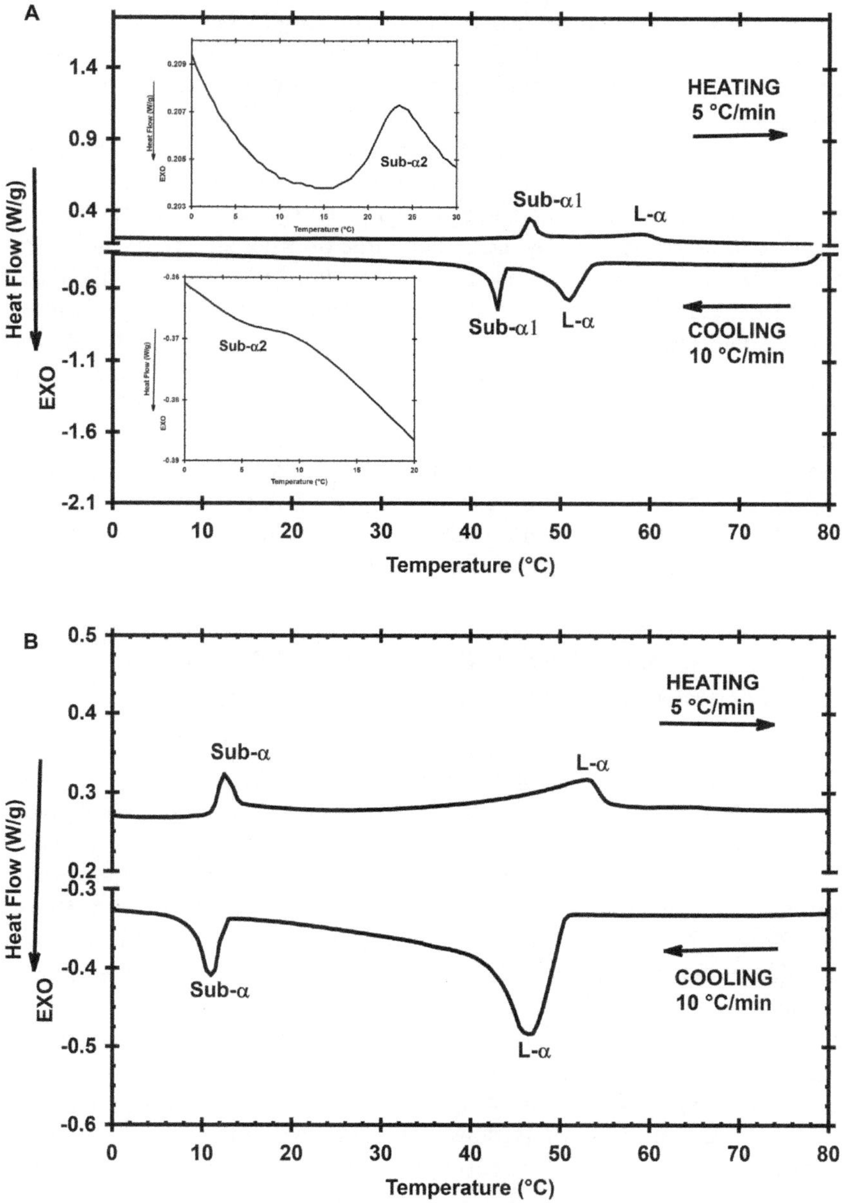

Figure 8.3 Cooling (A) and heating (B) thermograms for pure monoglycerides (>99%) from C14 up to C22. To observe the sub-α2 transition occurring in C16, C18, C20, and C20 the panels C and D correspond to augmentations of the corresponding thermograms for the interval between −20 °C and 40 °C.

well established. According to the initial studies of Lutton[62] the sub-α1 ⇄ sub-α2 phase transition is present just in pure MGs with C18 fatty acids or longer (Figure 8.3A).[62] Based on the results obtained by our group, we suggested that the sub-α1 ⇄ sub-α2 transition could be associated with a rotator phase remnant after the sub-α1 phase crystallization in MGs with an 18 carbon alkyl chain or longer.[25] However, additional experiments showed the presence of the sub-α1 ⇄ sub-α2 transition in MGs with fatty acids shorter than C18 (*i.e.*, 1-mono-palmitoyl-glycerol). It seems that the detection of this transition by DSC depends on the signal to noise ratio achieved by the calorimeter used. Additionally, MGs with an alkyl chain shorter than C16 (*i.e.*, C14) go through polymorph changes (*e.g.*, formation of the β polymorph) at temperatures lower than the Lα → sub-α1 transition. The development of polymorphism during cooling might make it difficult to detect the sub-α1 ⇄ sub-α2 transition. In any case, when cooling–heating cycles are applied to mixtures of MGs like those present in commercial food-grade MGs (*e.g.*, mixtures of 1-mono-stearoyl-glycerol and 1-mono-palmitoyl-glycerol), these systems do not show the sub-α1 ⇄ sub-α2 transition (Figure 8.3B). Here, the development of a mixed lamellar structure during cooling might make the achievement of higher structural order difficult.

8.2.2 Mesophases, Crystal Phases, and Polymorphic Transitions of Monoglycerides in Oleogels

Within the context of the previous discussion, Figure 8.4 shows the cooling and heating thermograms for the following pure MGs: 1-mono-behenoyl-glycerol (C22), 1-mono-araquiloyl-glycerol (C20), 1-mono-stearoyl-glycerol (C18), 1-mono-palmitoyl-glycerol (C16), and 1-mono-myristoyl-glycerol (C14) (Toro-Vazquez J. F. and Charó-Alvarado M. E., unpublished results). Upon cooling from the isotropic phase (90 °C) the corresponding thermograms for pure C16 to C22 showed the exotherms associated with their initial molecular self-assembly (Lα mesophase) and, as cooling proceeded, the subsequent sub-α1 and sub-α2 crystallization (Figure 8.4A and C). The cooling of C14 showed just the exotherms associated with the initial molecular self-assembly (Lα) and the crystallization of the sub-α phase (Figure 8.4A). We observed the same calorimetric behavior during cooling of 8% C14 to C20 solutions in safflower oil. The peak temperature of the exotherms associated with the Lα, sub-α, or sub-α1 phases for pure C14 to C22 and those corresponding to 8% solutions in safflower oil, showed a significant direct linear regression with the number of carbons of the alkyl chain ($R^2 > 0.96$, $P < 0.001$). This behavior showed that the MGs initial molecular self-assembly and the crystallization of the sub-α or sub-α1 phase occurred at higher temperature as the length (*i.e.*, number of carbons) of the esterified fatty acid increased. Our studies had previously showed that the solubility limit temperature of different gelator molecules is directly associated with the gelator crystallization temperature in vegetable oils (*i.e.*, the

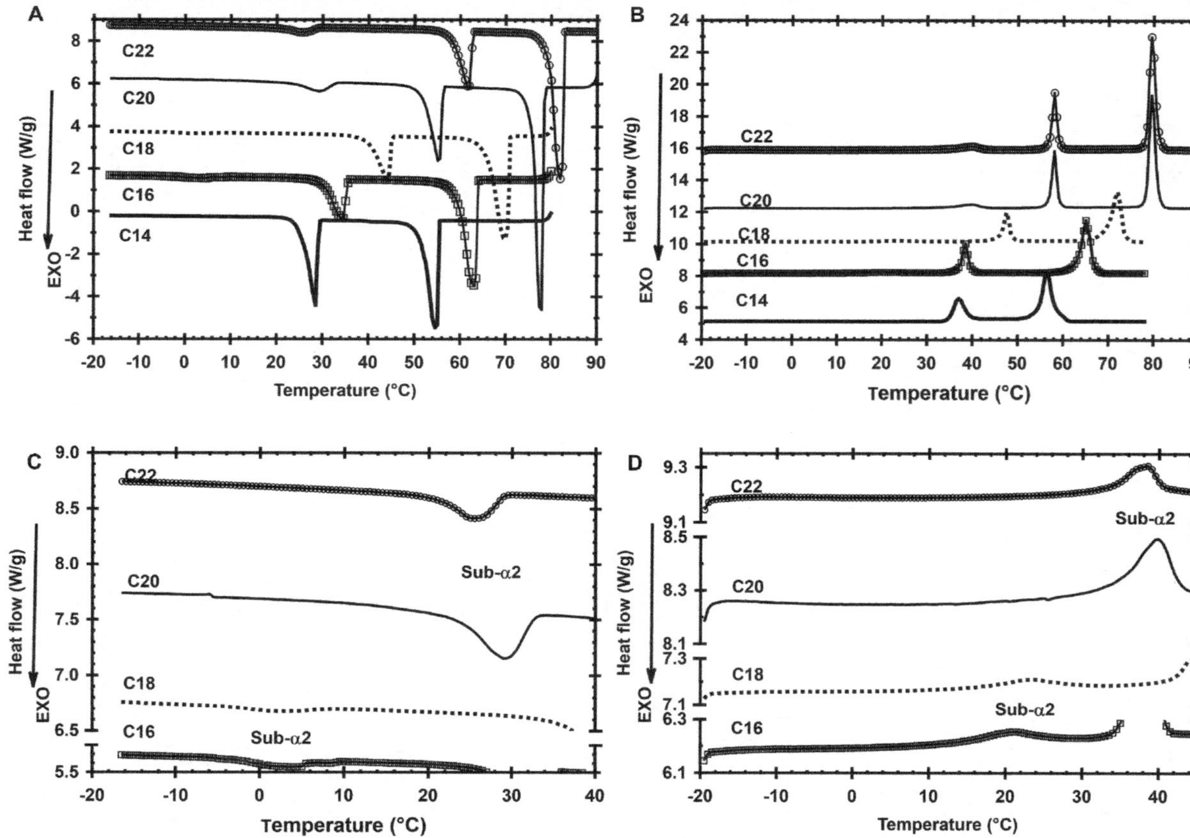

Figure 8.4 Cooling (10 °C per min; A) and heating (5 °C per min; B) thermograms for pure 1-mono-behenoyl-glycerol (C22), 1-mono-araquiloyl-glycerol (C20), 1-mono-stearoyl-glycerol (C18), 1-mono-palmitoyl-glycerol (C16), and 1-mono-myristoyl-glycerol (C14). Panels C (cooling) and D (heating) are magnifications (−20 °C to 40 °C) of the thermograms to show the crystallization and melting of the sub-α2 phase in C16 to C22. Unpublished results from Toro-Vazquez J. F. and Charó-Alvarado M. E.

temperature at the maximum of the exotherm associated with the onset for molecular self-assembly).[18] Therefore, although the polarity of the hydrophilic head thermodynamically drives the molecular self-assembly of MGs in the vegetable oil, an additional driving factor is the alkyl chain solubility in the oil. Thus, the MG's solubility temperature in vegetable oils is associated with the peak temperature for the Lα, and the sub-α/sub-α1 exotherms. In contrast, the temperature for the sub-α1 ⇄ sub-α2 transition for C16 to C22 did not show any relationship with the number of carbons in the MG alkyl chain. The sub-α1 ⇄ sub-α2 transition for C20 and C22 occurred between 26 °C and 30 °C, and for C16 and C18 between 0 °C and 6 °C, with a significantly smaller heat of crystallization for these last two MGs. As previously discussed, these results indicated that the sub-α1 ⇄ sub-α2 transition is probably associated with the crystallization of a rotator phase remnant in the alkyl chain after the sub-α1 phase crystallization.[25] The corresponding heating thermograms of the crystallized pure (Figure 8.4B and D) and 8% vegetable oil solutions of C16 to C22 showed the endotherms associated with sub-α2, sub-α1, and Lα melting. In agreement with the crystallization thermogram for the pure C14, the corresponding heating thermogram showed just the endotherms associated with Lα and sub-α phase melting (Figure 8.4B and D). The 8% C14 vegetable oil solution observed a similar melting behavior.

For a given MG–vegetable oil system, the temperature for the Lα phase formation and melting are concentration dependent occurring at lower temperature as the MG concentration decreases. In contrast, the temperature for sub-α phase crystallization and melting is independent of the MG concentration (Figure 8.5; Toro-Vazquez J. F. and Charó-Alvarado M. E., unpublished results). As shown in Figure 8.4A, the Lα → sub-α or Lα → sub-α1 transition depends on the number of carbons of the fatty acid esterified to the MG. During storage of oleogels or after cooling of MG–vegetable oil solutions at temperatures lower than the sub-α phase crystallization, the MGs tend to develop a more stable polymorph, *i.e.*, the β phase. The β polymorphic transition results because, tentatively, in the sub-α phase the repulsive forces throughout the bilayer are smaller than the van der Waals forces along the lamella. In consequence, as a function of time there is a microstructural densification in the MG crystals (*i.e.*, an additional tilt of the alkyl chains) that results in an increase in the main diffraction peak from 4.17 Å (orthorhombic sub-α phase) to 4.55–4.65 Å (triclinic β polymorph).[41,43] As suggested by Chen and Terentjev,[41] this is an indication of the formation of intermolecular hydrogen bonding between –OH of different MGs derived from an enantiomeric segregation of the D and L monoglycerides.[41,49] The enantiomeric segregation of the D and L monoglycerides is similar to the one occurring in MG–water solutions during the development of the coagel.[71] It is important to note that the formation of the β polymorph and subsequent β crystal agglomeration occurring during storage of MG oleogels, results in a significant modification of the three-dimensional crystal network that results in a deleterious effect of the

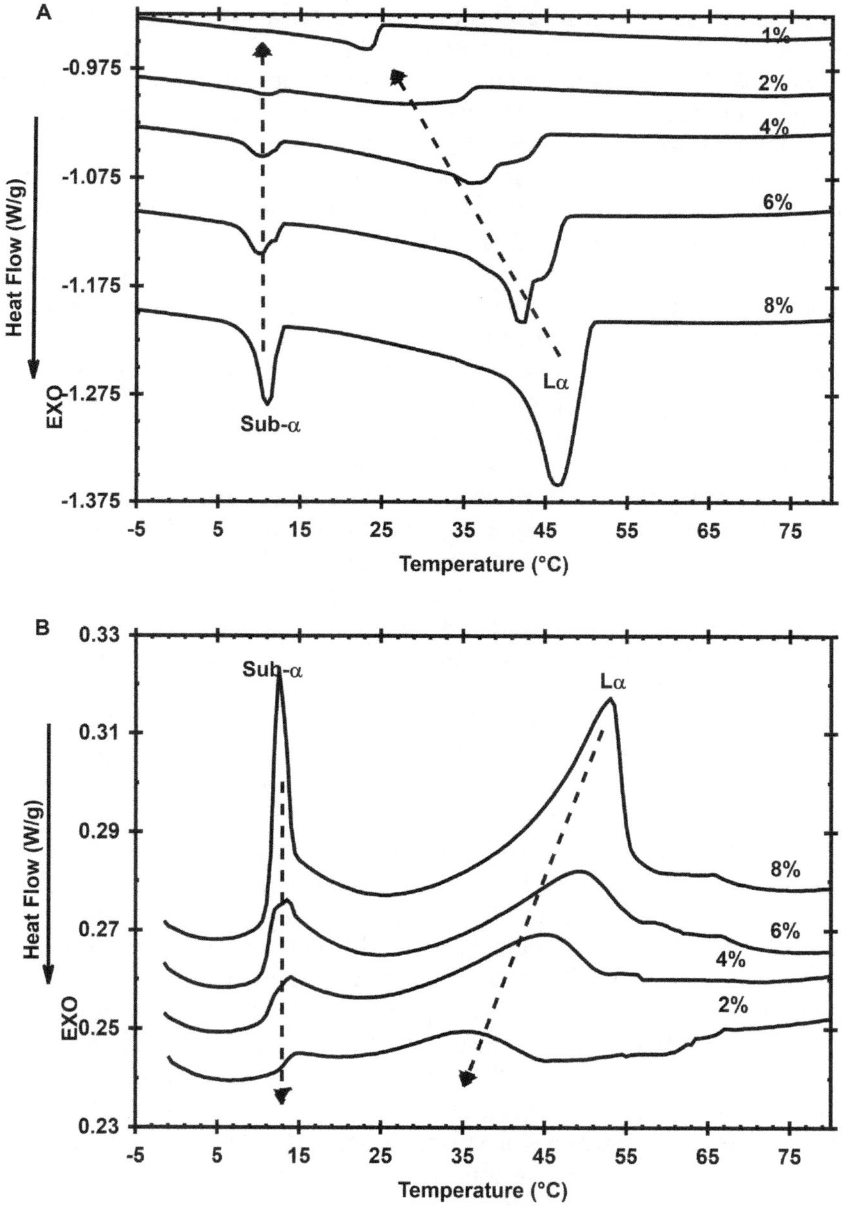

Figure 8.5 Cooling (10 °C per min; A) and heating (5 °C per min; B) thermograms of oleogels formulated in safflower oil high in triolein with 1% to 8% of a commercial MG mixture (37.66% of C18 and 54.02% of C16). The arrows show that the Lα phase formation/melting is concentration dependent and the sub-α phase crystallization/melting is concentration independent. Unpublished results from Toro-Vazquez J. F. and Charó-Alvarado M. E.

thermo-mechanical and oil-binding properties of the MG oleogels.[41,43,46] This series of phenomena restricts the commercial use of MG oleogels in food systems. It is important to note that the Lα and sub-α states are metastable phases,[41] that as a function of time and temperature conditions achieve a thermodynamically stable state, which is the β polymorph. Consequently, the search for mechanisms that avoid or at least limit the formation of the β polymorph during the formation and storage of MG oleogels is of utmost importance. To follow the β phase formation as a function of processing (*i.e.*, time–temperature, shearing, MG concentration, MG composition) and/or storage conditions, and its effects on the microstructure and thermo-mechanical properties of MG oleogels, the use of a combination of analytical techniques is essential. Most of the studies include at least DSC, oscillatory rheological measurements, X-ray diffraction analysis, and microscopy.

Within the previous context, phase diagrams are used to describe the phases (*i.e.*, Lα, sub-α1, sub-α2, β) occurring and coexisting during molecular self-assembly and subsequent crystallization of pure gelators, gelator solutions, or organogels (*i.e.*, MGs oleogels). This, as a function of temperature and gelator concentration, assuming sufficient time and energy to achieve thermodynamic equilibrium conditions is provided.[72] In the particular case of MG oleogels, to evaluate their practical use is important to establish the relationship between the MG phase diagrams in vegetable oils and the corresponding oleogel rheological properties. The viscoelastic properties of MG oleogels, not present in the MG–water system, are essential in the production of *trans*-free substitutes for food systems. As long as the β phase formation is limited or avoided, the formation of stable MG oleogels can be achieved. Within this framework, our group determined the phase diagram (5 °C) of oleogels developed in safflower oil high in triolein with 0.5% to 8% (wt/wt) of C18 (93.5% 1-mono-stearoyl-glycerol) and with a commercial food-grade MG (37.7% C18 and 54.0% C16). An additional positive point of this study is that the evolution of the oleogel's microstructure as a function of MG concentration was associated with the corresponding crystal phase present as established by X-ray diffraction. The microstructure developed by the MG–vegetable oil system as temperature decreased was monitored by plotting the difference in elasticity (G') between subsequent time–temperature conditions [*i.e.*, d(G')/d(time)].[25] This study established that the Lα mesophase provides a limited structure to the vegetable oil resulting just in sol systems. However, the sub-α phase, particularly the one developed by the mixture of MGs (*i.e.*, the commercial MG), structured the vegetable oil through the uptake and retention of oil within the sub-α microstructure. The overall result was that, for a similar MG concentration, the oleogels formulated with the commercial MG achieved higher G' and solid fat content (SFC) than the oleogels formed with pure C18. The difference between the commercial and pure MG oleogel's elasticity (*i.e.*, G') and SFC was greater as the MG concentration increased, particularly at MG concentrations of 2% and above (Figure 8.6).[25] The difference between the oleogels' rheology was partially explained considering the higher SFC achieved by the commercial

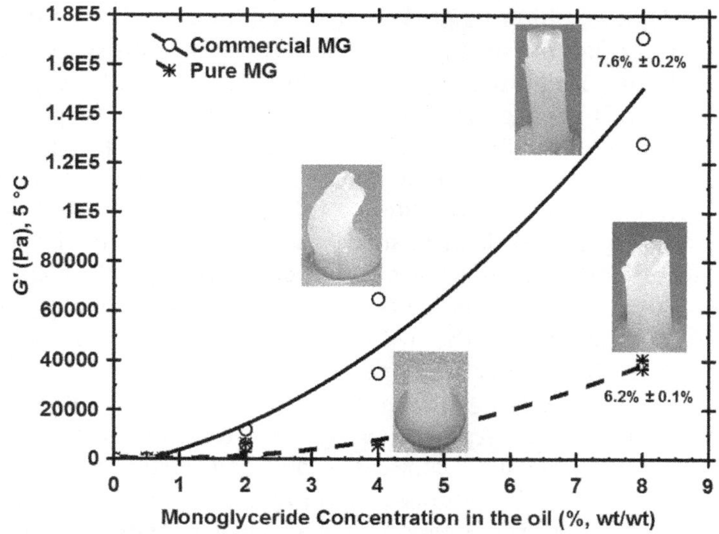

Figure 8.6 Elasticity (G') of oleogels (5 °C) developed in high oleic safflower oil with a commercial MG mixture (37.66% of C18 and 54.02% of C16) or pure MG (93.5% of C18). The inserts show photographs of the oleogels developed with 4% and 8% of the corresponding MG. The solid content of the 8% MG oleogels is indicated.
Reproduced from ref. 25 with permission from Elsevier, Copyright 2014.

MG, but mainly because the mixed sub-α structure developed by C18 and C16 favored the incorporation and retention of higher amounts of oil. This is in contrast with the behavior observed for the sub-α1 and sub-α2 phases present in the oleogel developed just by the pure C18.[25] An important observation of this study was that, in the commercial MG oleogel, the transition to the β polymorph occurred within 3 hours of storage (5 °C). In contrast, in the pure C18 oleogel the formation of the β polymorph did not occur even up to 48 hours of storage (5 °C). As already mentioned, the presence of β crystals developed either through a polymorphic transition from the Lα phase or during storage from the sub-α phase is associated with the collapse of the oleogel three-dimensional crystal network and the subsequent oil phase separation.[41,42] Evidently, it is important to investigate the mechanisms that limit the development of the β polymorph and their subsequent agglomeration during storage.

8.3 Lecithin and Ethylcellulose as Gelators that Limit β Phase Formation

8.3.1 Molecular Self-assembly of Lecithin in Vegetable Oil

Lecithin (LC), as a MG, is another amphiphilic molecule also capable of developing organogels in organic solvents and oleogels in vegetable oils.

LC is a family of amphiphilic phospholipids composed of two alkyl tails of fatty acids esterified to the primary and secondary hydroxyl groups of a glycerol molecule. The remaining primary hydroxyl group of the glycerol is, usually, esterified to phosphocholine given origin to phosphatidylcholine (2-diacyl-*sn*-glycerol-3-phosphocholine). Phosphatidylcholine (PC) is the major component in lecithin. Commonly, PC and LC are used interchangeably, although, evidently the two are different. The major sources of LC are sunflower seeds, eggs, and particularly soybeans. As MGs, we generally depict LC and all phospholipids, as molecules with a polar "head," but in contrast to MGs, the LC has two hydrophobic or alkyl "tails".[73] Additionally, the polar "head" of LC is larger than the hydrophilic "head" of MGs. As with other surfactants, the "head" area, the "tail" length, and the "tail" volume determine the critical packing parameter (CPP) of LC and MG molecules. Theoretically, the CPP determines, to a great extent, the type of structure developed by the surfactant.[74] However, additional factors like the solvent characteristics (*i.e.*, polar, apolar), temperature, and alkyl chain unsaturation also affect the type of molecular aggregates developed by surfactants.[75]

The charged functional groups of PC confer zwitterionic characteristics to the molecule. Consequently, in organic solvents like vegetable and mineral oils, polar interactions thermodynamically drive the onset of LC's molecular self-assembly developing inverse micelles with spheroidal or elliptic shapes. If sufficient LC concentration is present and small quantities of water or other primer molecules (*i.e.*, citric acid) are present, LC might develop wormlike structures that provide viscoelastic properties to the system.[76] This self-assembly behavior of LC was originally reported by Scartazzini and Luisi[77] in particular organic solvent–LC systems, that upon addition of water developed organogels. In vegetable oils the LC can develop oleogels either alone,[23] or through the interaction with other molecules like vegetable waxes[78] and sitosterol,[79] and in the presence of water with fatty acids[80] or ceramide.[81] Thus, LC might have synergistic interactions with the other molecules developing oleogels at lower LC concentration but achieving higher rheological properties than oleogels developed just by the LC.

Our group has studied the development of oleogels using commercial LC, MGs, and LC + MG mixtures, assessing the effect of water addition on the oleogels' thermo-mechanical properties and the β polymorph formation during oleogel storage.[43] It is well known that water has a strong affinity to the MGs' hydroxyl groups. We considered that in MG or in MG–LC oleogels, the presence of water bound to the hydroxyl groups in the hydrated inverse lamella might allow for sufficient molecular fluctuations of the glycerol hydroxyl groups. Tentatively, this could restrict the alkyl chains' alignment along the lamella, thus limiting their crystallization and the subsequent β phase formation during oleogel storage.[43] Within this context, we established that the MG and LC minimal gelling concentration in high oleic safflower oil after 24 h at 15 °C was 3.5% and 5%, respectively. The minimal gelling concentration is a gelling parameter established by the gelator solubility in the solvent (*i.e.*, the vegetable oil) as affected by time–temperature conditions (*i.e.*, gel setting temperature, cooling rate). These results showed that LC was more soluble in the vegetable oil than the MG, requiring higher concentration to develop a self-supporting structure (*e.g.*, an oleogel). Based on these observations, we evaluated the tentative MG–LC interactions during oleogel development, the oleogel's thermo-mechanical properties, and the β phase formation. In the first approach used, the systems were formulated using a constant MG concentration (2% wt/wt) while increasing the LC concentration from 0.25% to 2.5%. Consequently, with this methodology the total gelator mass (*i.e.*, MG + LC) in the oleogel increased from 2.25% up to 4.5%. This interval of MG + LC corresponded to a proportion of moles of MG per mole of LC from 15.3 down to 1.5.[43] It is important to point out that the MG and LC concentrations used were below their corresponding minimal gelling concentration in the vegetable oil. In a second approach used, we kept the MG + LC at either 2% or 4%, using in each case between 2 and 16 moles of MG per mole of LC (Bello-Santillán and Charó-Alonso, unpublished results). Figure 8.7A shows the G' and solid content of oleogels formulated with different moles of MG per mole of LC using a constant MG concentration (2%) and increasing concentrations of LC. Figure 8.7B and C show the G' and solid content of oleogels formulated with a constant MG + LC of 2% and 4%, respectively. The arrows indicate the G' of systems formulated just with 2% MG (Figure 8.7A and B) or 4% MG (Figure 8.7C). It is important to note that the 2% MG system was below the minimal gelling concentration in the vegetable oil, and thus this solution did not gel. Thus, except for the system with the highest LC concentration and, therefore, the lowest MG proportion (*i.e.*, 1.5 moles of MG per mole of LC; Figure 8.7A), independent of the methodology used to formulate the oleogels, at 15 °C we obtained well-structured oleogels where the elastic modulus was above the viscous modulus (*i.e.*, $G' > G''$; Figure 8.7). In the oleogels formulated with 2% MG plus increasing amounts of LC we achieved the highest G' at 15.3 moles of MG per mole of LC (Figure 8.7A).[43] These oleogels had a G' of 8415 Pa (±1638 Pa) and a solid content of 1.86% (±0.10%). In contrast, in the oleogels formulated with a constant total gelator mass of 2% we achieved the highest G' at 8.1 moles of MG per mole of LC, obtaining a G' of 2342 Pa (±40 Pa) and a solid content of 1.25% (±0.09%) (Figure 8.7B). Evidently, the increment in G' was not associated with the solid content of the system but with the oleogel's microstructure (Figure 8.8). Independent of the methodology used to formulate the oleogels, in each case we obtained the highest G' at the lowest solid content in the oleogels (Figure 8.7A and B). The photographs obtained by polarized light microscopy shown in Figure 8.8 correspond to the results shown in Figure 8.7B. The system just with 2% MG did not develop an oleogel and showed only large acicular spherulites associated with MG crystals in the β phase (Figure 8.8). In contrast, the oleogels developed with a mixture of MG and LC (MG–LC) were structured by acicular spherulites associated with MG crystallization, and small birefringent plate-shaped and rod-shaped crystals. The amount of the small

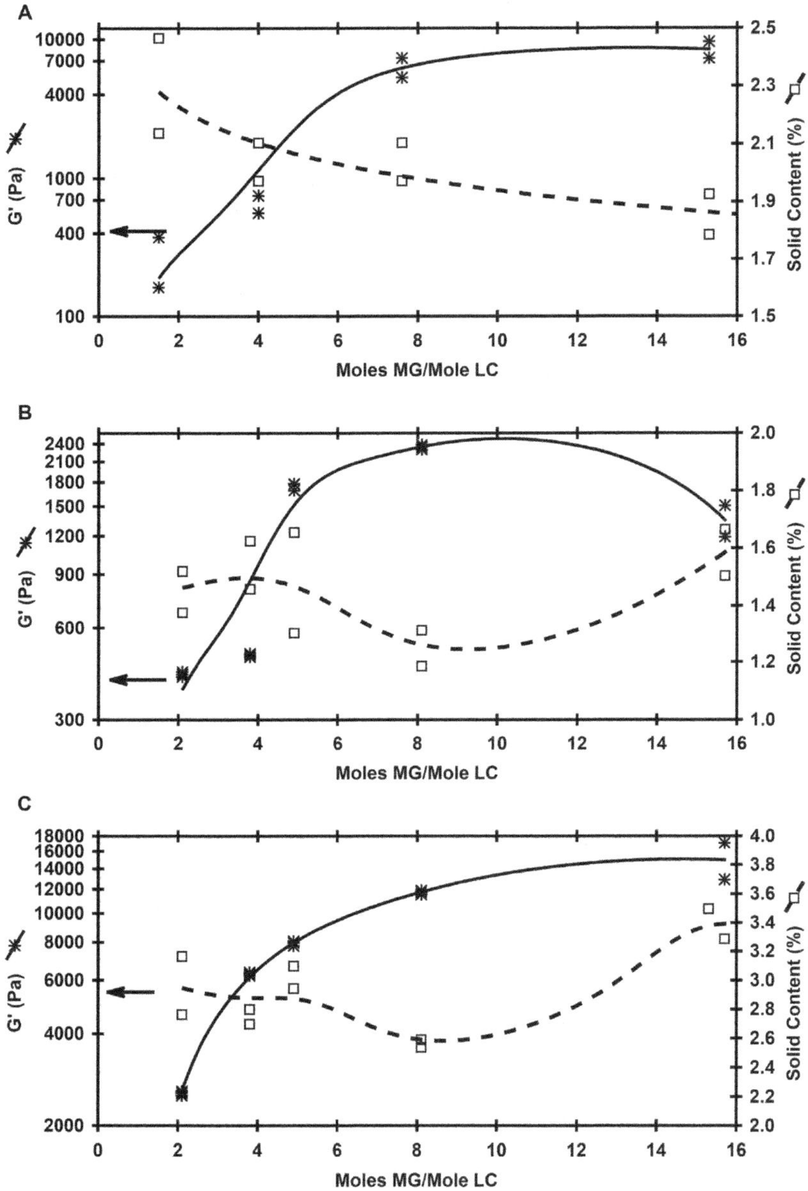

Figure 8.7 Elasticity (G') and percentage of solid content as a function of moles of MG per mole of LC of oleogels developed in safflower oil high in triolein. The measurements were done after 24 h at 15 °C. Panel A, oleogels developed using a constant MG concentration (2% wt/wt) and different LC concentrations between 0.25% and 2.5%. Adapted from ref. 43 with permission from Springer Nature, Copyright 2019. In panels B and C, the moles of MG per mole of LC in the oleogels were achieved keeping a constant total gelator mass (MG + LC) of 2% or 4%, respectively. Unpublished results from Bello-Santillán and Charó-Alonso.

Figure 8.8 Photographs of MG–LC oleogels corresponding to the rheological and solid content reported in Figure 9.7B. The total mass of gelators in the oleogels was 2% (wt/wt). The photographs were obtained by polarized light microscopy after 24 hours at 15 °C. The photographs indicate the corresponding moles of MG per mole of LC in the oleogels. Unpublished results from Bello-Santillán and Charó-Alonso.

birefringent plate-shaped and rod-shaped crystals increased while the large acicular spherulite MG crystals decreased as the LC proportion in the system decreased (*i.e.*, as the MG proportion increased in the oleogel; Figure 8.8). Additional studies at 4% total mass of gelators using the same moles of MG per mole of LC provided oleogels showing similar microstructural and rheological behavior (Figures 8.7C and 8.9) than the oleogels formulated with a constant total gelator mass of 2% (Figures 8.7B and 8.8). Evidently, in contrast with the oleogels developed with a total mass of gelator of 2%

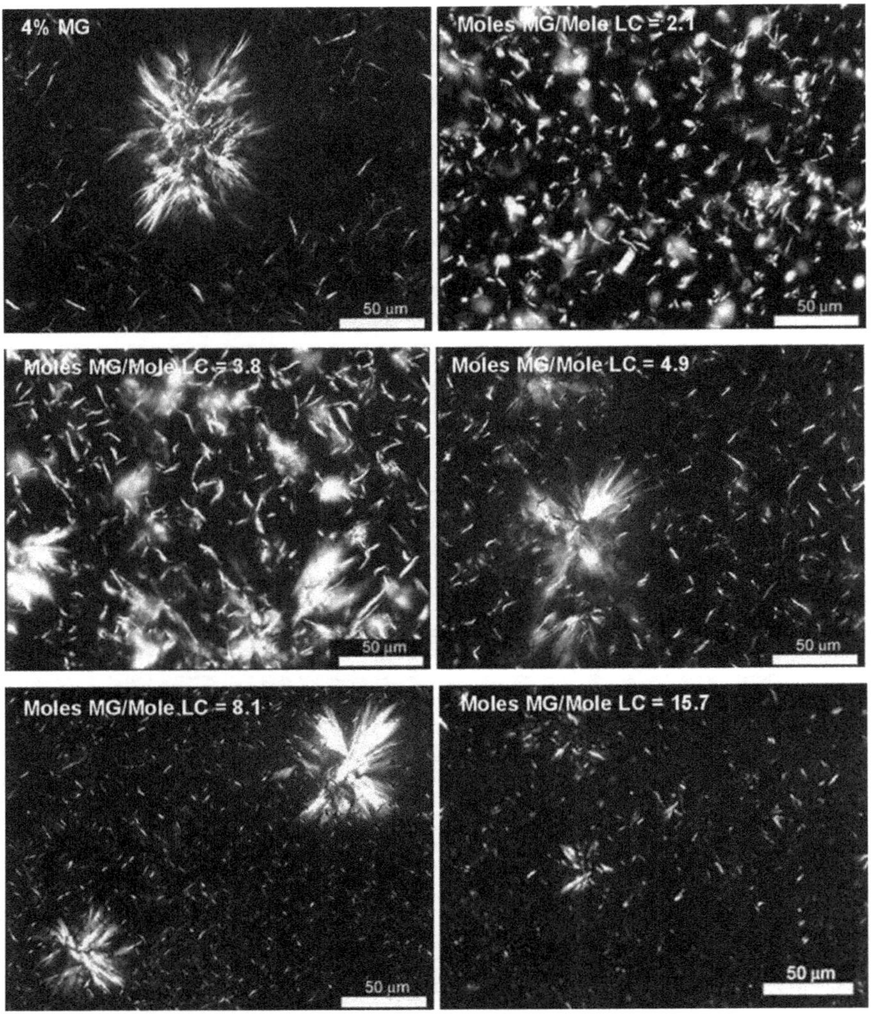

Figure 8.9 Photographs of MG–LC oleogels corresponding to the rheological and solid content reported in Figure 9.7C. The total mass of gelators in the oleogels was 4% (wt/wt). The photographs were obtained by polarized light microscopy after 24 hours at 15 °C. The photographs indicate the corresponding moles of MG per mole of LC in the oleogels. Unpublished results from Bello-Santillán and Charó-Alonso.

(Figure 8.7B), the higher G' achieved by the oleogels developed with 4% total gelator was associated with a higher solid content (Figure 8.7C) (Bello-Santillán and Charó-Alonso, unpublished results). This rheological behavior was associated with the presence of increasing amounts of small birefringent plate-shaped and rod-shaped crystals and a lower proportion of acicular spherulites. Previous DSC studies clearly showed that these small plate-shaped and rod-shaped crystals were associated with the development of MG–LC co-crystals.[43] Thus, at the highest moles of MG per mole of LC (*i.e.*, 15.7) the oleogel was structured just by small birefringent plate-shaped and rod-shaped crystals (Figure 8.9). We explained the rheological behavior of the MG–LC oleogels considering that the crystal network structured by co-crystals provided a larger solid surface to liquid phase ratio than the microstructure formed by large crystal spherulites. Thus, we proposed that the crystal network microstructure developed above 8.1 moles of MG per mole of LC physically trapped the vegetable oil more efficiently than the microstructure developed at lower moles of MG per mole of LC.[43] The overall result was that the oleogels formulated using more than 8.1 moles of MG per mole of LC had a self-standing structure of higher elastic properties that the oleogels developed at lower moles of MG per mole of LC (*i.e.*, at a higher proportion of LC). We concluded that there is a synergistic interaction between MG and LC that resulted in the formation of MG–LC co-crystals. At particular MG to LC proportions the co-crystals develop a crystal network with high oil binding efficiency and high elasticity.[43] This effect seems to be more evident in oleogels formulated with a MG + LC above 2% (Figures 8.7A and C and 8.9). The most relevant results of all these studies were that upon addition of just 10% of water/total mass of gelator, we increased the G' of oleogel formulated with 15.3 moles of MG per mole of LC from 3446 Pa (±0.0 Pa) up to 6420.1 Pa (±412.9 Pa).[43] Additionally, through DSC and X-ray studies we observed that in the oleogels with 10% of water/total mass of gelator, the development of the β polymorph was delayed. Thus, even after 12 months of storage at 15 °C the oleogels formulated with 15.3 moles of MG per mole of LC plus the addition of 10% of water/total mass of gelator, showed the presence of the Lα phase and limited formation of the β polymorph (Figure 8.10). In contrast, without the addition of water the same type of oleogels showed the characteristic melting profile and diffraction peaks of the β polymorph (4.55 Å to 4.65 Å) just after 24 h of storage, and particularly after 12 months of storage (Figure 8.10B).[43] As already discussed, the MG oleogels are metastable and during storage the Lα to β polymorph transition and subsequent β crystal agglomeration occur. These phenomena result in oil phase separation and deterioration of the mechanical and oil-binding properties of the MG oleogel.[41,42,46] Within this context, all the MG–LC oleogels, particularly those developed with the addition of 10% water, maintained good rheological properties and did not show oil phase separation even after 12 months of storage at 15 °C.[43] The increase in elasticity and stability to the β polymorph transition was more evident in MG-PL oleogels developed in mineral oil either with or without the addition of water.[43] As a corollary of these results we conclude that by using MG–LC mixtures we can obtain oleogels at lower gelator concentrations, with higher elasticity and more stability to oil phase separation than with oleogels developed just with MG. Additionally, the results obtained by our group showed that the polarity of the liquid phase is a useful variable to engineer the rheological properties and extend the storage stability of MG–LC oleogels. Consequently, through the synergistic interaction of LC and MGs we can extend and diversify the use of MG oleogels in food products.

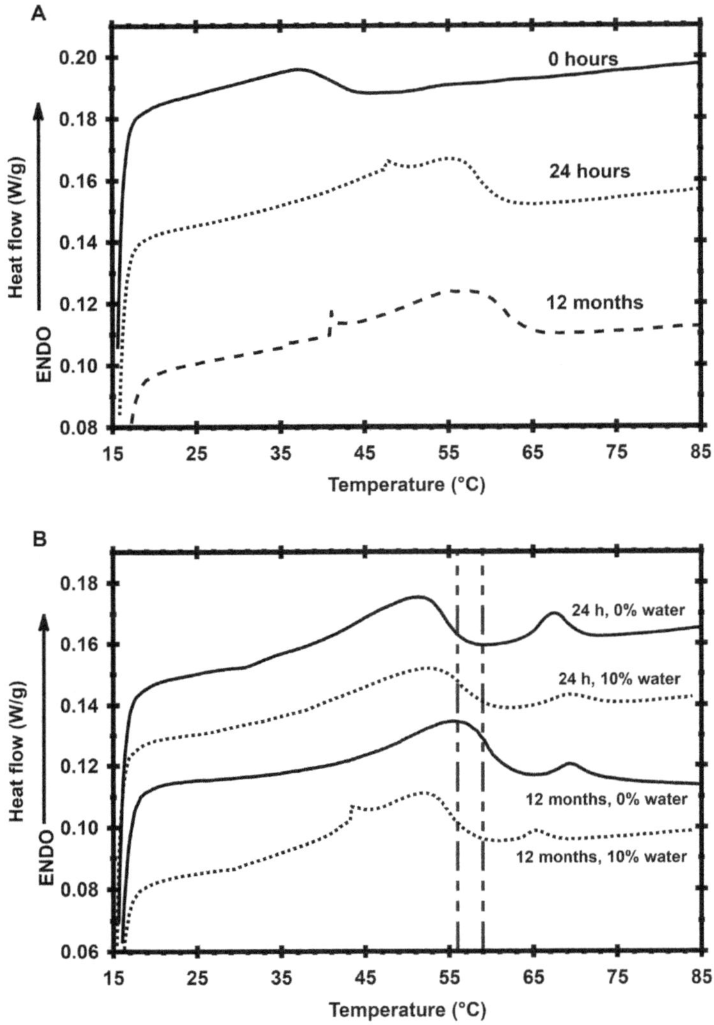

Figure 8.10 Melting thermograms of the 2% MG oleogel (A) and the 15.3 moles of MG per mole of LC oleogel (B). The thermograms were obtained after 0 h, 24 h, and 12 months of storage at 15 °C. The dotted lines indicate the melting temperature interval for the β polymorph present in the 2% MG solution in vegetable oil after 24 h or 12 months of storage at 15 °C (A). Adapted from ref. 43 with permission from Springer Nature, Copyright 2019.

8.3.2 Ethylcellulose Effect on MG Crystallization in Vegetable Oils

Ethylcellulose (EC) is a linear polysaccharide derived from cellulose, and therefore, formed by 1,4-β-D-glucose units. EC is produced through the ethoxylation of the –OH groups present in each glucose monomer at carbons

2, 3, or 6. Through the extent of ethoxylation, known as the substitution degree (SD), it is possible to establish EC's solubility profile. Thus, EC with a SD between 1.0 and 1.5 is water-soluble, while solubility in organic solvents, including vegetable oils, is achieved with a SD between 2.4 and 2.5.[82] We commonly characterize EC by its SD and by its molecular weight. Within this context, using EC soluble in vegetable oil Zetzl *et al.*[83] showed that oleogels formulated with 10% EC were stronger as the molecular weight and EC concentration increased.

EC, chitin, and chitosan are the only polymers known that can gel edible oils.[83–86] However, the self-assembly behavior of EC in vegetable oils is the most studied.[46,83,85,87] In EC vegetable oil solutions surfactant molecules like MGs, sorbitan esters, and fatty acids act as plasticizers by hydrogen-bonding with the free (*i.e.*, not substituted) hydroxyl groups on the polymer strands.[46,87–89] Through this process, the surfactant becomes embedded between polymer strands limiting the formation of junction zones between the polymer strands, increasing the free volume of the system, modifying the gelation temperature and the thermo-mechanical properties of the EC oleogel. Within this context, our group studied the effect of EC on MG self-assembly, measuring the thermo-mechanical and oil-binding properties of EC–MG oleogels and, particularly, the EC effect on the MG polymorphic transitions occurring during storage of EC–MG oleogels (14 days at 15 °C). This is in comparison with the behavior observed by the oleogels developed just with 2% and 8% MGs.[46] For this study, we used two types of distilled MG (>90% MG content) with the manufacturer codes Alphadim 90 SKB (SKB) and Alphadim 90 PKB (PKB) (Caravan Ingredients, Lenexa, KS). The MG composition for SKB was ≈79% of C18 and ≈2% of C16, while for PKB was ≈47% of C18 and ≈47% of C16.[46] It is important to note that the 2% MG concentration was below its minimal gelling concentration in the vegetable oil, while the 8% MG was well above the minimal gelling concentration.[25,43] The oleogels were developed just with 2% or 8% MG (*i.e.*, 0% EC) and with the addition of 6% EC (SD of 48–49.5%). At this concentration the EC was below its minimum concentration required for gel formation. The oleogels developed were evaluated measuring their melting thermograms by DSC, X-ray diffraction, oil loss as a function of storage time at 15 °C, and the oleogels' rigidity through complex modulus measurement (G^*; Figure 8.11).[46] Within this context, the results in Figure 8.11A show that when the MGs were used at a concentration below their minimal gelling concentration (*i.e.*, 2%) the G^* of the SKB–canola oil and the PKB–canola oil were lower than the G^* of the EC–vegetable oil system. However, the opposite occurred when we used the MGs at a concentration well above the minimal gelling concentration (*i.e.*, 8%), particularly with the MG that had a predominant concentration of C18 (*i.e.*, SKB; Figure 8.11B). We observed a similar behavior in the EC–MG oleogels, where, independent of the MG concentration, the EC–SKB oleogels achieved the highest G^*, particularly at 8% MG (Figure 8.11). Evidently, this showed that the EC and the MGs had a synergistic interaction that resulted in well-structured oleogels. Evidently,

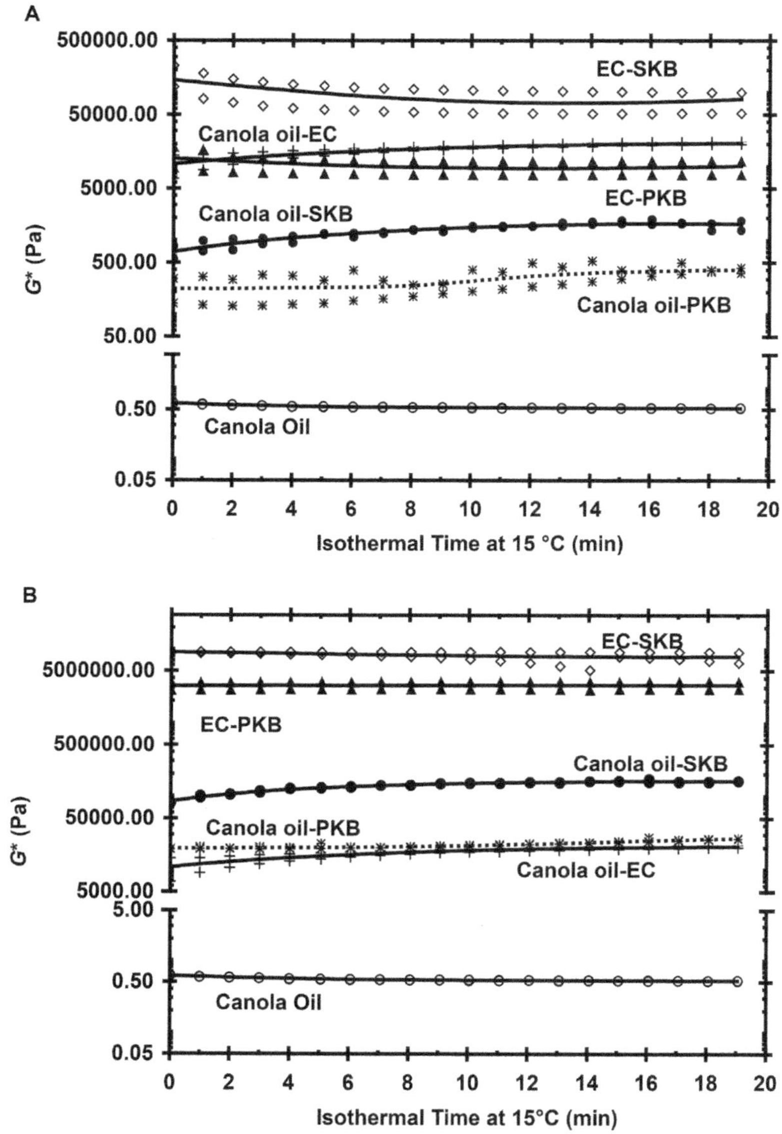

Figure 8.11 Complex modulus (G^*) as a function of storage time (15 °C) for the oleogels formulated in canola oil with monoglyceride types SKB or PKB in a mixture with 6% EC (EC–SKB, EC–PKB). The G^* for the canola oil and the oleogels formulated just with the monoglycerides (canola oil–SKB, canola oil–PKB) or with 6% EC (canola oil–EC) are also included. In panel A the monoglyceride concentration used was 2% and in panel B it was 8%.
Reproduced from ref. 46 with permission from Elsevier, Copyright 2021.

the interaction between the EC and the MG is dependent on both MG composition and MG concentration. These results were explained considering that the EC–MG interaction is at two levels. Initially, the hydrophilic "head" (*i.e.*, –OH groups) of MGs develops hydrogen-bonding with the free –OH groups of the EC strands. This results in a plasticizing effect on the EC and a decrease in the effective MG concentration in the oil phase. Consequently, when comparing the crystallization behavior of canola oil–MG and the canola oil–EC–MG systems formulated at the same MG concentration, and crystallized using the same crystallization temperature (*i.e.*, 15 °C) and cooling rate (*i.e.*, 8 °C per min), the supercooling for MG crystallization was lower in the system with EC than in the system without EC. This is because in the canola oil–EC–MG system the effective MG concentration was lower than in the canola oil–MG system.[46] At low supercooling crystal growth occurs at a faster rate than nucleation, while at higher supercooling the opposite occurs.[89] The overall result is that in MG oleogels formulated with EC, the "free" MG developed larger crystals while in MG oleogels formulated without EC the MG developed a large number of smaller crystals. As an example of the previous discussion, Figure 8.12 shows

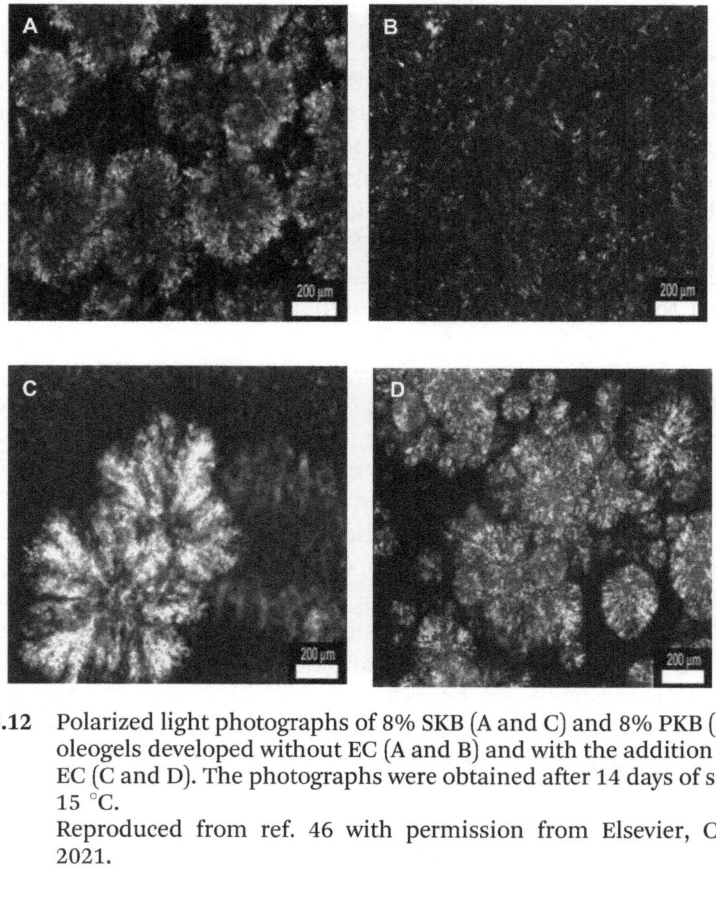

Figure 8.12 Polarized light photographs of 8% SKB (A and C) and 8% PKB (B and D) oleogels developed without EC (A and B) and with the addition of 6% of EC (C and D). The photographs were obtained after 14 days of storage at 15 °C.
Reproduced from ref. 46 with permission from Elsevier, Copyright 2021.

polarized light photographs of the oleogels formulated with 8% of SKB or 8% of PKB with or without 6% of EC. The photographs were obtained after 14 days of storage at 15 °C. In the analysis of the oleogel microstructure shown in Figure 8.12, the EC fibers were not observed. EC fibers and glass (from the cover slip) have similar refractive indexes under visible light, *i.e.*, 1.47 and 1.52, respectively.[90] This resulted in the impossibility of observing the entangled network of EC strands throughout the oleogels using polarized light or light microscopy. The crystallization of free MGs throughout the three-dimensional EC fiber network would act as an active filler resulting in oleogels with higher elasticity than in canola oil–EC and canola oil–MG oleogels (Figure 8.11). Additional results from this study obtained through X-ray diffraction and melting thermograms obtained in oleogels stored at 15 °C showed that, independent of the type of MG used, the EC increased the time at which the $L\alpha \rightarrow$sub-$\alpha \rightarrow \beta$ polymorph transition occurred. Despite the amorphous phase present in the oleogels associated with the vegetable oil, we detected diffraction peaks that allowed us to evaluate the EC effect on the MG crystallization behavior in the oleogels. Due to the higher number of crystals developed, these results were particularly evident in diffractograms for the 8% MG and 8% MG–EC oleogels. Thus, during the whole storage period the 8% SKB oleogels showed just the characteristic diffraction pattern of the sub-α phase (*i.e.*, diffraction peaks at 4.21 Å and 4.34 Å; Figure 8.13A). However, the corresponding melting thermogram did show the endotherm associated with the β phase. In contrast, during the first 7 days of storage at 15 °C the 8% SKB–EC oleogels showed that the MG remained mostly in the $L\alpha$ phase (peak at 4.11 Å), then crystallized into the sub-α phase (faint peak at 4.34 Å and a well-defined peak at 3.95 Å; Figure 8.13B). In contrast with the 8% SKB oleogels (*i.e.*, without EC), during the whole storage period the 8% SKB–EC oleogels did not show the melting profile nor the appearance of diffraction peaks associated with the presence of the β polymorph (Figure 8.13B).[46] The 8% PKB oleogels (without EC) at 0 days of storage showed the presence of the $L\alpha$ phase (diffraction peak at 4.23 Å; Figure 8.13C). However, after 2 days of storage the $L\alpha$ phase went to a β polymorphic transition (diffraction peaks at 4.37 Å, 4.47 Å, and 4.56 Å; Figure 8.13D). The melting thermograms of the 8% PKB oleogels confirmed these results.[46] In contrast, the melting thermogram of the 8% PKB–EC oleogels showed the development of the β polymorph after 2 days of storage. However, the X-ray diffraction analysis did not show evidence of the β phase in the oleogels during the whole storage period (Figure 8.13D). The authors explained these results considering that the amount of β crystals present in the 8% PKB–EC oleogels was too low, and the amorphous phase present (*i.e.*, vegetable oil) overshadowed the crystals' diffraction pattern.[46] We concluded that, in the MG organogels the EC slowed the rate for the sub-α to β polymorphic transition. As already discussed, this polymorphic transition is directly associated with the collapse of the oleogel three-dimensional crystal network. Thus, the study also included the measurement of the oil loss from the MG and MG–EC oleogels during the 14 days of storage at 15 °C.[46]

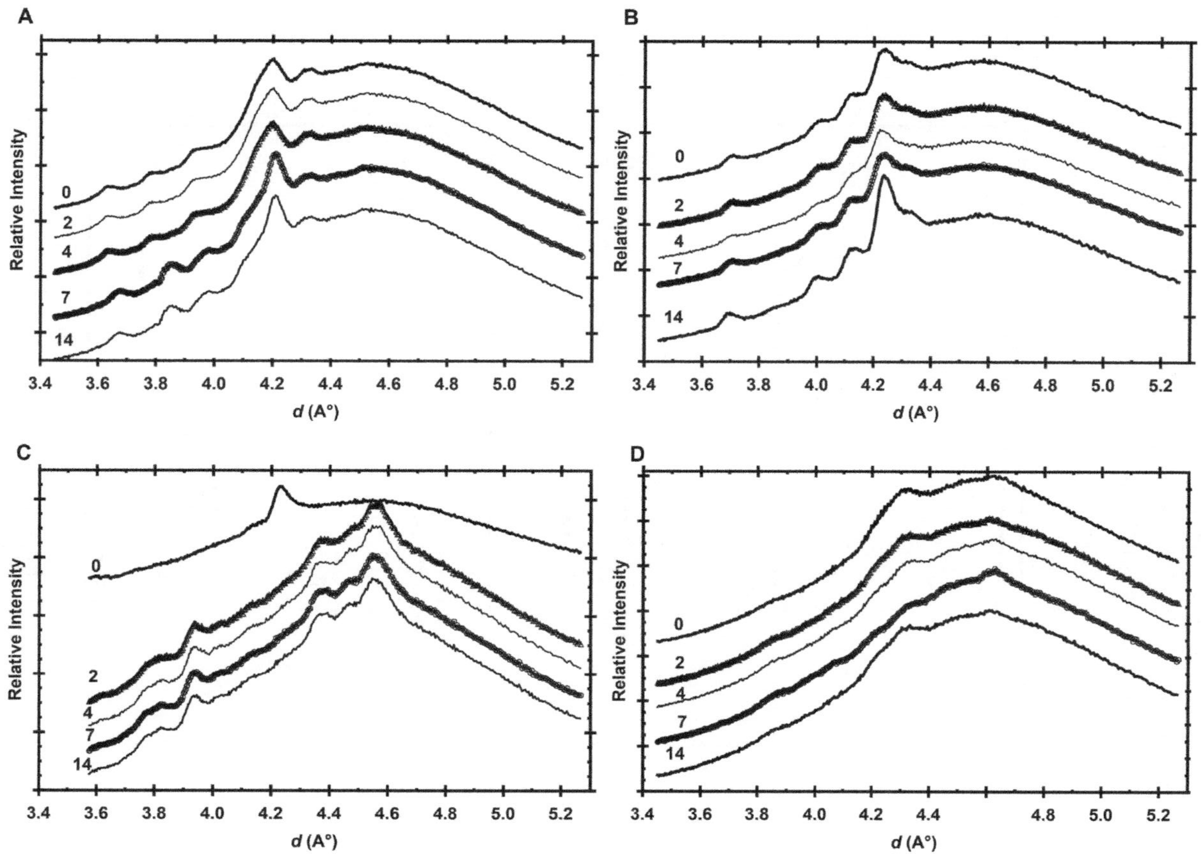

Figure 8.13 X-ray diffractograms of 8% SKB (A), 8% PKB (C), 8% SKB-6% EC (B), and 8% PKB-6% EC (D) oleogels. The diffractograms were obtained as a function of storage time (0 to 14 days) at 15 °C.
Reproduced from ref. 46 with permission from Elsevier, Copyright 2021.

The results obtained indicate that the EC decreased significantly the oleogels' oil loss, particularly in those formulated with 8% SKB (Figure 8.14). This study provided three major conclusions:[46]

1. The MG and EC had a synergistic interaction that resulted in oleogels with higher rheological properties and lower oil loss than the ones observed in MG oleogels without EC.

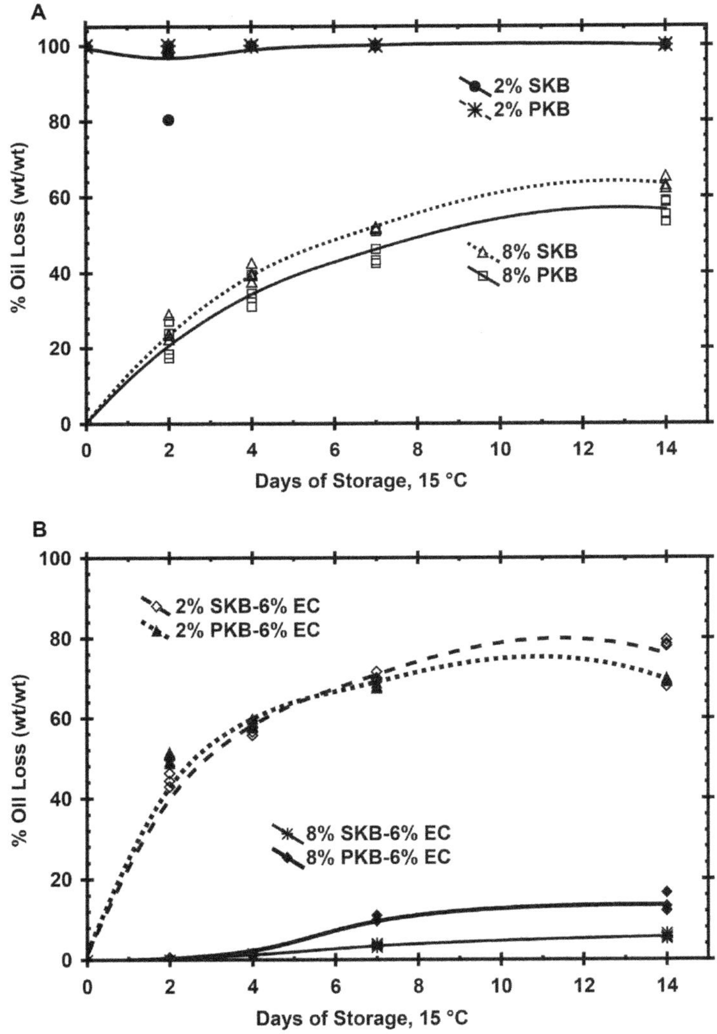

Figure 8.14 Percentage of oil loss (wt/wt) as a function of storage time (15 °C) for the 2% and 8% (wt/wt) SKB and PKB oleogels developed without EC (A) and with 6% EC (B).
Reproduced from ref. 46 with permission from Elsevier, Copyright 2021.

2. The EC slowed the molecular mobility and, therefore, the sub-α to β polymorphic transition and the subsequent β crystal agglomeration.
3. The EC–MG interaction, and therefore the higher rheological properties, the sub-α to β polymorphic transition, and the oleogel's oil loss depended on the MG concentration and composition. Thus, in the oleogels formulated with 8% SKB, the MG constituted mainly by C18, achieved the highest G^*, the sub-α to β polymorphic transition occurred slower, and the oleogels had the lowest oil loss during storage. This was in comparison with the oleogels formulated with 2% MG, particularly with the MG constituted by a mixture of C18 and C16 (*i.e.*, PKB).

8.4 Application of Oleogels in Food Systems and Conclusions

Most of the interest in oleogels is associated with the current challenge faced by the food industry to produce *trans*-free food products low in saturated fatty acids, but still having the functional and, particularly, the nutritional properties demanded by today's consumers. Some nutritional demands of consumers are associated with the consumption of vegetable and marine oils high in mono- and poly-unsaturated fatty acids. These edible oils do not have the physical and chemical properties that allow their incorporation into food systems, without having a deleterious effect on the product's texture and oxidative stability. Within this framework, we have to look to molecular self-assembly, and particularly to oleogels, with a two-fold perspective. First, as an alternative to develop structured edible oils that mimic the functional properties of saturated and *trans* fats. Secondly, as a process to develop new systems that provide foods with innovative and tailor-made functional and nutritional properties (*i.e.*, controlled and sustained release of flavor and volatile compounds, foods containing bioactive ingredients for health promotion or illness prevention, and edible food packaging). The main functional role of saturated and *trans* fats in food products is providing structure, and subsequently texture (hardness, spreadability, changes of consistency as a function of temperature). Additionally, saturated and *trans* fats provide the matrix for the incorporation of other phases (*i.e.*, air, water) to the food system, which results in significant textural changes. Consequently, the use of oleogels in food formulation largely depends on supplying these functional properties, achieving similar organoleptic properties and shelf-life than the products formulated with saturated and *trans* fats. Assuming we successfully develop functional oleogels, economic aspects (*i.e.*, gelator cost, availability) and food regulation issues must be additionally taken into account before the food industry considers their use. From our perspective, among the great variety of food-grade oleogelators studied by different research groups worldwide, vegetable waxes (*i.e.*, rice bran wax, candelilla wax, and sunflower wax) and MGs are the closest to achieving all these characteristics and have better functional properties to be used as fat mimics.

Thus, for margarine and spread applications, vegetable wax-based oleogels are the most suitable options.[10,91] However, because of the susceptibility of waxes to crystal aggregation during storage these vegetable wax-based oleogel products may not be stable over long storage periods. Additionally, vegetable wax imparts a waxy flavor to the food product, even at the low concentrations used to develop oleogels (*i.e.*, 1% to 5%).[16] Regarding the use of MG oleogels in food systems, as already discussed, their major drawback is the polymorphic change that occurs during storage resulting in the formation of β crystals and the subsequent crystal agglomeration. This polymorphic transition produces severe changes in the oleogel's microstructure that results in a deleterious effect in the rheology of the system and, eventually, oil phase separation. However, the research done by our group[23,25,43,46] has shown that through the development of mixed oleogels with LC (MG–LC) or with EC (MG–EC), we can delay the sub-α to β polymorphic transition and the subsequent oil phase separation. For the MG–LC systems, our results show that these gelators can develop a co-crystal that structure the oleogels' crystal network that physically confines the vegetable oil. All these phenomena were more evident in MG–LC oleogels developed with moles of MG per mole of LC above 8.1, using a total gelator concentration (MG + LC) of 4% and the addition of just 10% of water/total mass of gelator. Thus, the oleogels formulated with 15.3 moles of MG per mole of LC and the addition of 10% of water/total mass of gelator, showed crystals in the Lα phase and limited formation of the β polymorph even after 12 months of storage at 15 °C. In contrast, without the addition of water the same type of MG–LC oleogels developed the β polymorph just after 24 hours of storage, followed by severe oil phase separation, particularly after 12 months of storage. In the same way, EC showed a synergistic interaction with commercial MGs (*i.e.*, SKB and PKB) that resulted in MG–EC oleogels with higher rheological properties, lower oil loss, and slower sub-α to β polymorphic transition than the ones observed in MG oleogels. The results showed that the MG–EC interaction depended on MG concentration and composition. Thus, the MG–EC oleogels formulated essentially by 1-mono-stearoyl-glycerol (C18) achieved the highest rheological properties, the sub-α to β polymorphic transition occurred at a slower rate and had the lowest oil loss during storage. This was in comparison with the oleogels formulated with 2% MG, particularly with those formulated with a MG constituted by a mixture of C18 and 1-mono-palmitoyl-glycerol (C16). Then, using MG–LC and MG–EC mixtures we can obtain oleogels at a lower gelator concentration, with higher elasticity and more stability to oil phase separation than with oleogels developed just with MG. Consequently, through the synergistic interaction of LC and EC with MGs we can extend and diversify the use of MG oleogels in food products.

Acknowledgements

The results regarding the lecithin effect on monoglyceride self-assembly originated in Dr J. F. Toro-Vazquez laboratory were supported by Consejo Nacional de Ciencia y Tecnología (CONACYT) through the grant CB-280981-2018.

References

1. C. A. Martin, M. C. Milinsk, J. V Visentainer, M. Matsushita and N. E. De-Souza, *J. Am. Oil Chem. Soc.*, 2007, **79**, 343–350.
2. A. Ascherio, M. B. Katan, P. L. Zock, M. J. Stampfer and W. C. Willett, *N. Engl. J. Med.*, 1999, **340**, 1994–1998.
3. F. B. Hu, M. J. Stampfer, J. E. Manson, E. Rimm, G. A. Colditz, B. A. Rosner, C. H. Hennekens and W. C. Willett, *N. Engl. J. Med.*, 1997, **337**, 1491–1499.
4. U. Risérus, W. C. Willett and F. B. Hu, *Prog. Lipid Res.*, 2008, **48**, 44–51.
5. R. Micha and D. Mozaffarian, *Nat. Rev. Endocrinol.*, 2009, **5**, 335–344.
6. Ten threats to global health in 2019, https://www.who.int/news-room/spotlight/ten-threats-to-global-health-in-2019, (accessed 30 January 2021).
7. I. A. Brouwer, *Nutr. Bull.*, 2020, **45**, 271–280.
8. 57th Directing Council, 71st Session of the Regional Committee of WHO for the Americas - PAHO/WHO | Pan American Health Organization, https://www.paho.org/en/governing-bodies/directing-council/57th-directing-council-71st-session-regional-committee-who, (accessed 30 January 2021).
9. Global Trans Fat Elimination: Launch of WHO's 2020 Progress Report | NCD Alliance, https://ncdalliance.org/news-events/news/global-trans-fat-elimination-launch-of-who's-2020-progress-report, (accessed 30 January 2021).
10. A. R. Patel and K. Dewettinck, *Food Funct.*, 2016, **7**, 20.
11. C. Park and F. Maleky, *Front. Sustainable Food Syst.*, 2020, **4**, 1–8.
12. A. J. Martins, A. A. Vicente, R. L. Cunha and M. A. Cerqueira, *Food Funct.*, 2018, **9**, 758–773.
13. L. S. K. Dassanayake, D. R. Kodali, S. Ueno and K. Sato, *J. Am. Oil Chem. Soc.*, 2009, **86**, 1163–1173.
14. C. Diem Doan, C. Ming, M. De Vrieze, F. Lynen, S. Danthine, A. Brown, K. Dewettinck and A. R. Patel, *Food Chem.*, 2017, **214**, 717–725.
15. F. M. Alvarez-Mitre, J. A. Morales-Rueda, E. Dibildox-Alvarado, M. A. Charó-Alonso and J. F. Toro-Vazquez, *Food Res. Int.*, 2012, **49**, 580–587.
16. A. I. Blake, J. F. Toro-Vazquez and H.-S. Hwang, *Edible Oleogels*, Elsevier, 2018, pp. 133–171.
17. H.-S. Hwang, S. Kim, M. Singh, J. K. Winkler-Moser and S. X. Liu, *J. Am. Oil Chem. Soc.*, 2012, **89**, 639–647.
18. J. F. Toro-Vazquez, J. Morales-Rueda, A. Torres-Martínez, M. A. Charó-Alonso, V. A. Mallia and R. G. Weiss, *Langmuir*, 2013, **29**, 7642–7654.
19. A. Bot and W. G. M. Agterof, *J. Am. Oil Chem. Soc.*, 2006, **83**, 513–521.
20. A. Bot, R. Den Adel and E. C. Roijers, *J. Am. Oil Chem. Soc.*, 2008, **85**, 1127–1134.
21. R. Den Adel, P. C. M. Heussen and A. Bot, *J. Phys.: Conf. Ser.*, 2010, **247**, 1–12.

22. L. Han, L. Li, B. Li, L. Zhao, G. Q. Liu, X. Liu and X. Wang, *J. Am. Oil Chem. Soc.*, 2014, **91**, 1783–1792.
23. M. Martínez-Ávila, A. De la Peña-Gil, F. M. Álvarez-Mitre, M. A. Charó-Alonso and J. F. Toro-Vazquez, *J. Am. Oil Chem. Soc.*, 2019, **96**, 273–289.
24. C. H. Chen, I. Van Damme and E. M. Terentjev, *Soft Matter*, 2009, **5**, 432–439.
25. A. López-Martínez, J. A. Morales-Rueda, E. Dibildox-Alvarado, M. A. Charó-Alonso, A. G. Marangoni and J. F. Toro-Vazquez, *Food Res. Int.*, 2014, **64**, 946–957.
26. F. G. Gandolfo, A. Bot and E. Flöter, *J. Am. Oil Chem. Soc.*, 2004, **81**, 1–6.
27. J. A. Morales-Rueda, E. Dibildox-Alvarado, M. A. Charó-Alonso, R. G. Weiss and J. F. Toro-Vazquez, *Eur. J. Lipid Sci. Technol.*, 2009, **111**, 207–215.
28. R. S. H. Lam and M. A. Rogers, *Cryst. Growth Des.*, 2011, **11**, 3593–3599.
29. S. S. Sagiri, B. Behera, R. R. Rafanan, C. Bhattacharya, K. Pal, I. Banerjee and D. Rousseau, *Soft Mater.*, 2014, **12**, 47–72.
30. J. F. Toro-Vazquez, J. A. Morales-Rueda, E. Dibildox-Alvarado, M. Charó-Alonso, M. Alonzo-Macias and M. M. González-Chávez, *J. Am. Oil Chem. Soc.*, 2007, **84**, 989–1000.
31. J. F. Toro-Vazquez, R. Mauricio-Pérez, M. M. González-Chávez, M. Sánchez-Becerril, J. J. Ornelas-Paz and J. D. Pérez-Martínez, *Food Res. Int.*, 2013, **54**, 1360–1368.
32. M. Chopin-Doroteo, J. A. Morales-Rueda, E. Dibildox-Alvarado, M. A. Charó-Alonso, A. de la Peña-Gil and J. F. Toro-Vazquez, *Food Biophys.*, 2011, **6**, 359–376.
33. A. R. Patel, M. Babaahmadi, A. Lesaffer and K. Dewettinck, *J. Agric. Food Chem.*, 2015, **63**, 4862–4869.
34. M. A. Rogers and A. G. Marangoni, *Cryst. Growth Des.*, 2008, **8**, 4596–4601.
35. S. Abraham, Y. Lan, R. S. H. Lam, D. A. S. Grahame, J. Jae, H. Kim, R. G. Weiss and M. A. Rogers, *Langmuir*, 2012, **28**, 4955–4964.
36. E. Co and A. G. Marangoni, *J. Am. Oil Chem. Soc.*, 2013, **90**, 529–544.
37. U.S. Food and Drug Administration, *CFR - Code of Federal Regulations Title 21*, Chapter I, Subchapter B, Part 178 Indirect Food Additives: Adjuvants, Production aids, and Sanitizers.
38. D. J. Abdallah and R. G. Weiss, *Langmuir*, 2000, **16**, 352–355.
39. A. De La Peña-Gil, J. F. Toro-Vazquez and M. A. Rogers, *Food Biophys.*, 2016, **11**, 283–291.
40. N. K. Ojijo, I. Neeman, S. Eger and E. Shimoni, *J. Sci. Food Agric.*, 2004, **84**, 1585–1593.
41. C. H. Chen and E. M. Terentjev, *Langmuir*, 2009, **25**, 6717–6724.
42. C. H. Chen and E. M. Terentjev, *Edible Oleogels*, Elsevier, 2018, pp. 103–131.
43. M. Aguilar-Zárate, A. De La Peña-Gil, F. M. Álvarez-Mitre, M. A. Charó-Alonso and J. F. Toro-Vazquez, *Food Biophys.*, 2019, **14**, 326–345.
44. L. J. Han, L. Li, L. Zhao, B. Li, G. Q. Liu, X. Q. Liu and X. De Wang, *Food Res. Int.*, 2013, **53**, 42–48.

45. M. Pernetti, K. van Malssen, D. Kalnin and E. Flöter, *Food Hydrocolloids*, 2007, **21**, 855–861.
46. A. López-Martínez, M. A. Charó-Alonso, A. G. Marangoni and J. F. Toro-Vazquez, *Food Res. Int.*, 2015, **72**, 37–46.
47. A. K. Rodríguez-Hernández, J. D. Pérez-Martínez, J. A. Gallegos-Infante, J. F. Toro-Vazquez and J. J. Ornelas-Paz, *Carbohydr. Polym.*, 2021, **252**, 117–171.
48. G. Avendaño-Vásquez, A. De la Peña-Gil, M. E. Charó-Alvarado, M. A. Charó-Alonso and J. F. Toro-Vazquez, *Front. Sustainable Food Syst.*, 2020, **4**, 1–21.
49. C. H. Chen and E. M. Terentjev, *Edible Oleogels: Structure and Health Implications*, Elsevier, 2011, pp. 173–201.
50. U.S. Food and Drug Administration, *CFR - Code of Federal Regulations Title 21*, Chapter I, Subchapter B, Part 184 Direct Food Substances Affirmed as Generally Recognized as Safe, Subpart B Listing of Specific Substances Affirmed as GRAS.
51. B. J. F. Hudson, *Encyclopedia of Food Sciences and Nutrition*, Elsevier, 2003, pp. 2297–2300.
52. L. V. Fregolente, C. B. Batistella, R. M. Filho and M. R. W. Maciel, *Twenty-Seventh Symposium on Biotechnology for Fuels and Chemicals*, Humana Press, Totowa, NJ, USA, 2007, pp. 680–693.
53. N. J. Krog and F. Sparso, *Food Emulsions*, Marcel Dekker, New York, NY, USA, 2004, pp. 45–91.
54. S. P. J. N. Senanayake and F. Shahidi, *Bailey's Industrial Oil and Fat Products*, ed. F. Shahidi, Wiley, 2005, pp. 555–584.
55. T. L. G. Carlson, K. Larsson, N. Dinh-Nguyen and N. Krog, *Starch-Stärke*, 1979, **31**, 222–224.
56. K. Larsson, *Food Struct.*, 1982, **1**, 52–56.
57. L. Sagalowicz, M. E. Leser, H. J. Watzke and M. Michel, *Trends Food Sci. Technol.*, 2006, **17**, 204–214.
58. G. Cassin, C. De Costa, J. P. M. Van Duynhoven and W. G. M. Agterof, *Langmuir*, 1998, **14**, 5757–5763.
59. A. G. Marangoni, S. H. J. Idziak, C. Vega, H. Batte, M. Ollivon, P. S. Jantzi and J. W. E. Rush, *Soft Matter*, 2007, **3**, 183–187.
60. A. G. Marangoni, S. H. J. Idziak and J. W. E. Rush, *Food Biophys.*, 2008, **3**, 241–245.
61. T. Maruyama, I. Niiya, M. Okada and T. Matsumoto, *J. Jpn. Oil Chem. Soc.*, 1979, **28**, 486–490.
62. E. S. Lutton, *J. Am. Oil Chem. Soc.*, 1971, **48**, 778–781.
63. N. Krog and K. Larsson, *Chem. Phys. Lipids*, 1968, **2**, 129–143.
64. R. Mezzenga, C. Meyer, C. Servais, A. I. Romoscanu, L. Sagalowicz and R. C. Hayward, *Langmuir*, 2005, **21**, 3322–3333.
65. L. K. Shrestha, T. Sato, D. P. Acharya, T. Iwanaga, K. Aramaki and H. Kunieda, *J. Phys. Chem. B*, 2006, **110**, 12266–12273.
66. N. K. O. Ojijo, E. Kesselman, V. Shuster, S. Eichler, S. Eger, I. Neeman and E. Shimoni, *Food Res. Int.*, 2004, **37**, 385–393.

67. A. H. Poghosyan, L. H. Arsenyan and A. A. Shahinyan, *Langmuir*, 2012, **29**, 29–37.
68. P. G. Petrov, S. V. Ahir and E. M. Terentjev, *Langmuir*, 2002, **18**, 9133–9139.
69. J. P. M. Van Duynhoven, I. Broekmann, A. Sein, G. M. P. Van Kempen, G. J. W. Goudappel and W. S. Veeman, *J. Colloid Interface Sci.*, 2005, **285**, 703–710.
70. C. H. Chen and E. M. Terentjev, *Langmuir*, 2010, **26**, 3095–3105.
71. B. A. C. Horta and P. H. Hünenberger, *J. Am. Chem. Soc.*, 2011, **133**, 8464–8466.
72. J. F. Toro-Vazquez and J. D. Pérez-Martínez, *Molecular Gels: Structure and Dinamics*, The Royal Society of Chemistry, 2018, pp. 57–87.
73. A. López-Martínez, J. A. Morales-Rueda, E. Dibildox-Alvarado, M. A. Charó-Alonso, A. G. Marangoni and J. F. Toro-Vazquez, *Food Res. Int.*, 2014, **64**, 946–957.
74. R. Pichot, R. L. Watson and I. T. Norton, *Int. J. Mol. Sci.*, 2013, **14**, 11767–11794.
75. J. Israelachvili and I. Ladyzhinski, *NATO Sci. Ser. II*, 2004, **160**, 1–28.
76. P. Walde, A. M. Giuliani, C. A. Boicelli and P. L. Luisi, *Chem. Phys. Lipids*, 1990, **53**, 265–288.
77. P. K. Okuro, A. A. Malfatti-Gasperini, L. H. Fasolin, A. A. Vicente and R. L. Cunha, *J. Surfactants Deterg.*, 2020, **23**, 725–735.
78. R. Scartazzini and P. L. Luisi, *J. Phys. Chem.*, 1988, **92**, 829–833.
79. P. K. Okuro, I. Tavernier, M. D. B. Sintang, A. G. Skirtach, A. A. Vicente, K. Dewettinck and R. L. Cunha, *Food Funct.*, 2018, **9**, 1755–1767.
80. L. Han, L. Li, B. Li, L. Zhao, G.-Q. Liu, X. Liu and X. Wang, *J. Am. Oil Chem. Soc.*, 2014, **91**, 1783–1792.
81. N. Gaudino, S. M. Ghazani, S. Clark, A. G. Marangoni and N. C. Acevedo, *Food Res. Int.*, 2019, **116**, 79–89.
82. S. Guo, M. Lv, Y. Chen, T. Hou, Y. Zhang, Z. Huang, Y. Cao, M. Rogers and Y. Lan, *Food Funct.*, 2020, **11**, 2048–2057.
83. A. K. Zetzl, A. G. Marangoni and S. Barbut, *Food Funct.*, 2012, **3**, 327–337.
84. W. Koch, *Ind. Eng. Chem.*, 1937, **29**, 687–690.
85. M. Davidovich-Pinhas, S. Barbut and A. G. Marangoni, *Carbohydr. Polym.*, 2015, **117**, 869–878.
86. J. M. Milani and M. H. Naeli, *Handbook of Chitin and Chitosan*, Elsevier, 2020, pp. 383–406.
87. M. Davidovich-Pinhas, S. Barbut and A. G. Marangoni, *Carbohydr. Polym.*, 2015, **127**, 355–362.
88. A. Haj Eisa, S. Laufer, J. Rosen-Kligvasser and M. Davidovich-Pinhas, *Eur. J. Lipid Sci. Technol.*, 2020, **122**, 1–10.
89. A. G. Marangoni, *Fat Crystal Networks*, ed. A. G. Marangoni, Marcel Dekker, New York, NY, USA, 2005, pp. 21–82.
90. L. W. McKeen, *Film Properties of Plastics and Elastomers*, Elsevier, 2012, pp. 353–378.
91. H.-S. Hwang, M. Singh, E. L. Bakota, J. K. Winkler-Moser, S. Kim and S. X. Liu, *J. Am. Oil Chem. Soc.*, 2013, **90**, 1705–1712.

CHAPTER 9

Functionality of Natural Waxes in Hybrid Fat Crystal Networks

FILIP VAN BOCKSTAELE,*[a,b] MARIE ROMANUS,[a,b] IVANA A. PENAGOS[a,b] AND KOEN DEWETTINCK[a,b]

[a] Ghent University, Department of Food Technology, Safety and Health, Food Structure and Function (FS&F) Research Group, Coupure Links 653, Ghent, Belgium; [b] Vandemoortele Centre 'Lipid Science and Technology', Coupure Links 653, Ghent, Belgium
*Email: filip.vanbockstaele@ugent.be

9.1 Introduction

Saturated and *trans* fatty acids greatly contribute to the properties and functionality of solid fat products like shortenings, fat spreads, and margarines. More precisely, they form a fat crystal network that entraps the liquid oil and prevents its exudation. Although saturated and *trans* fatty acids are important structurants, their intake should be restricted from a nutritional perspective due to health concerns.[1,2]

Trans fats are known to increase the risk of cardiovascular health diseases, the most important cause of death in the European Union (EU) and in the United States of America (US).[3,4] Current EU legislation declares that the amount of *trans* fats in food cannot exceed 2 grams per 100 grams of fat, excluding naturally occurring *trans* fats.[5] Also, the World Health Organization (WHO) advises against the intake of industrial *trans* fats.[6]

An obvious alternative to *trans* fats is to replace them with saturated fatty acids (SFAs). However, strong evidence supports that partial replacement of SFAs with oils rich in *cis*-polyunsaturated fatty acids reduces the risk of

Food Chemistry, Function and Analysis No. 31
Development of *Trans*-free Lipid Systems and their Use in Food Products
Edited by Jorge F. Toro-Vazquez

Published by the Royal Society of Chemistry, www.rsc.org

cardiovascular health diseases.[1,7] Consequently, various organizations advise the consumption of less SFAs. For instance, based on the advice of the Food and Agriculture Organization of the United Nations (FAO), the WHO recommends reducing the intake of SFAs to less than 10% of the total energy intake.[8,9]

A major problem when replacing *trans* fats and SFAs with unsaturated oil is that there is usually a loss in structure. Hence, in order to obtain a semi-solid behavior on a macroscopic scale, an alternative oil structuring technique is needed. Within this context, oleogelation is one of the most studied approaches. A classic oleogel is a bicontinuous colloidal system that consists of a micro-heterogeneous solid (*i.e.*, gelator) and an edible liquid oil. The gelator entraps the liquid oil in a three-dimensional matrix of randomly entangled fiber-like or plate-like structures to form a solid-like material.[1,10–12] Gelators are surface-active and are composed of two parts, a lipophilic part and an interacting part of functional groups, which induce the formation of supramolecular networks of self-assembled molecules.[13] Not only should a gelator be economical at the gelling concentration and therefore have efficient gelling capacities, but the formed oleogel should also exhibit similar material properties as the fat material that it is supposed to replace. Additionally, gelators should be readily available and approved for use in food, *i.e.*, food-grade. Considering the previous aspects, several compounds might be used as oleogelators, including monoglycerides, phytosterols, fatty acids, and waxes.[1,12]

9.2 Waxes from Vegetal and Insect Sources

Natural waxes from vegetal and insect sources are organogelators that have been studied intensively during the last 10 to 15 years. Overall, natural waxes, particularly vegetal waxes, are efficient gelators of vegetable oils at concentrations lower than 10 wt%. The resulting oleogels usually have functional properties suitable for their use in edible and cosmetic products. Their functionality is often linked to their chemical composition. Some vegetable waxes have a relatively low price and are commercially abundant, although this does not apply to all of them.[14–17] A few plant waxes, like sunflower wax, are by-products of agricultural and industrial processes, so new applications might increase their market value.[18,19]

From a chemical perspective, a wax has been described as an ester of a monohydric long chain alcohol with a long chain acid.[20] However, when referring to natural waxes, the terminology "wax" is used to describe the complex mixtures of long chain compounds that cover the surfaces of several organisms. The composition of these mixtures varies among organisms and contains different chemical substances such as hydrocarbons, free fatty alcohols, free fatty acids, and wax esters (Scheme 9.1). Ketones, mono-, di-, triacylglycerols, and sterol esters can also be present in various concentrations.[13,20,21]

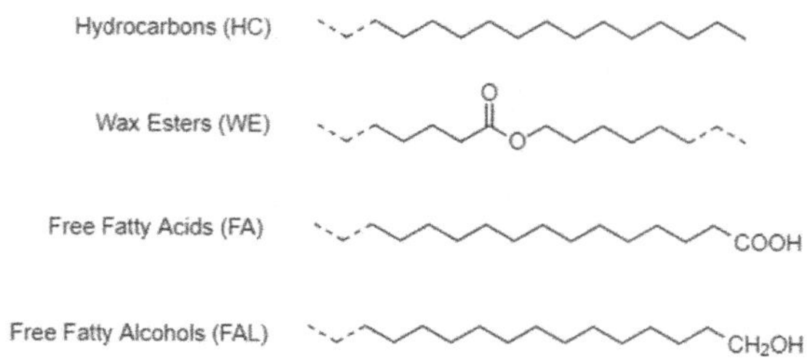

Scheme 9.1 Structures of the main chemical components in natural waxes.

The exact composition differs between waxes of different sources. For example, berry wax has different chemical constituents than candelilla wax and beeswax. In addition, there are chemical differences when comparing the same type of wax from different suppliers. Moreover, minor components and impurities considerably affect the gelation property of natural waxes.[15] Table 9.1 provides a summary of several physicochemical properties of the most commonly studied waxes obtained from insect or vegetable sources: beeswax (BW), berry wax (BEW), candelilla wax (CW), carnauba wax (CRW), fruit wax (FW), rice bran wax (RBW), shellac wax (SHW), and sunflower wax (SW).

The chemical composition also controls the melting and crystallization behavior of the waxes, which can be studied by differential scanning calorimetry (DSC). The crystallization and melting thermograms of some commercial waxes are presented in Figure 9.1.

CRW shows the highest melting point (83 °C), followed by SW (78 °C). White beeswax (WBW), yellow beeswax (YBW), and CW have melting temperatures of 65 °C, 65 °C, and 66 °C, respectively, and BEW has the lowest melting temperature (46 °C). The onset temperature of crystallization of the waxes decreases in the order of CRW > SW > CW > YBW > WBW > BEW, following the same order as the melting temperature. In general, waxes with a higher melting point also have a larger crystallization and melting enthalpy. It can also be noticed that the crystallization and melting thermograms of SW show a single-peak, while all other neat waxes show multiple peaks. These multiple peaks are caused by the complex chemical composition (Table 9.1). In contrast, SW mainly consists of wax esters, resulting in a unique peak in the crystallization thermogram.

The functionality of natural waxes in oleogels is determined by their chemical composition.[21] The critical gelling concentration and gel strength both depend on the waxes' chemical constituents. It has been found that the higher the portion of wax esters in the wax, the stronger and more brittle oleogels become. Furthermore, hydrocarbons and free fatty acids are reported to be responsible for consistent and stable oleogels, with higher critical stress and higher consistency index. Short-chain free fatty acids ($C < 16$) lower the crystallization temperature of natural waxes, which could limit the use in food products, while very long-chain free fatty alcohols ($C > 30$) hinder gelation.[21]

Table 9.1 Chemical composition and physical properties of natural waxes commonly used in the development of oleogels.

	Chemical composition[a] (%)				Physical properties		
	WE	FA	FAL	HC	CGC (%)	Tm (°C)	References
BEW[bc]	0–1	94–96	3–5	0–1	1–6	—	21, 22
CW[c]	15–39	7–11	1–15	42–75	1–2	60–73	15, 18, 21, 23–25
CRW[c]	54–85	3–10	2–38	0–3	2–5	80–85	15, 18, 21, 22, 26
FW[c]	0–1	29–43	55–70	0–1	5–>10	39–40	13, 15, 22
RBW[c]	92–97	0–8	0–1	0–1	0.5–5	77–83	13, 15, 18, 21, 27, 28
SW[c]	96–100	0–16	0–13	0–7	0.5–1	74–77	15, 18, 21, 29
BW[d]	57–71	8–18	1–12	12–28	1–3	—	15, 21, 22, 30–33
SHW[d]	—	—	—	—	2–5	—	15, 34

[a]Wax esters (WE), fatty acids (FA), free fatty alcohols (FAL), hydrocarbons (HC), critical gelling concentration (CGC), and melting temperature (Tm).
[b]Berry wax (BEW), candelilla wax (CW), carnauba wax (CRW), fruit wax (FW), rice bran wax (RBW), sunflower wax (SW), beeswax (BW), and shellac wax (SHW).
[c]Waxes from vegetable sources.
[d]Waxes from insect sources.

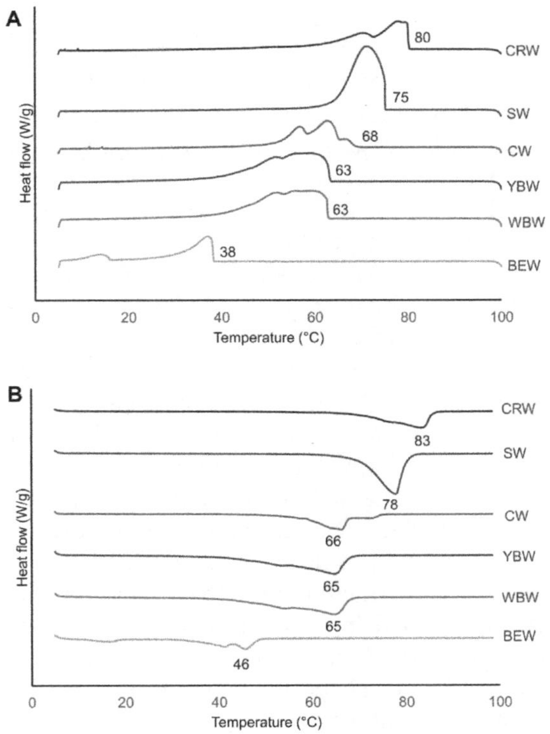

Figure 9.1 Crystallization (A) and melting (B) thermograms of neat waxes obtained by differential scanning calorimetry at a cooling or heating rate of 5 °C min^{-1}. For each crystallization thermogram, the numbers close to the baseline are the onset crystallization temperatures. For each melting thermogram, the numbers below the peak are the peak melting temperatures.[35] Yellow beeswax, YBW; white beeswax, WBW; berry wax, BEW; candelilla wax, CW; carnauba wax, CRW; sunflower wax, SW.

9.2.1 Waxes from Vegetable Sources

BEW is extracted from the berry fruit peels of the *Rhus verniciflua* tree and is commonly used in natural cosmetics.[36] BEW is not yet approved for use in food products. The wax consists of more than 90% of short chain fatty acids, including C16 and C18, with approximately 80% of palmitic acid. Hence, it has a low melting point, especially in comparison to other waxes. BEW in rice bran oil (RBO) crystallizes in small platelets due to the predominant short chain fatty acids.[21]

CW is obtained from the leaves of the candelilla plant, *Euphorbia antisyphilitica*, and further purified. The substance is identified in the CAS registry with the code 8006-44-8. The use of CW as a food additive is approved in EU as a glazing agent under the E-number E902. E-numbers are codes used in the European Union to identify food additives. They are used in legislation and often found in food packaging. On the other hand, the FDA (US Food and Drug Administration) has also approved the use of CW, attributing a GRAS status (generally recognized as safe) under the code of federal regulations 21CFR184.1976. CW has thermoreversible properties and consists of a large amount of hentriacontane, a self-assembling *n*-alkane, mainly supplemented by other *n*-alkanes, wax esters, and fatty acids.[21,25,37] CW forms small grain-like crystals (3 to 5 µm in length) in canola oil and has great oil binding capacity.[18]

CRW is obtained from the leaf buds and leaves of the carnauba palm *Copernicia cerifera* and further purified. CRW has the highest melting point of the commercial vegetable waxes (83–86 °C).[20] The difference in the maturity of the plant results in four types of CRW and explains the heterogeneous molecular composition.[38] CRW can be used as a food additive. The European Food Safety Authority (EFSA) restricts its use to a maximum concentration of 200 $mg\,kg^{-1}$ under the E-number E903. Identified with the CAS number 008-015-869, CRW is recognized as GRAS in the US (21CFR184.1978). Common applications are cosmetics and edible coatings for fruits and vegetables.[37] The main components of CRW are wax esters and fatty alcohols, which lead to the formation of dendritic crystals with a diameter between 10 and 20 µm in CRW oleogels.[18,21]

FW or bayberry wax is extracted from the berries of the *Myrica cerifera* shrub. The berries are boiled in water to melt the wax which is allowed to float to the surface. Next, the wax can be skimmed off and purified. Since it mainly consists of TAGs, it can be considered as a vegetable tallow. It is primarily used in candles but also in soap, cosmetics, crayons, and leather polish.[39] Its use as a food additive has not been approved in the EU or in the US. In oleogels, a concentration of at least 5 wt% is needed to form a gel, which is related to the large amount of low-melting short-chain FAs and FALs. FW forms large branched dendritic or spherulite crystals with a diameter of 30 to 50 µm that result in rheologically unstable gels.[13,24,40]

RBW is derived from rice bran, a by-product of rice processing. It has not been approved as a food additive in the EU. In the US, its use is allowed under restricted conditions (21CFR172.890). RBW is mainly constituted of wax esters which result in a relatively high melting temperature. In oleogels RBW crystallizes developing long (20–50 µm) needle-like and fibrous crystals[18,37] or big dendritic or spherulite crystals which interconnect.[21] In contrast to the minimum gelling concentration of 5% that Doan *et al.* (2017) found, other authors describe minimum gelling concentrations of 0.5–1%, which could be explained by the formation of different kinds of crystals.[15,18,37]

SW is obtained as a by-product of sunflower oil processing. The use of SW as a food additive is not allowed, nonetheless the wax is often used in cosmetics. SW is constituted mostly of wax esters (>90%) and forms fibrous crystals. In oleogels, just 0.5 wt% of SW is required to form a gel.[15,18,36,41]

9.2.2 Waxes Obtained from Insects

YBW and WBW are obtained from the honeycomb of the bee *Apis mellifera* and further purified. YBW is additionally bleached to obtain WBW. Both types of BW are approved as food additives in the EU with E-number E901 (EU Commission Regulation 1333/2008). The FDA also approved both YBW and WBW in the code of federal regulations (21CFR184.1973) under the CAS Number 8012-89-3. BW is used in the food industry as a glazing agent, coating, and carrier for colors.[42] The two most prominent compounds are wax esters and *n*-alkanes, although fatty acids and fatty acid alcohols are also present.[21,32] In RBO and olive oil, BW forms small needle-like crystals or platelets and is structurally similar to a β′ form.[41,43]

SHW is obtained after the purification of the secretion of lac insects, *Laccifer lacca*. SHW is commonly used in the food industry. In the EU, its use is regulated under the E-number E904 (EU Commission Regulation 1333/2008), while in the US the FDA approved shellac as an indirect food additive (21CFR175.300). Furthermore, SHW is also used in the pharmaceuticals and cosmetics industries.[44] Wax esters are the main component of shellac wax. The formed oleogels were reported to be more pliable and plastic, which was attributed to the high proportion of free alcohols, although this is still under debate. A minimum concentration of 2% is required to develop an oleogel.[44,45]

9.2.3 Legislation

Although natural waxes are studied widely for food applications, their use is often restricted by legislation. In the EU, EFSA carries out risk assessments taking into account exposure, biological, and toxicological data of the components. An overview of the applications currently allowed by EU legislation is listed in Table 9.2. Overall, natural waxes are mainly used as glazing agents at the *quantum satis* level. This means that a maximum dose has not been specified. Therefore, provided that substances are added following good manufacturing practices, they can be used to the necessary level for their intended purpose. For the use of CRW, maximum levels have been established.

In the US, the FDA regulates waxes as food additives. YBW, WBW, CRW, and CW are in the list of *direct food substances affirmed as generally recognized as safe*. Hence, provided that their addition is in line with good manufacturing practices, they are regarded as safe. YBW, WBW, CRW, CW, and SHW are also allowed as indirect food additives. RBW is allowed for specific applications.

There are only a few toxicology reports on waxes. As listed in Table 9.1, the main components of waxes are HC, WE, FAL, and FA. HC are poorly absorbed after ingestion. The absorbed amount, around 3%, is distributed in the organs and tissues and metabolized in the liver to acidic compounds which are then excreted in the urine. The remaining residues in organs and tissues return to near background levels within six months after exposure.[47,48] In parallel, the absorption of WE is generally considered as low. When partly hydrolyzed in the intestinal tract, their corresponding alcohols and acids are absorbed and incorporated into the normal metabolic

Table 9.2 Allowed use of waxes as food additives in the EU as of May 2021.[46] [a]

	BW (E901)	CW (E902)	CRW (E903)	SHW (E904)	MCW[b] (E905)	OPW[c] (E914)
Edible ices – only prepacked wafers containing ice cream	q.s.[d]	—	—	—	—	—
Entire fresh fruit and vegetables – only for the surface treatment of fruit: citrus fruit, melons, apples, pears, peaches, pineapples, bananas, mangoes, avocados and pomegranates and as glazing agent on nuts	q.s.	—	200[e]	q.s.	—	—
Entire fresh fruit and vegetables – only surface treatment of citrus fruit, melons, apples, pears, peaches and pineapples and glazing agent on nuts	—	q.s.	—	—	q.s.	q.s.
Cocoa and Chocolate products as covered by Directive 2000/36/EC – as glazing agent only	q.s.	q.s.	500	q.s.	—	—
Other confectionery including breath freshening microsweets – as glazing agent only	q.s.	q.s.	500	q.s.	q.s.	—
Chewing gum – as glazing agent only	q.s.	q.s.	1200	q.s.	q.s.	—
Decorations, coatings and fillings, except fruit-based fillings covered by category 4.2.4 – as glazing agent only	q.s.	q.s.	200	q.s.	q.s.	—
Fine bakery wares – only as glazing agents only for small products of fine bakery wares coated with chocolate	q.s.	q.s.	200	q.s.	—	—
Coffee, coffee extracts – only coffee beans, as glazing agent	q.s.	q.s.	200	q.s.	—	—
Potato-, cereal-, flour- or starch-based snacks – as glazing agent only	q.s.	q.s.	200	q.s.	—	—
Processed nuts – as glazing agent only	q.s.	q.s.	200	q.s.	—	—
Food supplements supplied in a solid form including capsules and tablets and similar forms, excluding chewable forms	q.s.	q.s.	200	q.s.	—	—
Food supplements supplied in a syrup-type or chewable form	q.s.	q.s.	200	q.s.	—	—
Processed eggs and egg products – only on the surface of unpeeled colored boiled eggs	—	—	3600	—	—	—
Processed eggs and egg products – only on the surface of unpeeled boiled eggs	—	—	—	q.s.	—	—

[a]—not allowed.
[b]Microcrytalline wax (MCW).
[c]Oxidized polyethylene wax (OPW).
[d]*Quantum satis* (q.s.).
[e]$mg\,l^{-1}$, or $mg\,kg^{-1}$ as corresponding.

pathways and eliminated.[42,49,50] FA and FAL are expected to follow the same metabolic pathway.[42]

The currently authorized uses of natural waxes as food additives are not of safety concern. EFSA has not established an Acceptable Daily Intake (ADI) value for CRW, CW, and BW because of the lack of long-term toxicity data. This means that the maximum amount of wax that can be ingested daily over a lifetime without any appreciable health risk has not been determined.[42,48,49] In contrast, the Joint FAO/WHO Expert Committee on Food Additives (JECFA) allocated an ADI of 7 mg $kgbw^{-1}$ day^{-1} for CRW. For BW, CW, and SHW the ADI has not been specified yet.[51]

Lastly, SW represents an interesting case. In fact, although SW is already consumed in human diets because it is present in sunflower oil (0.02–0.035 wt%) and sunflower seeds (about 0.1 wt%), it has not been listed as a GRAS ingredient. In addition, no toxicity of SW has been described, but some health benefits have been reported, such as a lowering effect on serum cholesterol.[52]

9.3 TAG–Wax Hybrid Fat Systems

9.3.1 Introduction

Oleogelation is a valuable approach to oil structuring, but often not a solution on its own. Hwang and co-workers[53] compared the firmness and melting behavior of commercial margarines with margarines made with different plant waxes and soybean oil. To achieve a firmness similar to that of commercial margarine, 10% of SW had to be added. However, high concentrations of wax are often associated with a waxy mouthfeel.[54] Trained evaluators scored margarines with 5% BW or 5% SW between 4 and 5 on a scale of 0 to 10. These results are most likely related to a detectable waxy mouthfeel.[55] Similar results were obtained for margarines made with 3.3% CW, where low sensory scores were attributed to a waxy taste.[56] Furthermore, oleogels have a lower solid content, spreadability, and plasticity, and are shear sensitive. Hence, when shear is applied, the solid-like structure transforms into a viscous liquid.[57,58] For instance, oleogels with CW are structured with intertwined platelets, which are destroyed when applying shear and do not recover afterwards. For certain food products, such as laminated doughs, structural recovery is one of the most relevant macroscopic properties, together with cohesion, consistency, and plasticity. Therefore, based on the shear sensitivity of wax oleogels, the high required concentrations of wax and the waxy sensory attributes, oleogels do not represent an alternative on their own.

A way to reduce the amount of *trans* and/or SFAs in margarines and still achieve the required properties is to combine an oleogel with a reduced amount of hardstock. This way, a hybrid TAG–wax fat system is created. Table 9.3 gives an overview of studies that combined waxes with saturated and unsaturated TAGs to produce hybrid fat systems for both shortenings and margarine applications.

Table 9.3 Hybrid fat systems combining waxes with triacylglycerol hardstock.

System	Type	Wax concentration (wt%)	Oil	Hardstock	Hardstock concentration (wt%)	References
Hybrid shortening – static cooling	BW, CRW	2–8	—	Palm kernel stearin	92–98	59
	BW, PW[a]	2.5	Pomegranate seed	Palm oil	50	60
	CW	0–3	Safflower	Tripalmitin (85%)	0–1	61
	CW	0–3	Safflower	Fully hydrogenated soybean oil	0–15	62
	CW	1–5	Soybean, Sunflower	Fully hydrogenated oil (HF)[b] Monoacylglycerols (MAG)	HF: 0–5 MAG: 0–4	63
	CW	0–5	Soybean	Cramble, palm, palm kernel and soybean hard fat	0–5	64
	CW	0–6	Soybean	Hydrogenated palm oil	0–6	65
	SW	0.1–0.25	—	Anhydrous milk fat	99.75–99.9	66
	SW	0.25–0.5	—	Anhydrous milk fat	99.50–99.75	21
Hybrid shortening – dynamic cooling	BW	10	Sunflower	Commercial cake shortening	20–85	67
	CW	3	Safflower	Tripalmitin (99%)	0–1	68
	CW	3–6	Canola	Commercial bakery shortening	40–70	69
	SW	1	Rice bran	Interesterified palm-based hardstock	15–25	70
Hybrid emulsion – dynamic cooling	CW	3.3	HOSO[c]	Interesterified fat	6.14	56
	CW	2	HOSO	Fully hydrogenated cramble oil	2	64
	SW	1–2.5	Rice bran	Interesterified palm-based hardstock	15–25	70

[a]Propolis wax (PW).
[b]Hard fat (HF).
[c]High oleic sunflower oil (HOSO).

The studies presented in Table 9.3 can be divided into three main categories based on their application and preparation method: (1) hybrid shortening systems with static cooling, (2) hybrid shortening systems with dynamic cooling, and (3) hybrid emulsions with dynamic cooling. In the first category, systems consisting exclusively of fats (*i.e.*, shortenings) are cooled down in the absence of shear. In the second category, shortenings are cooled down under dynamic conditions, combining shear and higher cooling rates. Lastly, in the third category, margarine-like systems (*i.e.*, W/O emulsions) are cooled under shear. Due to the increased cooling rate, dynamic conditions are often favorable for emulsion systems as they ensure a homogeneous emulsification of the water phase.

Table 9.3 suggests that most research has emphasized the use of CW and SW. Relatively small concentrations of the waxes are used, up to 10 wt%, since it is well-known that natural waxes form a gel at moderate concentrations and that high amounts lead to a waxy mouthfeel. Moreover, very different amounts and types of hardstock have been applied, ranging from 1 to 99.9 wt%. The lowest concentrations of hardstock TAG were applied in studies related to improving the rheological properties of the oleogel. The higher concentrations were mostly used in applications mimicking commercial shortenings or margarines.

9.3.1.1 Hybrid Shortenings

9.3.1.1.1 Static Cooling. For the formation of a statically cooled hybrid shortening system, different components are heated together and cooled down without applying any shear. A first effect of the addition of a vegetable wax to a hardstock is that the onset of crystallization of the hardstock shifts to higher temperatures. Different studies concluded that CW and SW crystallize first and induce the heterogeneous nucleation of the hardstock. This was demonstrated with anhydrous milk fat, tripalmitin, fully hydrogenated soybean oil, palm kernel stearin, and interesterified palm-based hardstock.[19,59,61,62,70] Ramírez-Gómez and co-workers[62] further indicated that the concentrations of the wax should be high enough to affect the crystallization temperature of the hardstock, in their case, a minimum of 2%. Figure 9.2 illustrates the crystallization thermograms of hybrid systems of SW and an interesterified palm-based hardstock (INT). The thermograms suggest that SW crystals appear earlier in the cooling process and serve as nucleation sites for the crystallization of the hardstock.[70]

The early crystallization of some natural waxes, like SW, impacts the crystal size and the morphology of the hardstock. Crystal size reduction has been reported recurrently,[19,62,70] most likely due to secondary nucleation processes. For example, an increased concentration of CW led to a crystal size reduction.[62] Since small crystals are preferred in edible lipid matrices to avoid a sandy mouthfeel and to provide the required texture and softness, a crystal size reduction is regarded as positive.[66] Figure 9.3 illustrates the

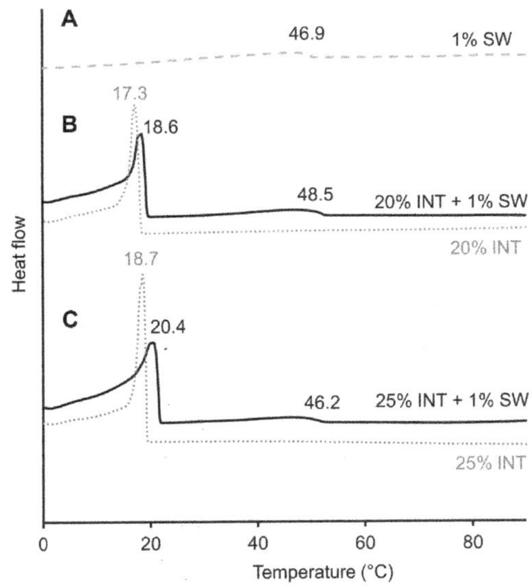

Figure 9.2 Crystallization thermograms of SW, interesterified palm-based hardstock (INT), and rice bran oil (RBO) systems. Dashed lines denote oleogel systems, dotted lines denote shortenings, and solid lines denote hybrid systems. (A) Oleogel made with 1 wt% SW in RBO. (B) Comparison between a hybrid system made with 20 wt% INT + 1 wt% SW in RBO and a shortening with 20% INT in RBO. (C) Comparison between a hybrid system made with 25 wt% INT + 1 wt% SW in RBO and a shortening with 25% INT in RBO. For each crystallization thermogram, the numbers above the peaks are the peak crystallization temperatures. The thermograms are obtained by differential scanning calorimetry at a cooling rate of 5 °C min^{-1}.
Adapted from ref. 70 with permission from John Wiley & Sons, Copyright © 2017 Society of Chemical Industry.

impact of SW and CW on the crystal size of a hardstock fat upon static cooling.

Several studies investigating the crystal morphology of hybrid wax-shortening systems have been carried out. Ramírez-Gómez and co-workers created mixtures of fully hydrogenated soybean oil (FHSO) and CW in vegetable oil. Upon crystallization of the melt, the systems developed mixed crystal networks containing FHSO crystals in the β form and CW crystals in orthorhombic packing.[62] It can be suggested that the heterogeneous nucleation induced by the CW crystals (formed at higher temperatures) largely modified the FHSO crystal microstructure. As the systems crystallized in a stable polymorphic form, the crystal habit did not change over 28 days of storage at 25 °C. The resulting hybrid system (FHSO–CW) showed a granular structure and a reduced crystal size in contrast to the spherulitic shapes in the systems without CW.[62]

In a different study, Martini and co-workers stated that, in the presence of BW and propolis wax (PW), palm oil crystallized in different morphologies.[19]

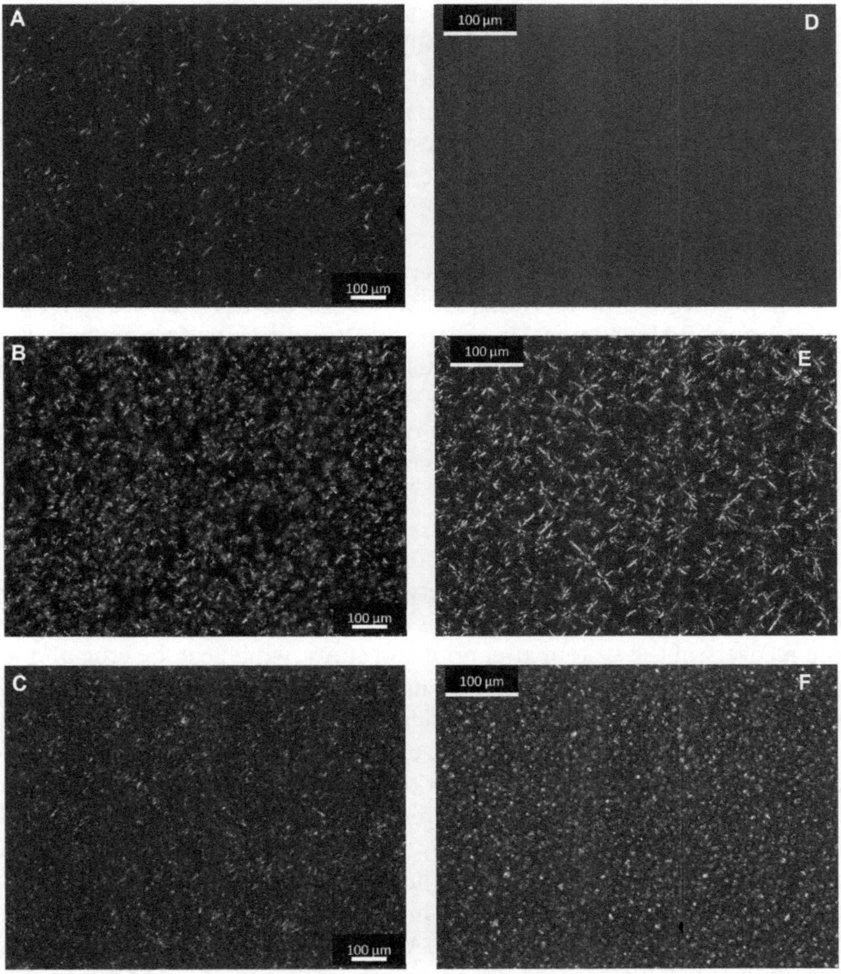

Figure 9.3 Impact of waxes on the crystal morphology of hardstock fats visualized by polarized light microscopy at 5 °C (A to C) and 15 °C (D to F). (A) 1% SW in RBO; (B) 20% INT in RBO; (C) 20% INT + 1% SW in RBO; (D) 1% CW in rapeseed oil; (E) 30% palm stearin in rapeseed oil; (F) 30% palm stearin + 1% CW. All samples were prepared under static conditions.[35,41]

Instead of the spherulitic crystal morphology of palm oil, the crystal morphology of a hybrid system of palm oil, BW, or PW and pomegranate seed oil was dominated by platelet-like crystals. These crystals were similar to the crystals observed in the pure oleogel samples.[60] On the other hand, it was observed that the crystal morphology did not change when replacing a small amount of CW by monoacylglycerols (MAG) and hardfat (HF).[63] A further increase of MAG and HF changed the crystal morphology to big spherulites formed by MAG, HF, or a combination of both, surrounded by a background of microplatelets formed by CW.

Results of both studies suggested that natural waxes drive the crystallization process and that fats co-crystallized. Several other authors reached similar conclusions. A study conducted by Toro-Vazquez *et al.* suggested that CW and tripalmitin might co-crystallize together and form mixed lamellar structures. This result was attributed to the similarity in structure of palmitic acid in tripalmitin and the aliphatic chains in hentriacontane, the most abundant hydrocarbon in CW.[61] Liu *et al.* (2019) investigated the impact of BW and CRW on the polymorphism of palm kernel oil.[59] CW addition significantly promoted the isothermal crystallization process of palm kernel stearin, showing a faster crystallization and a shorter time to get to the equilibrium solid fat content. In contrast, BW addition had no obvious effect on the crystallization rate. In both cases, BW and CW could participate in the crystallization of TAGs, inducing the formation of new hydrocarbon chain distances of 3.70 and 4.15 Å on the original basis and decreasing the lamellar distance (d (001)) and domain size (ξ) of the single crystallites. Hence, the addition of natural waxes leads to different morphologies of the fat crystals.

A key aspect to discuss is the effect on the macroscopic scale. Since waxes modify the lipid network through crystal size and morphology, this will impact the physicochemical properties on a macro level. Of special relevance is the recovery of the system after applying shear, which can be crucial in the production of certain foods, such as laminated doughs.[66] Overall, the rheological properties, such as elasticity and recovery of the system were enhanced by adding a hardstock to an oleogel.[61,62] A possible explanation for this improved elasticity is the growth of many hardstock crystals nucleated from the wax crystal surfaces. When enough CW was added to induce the nucleation of the hardstock, a grainy microstructure was created and the number of reversible crystal–crystal junction zones in the crystal network was increased leading to a higher recovery of the system after shear.[62] Another study showed that certain TAG combinations had a better interaction with CW.[64] TAGs formed by a combination of C16 and C18, and C18 and C22 obtained the best results: a small increase of G' (storage modulus), low oil loss values, and a decrease of the melting peak temperature. Shorter chain fatty acids, like C12 and C14, and C18, if used alone, did not improve the macroscopic properties.

In summary, in hybrid shortening systems with static cooling, the composition and concentration of each structuring component affect the crystal morphology and alter the physicochemical properties. Only if enough wax is added, the heterogeneous crystallization of the hardstock is induced and co-crystallization is obtained. The addition of more wax results in smaller crystals and has a positive influence on the texture. Altogether, hybrid systems have the possibility to achieve better elasticity, recovery, and oil loss value[60–64] compared to pure wax oleogels.

9.3.1.1.2 Dynamic Cooling. In contrast to the studies using static production, a limited number of publications described the production of

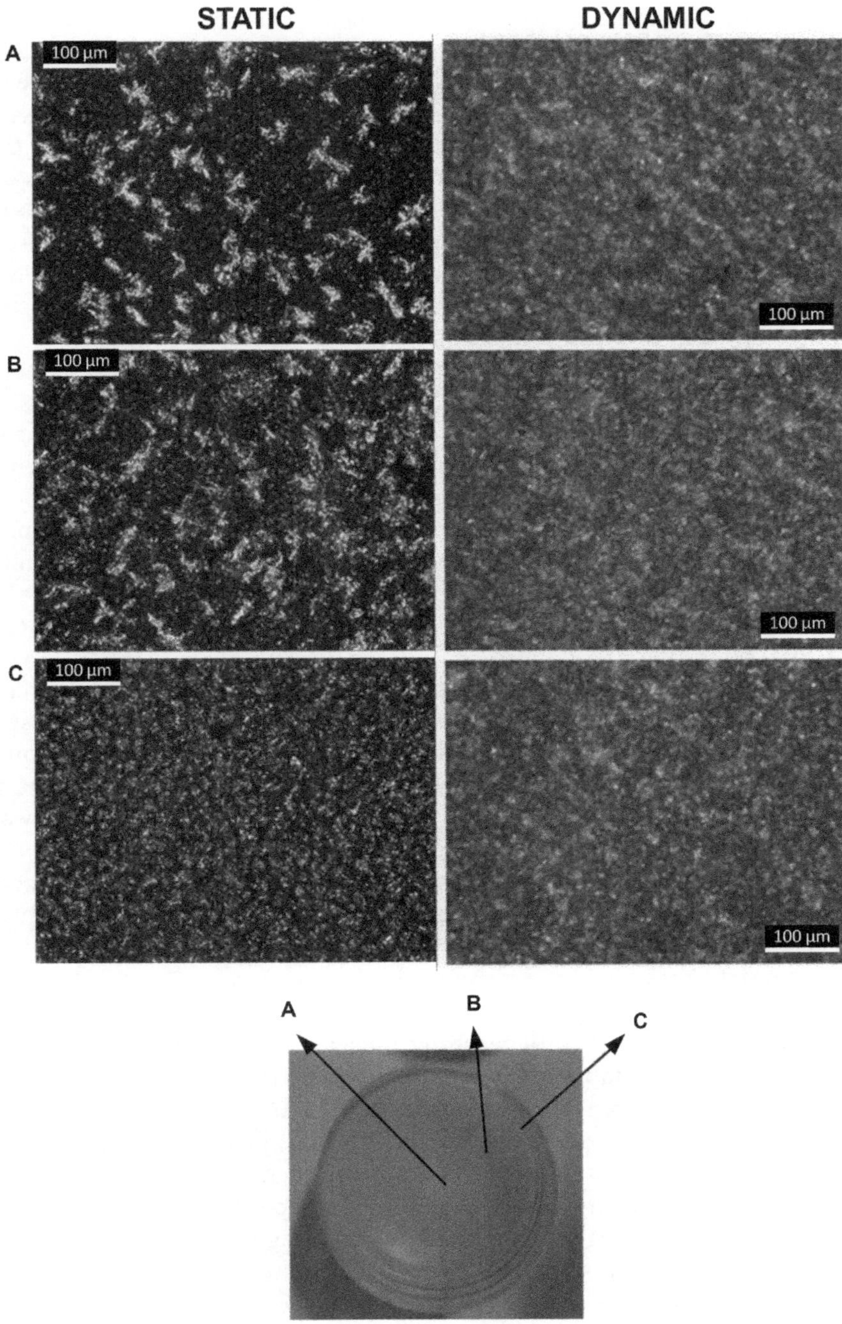

Figure 9.4 Polarized light micrographs of statically and dynamically produced hybrid systems (30% palm stearin and 6% CRW in rapeseed oil) at (A) the center, (B) half the radius, and (C) outside of the sample cup.[35]

hybrid shortenings using a dynamic method. Nonetheless, investigating dynamic conditions can be of great relevance as margarines and shortenings are industrially produced with simultaneous rapid cooling and shearing. In this way, a more uniform product is obtained and the processor has a greater degree of control over the final product.[71] The impact of the production method on a hybrid fat structure is nicely illustrated in Figure 9.4. A hybrid system consisting of 30% palm stearin and 6% CRW was produced under static (freezer to reach 15 °C) and dynamic conditions (bench-top crystallizer consisting of a surface scraped heat exchanger and pin worker unit and subsequent storage at 15 °C). Polarized light micrographs show the resulting crystal structures at different locations in the obtained samples. In the statically produced sample, crystal structures differ according to the sampling point. In fact, the cooling gradient results in smaller crystal structures near the cup wall and coarser structures in the center of the sample. Dynamically produced samples show a finer crystal morphology which is relatively consistent along the entire sample. In contrast, the statically produced sample shows a dual crystal morphology, presumably a combination of large CRW crystals surrounded by smaller palm stearin crystals. The dual crystal morphology is more pronounced in the center of the sample, where the lower cooling rate allows formation of larger CRW spherulites.

In a different study, wax oleogels were combined with a commercial shortening hardstock on a custom-made scraped surface heat exchanger and jacketed pin mixer.[67,69] These conditions resemble industrial operations, where pins are used to mechanically work the fat while it is crystallizing from the supercooled state to extend the plastic range of the product.[71] It was observed that the addition of shortening to a CW oleogel improved the shear sensitivity of the resulting hybrid fat system compared to the pure oleogel.[58] However, more in-depth physicochemical analysis on the hybrid fat samples was lacking.

Other researchers tried to implement shear on a lab scale. Chopin-Doroteo *et al.* applied shear during cooling of a CW oleogel and a CW hybrid system with 1% tripalmitin just until reaching nucleation (5 °C above the gelling temperature) or metastable conditions (10 °C above the gelling temperature). Then, the samples were statically cooled down to 15 °C.[68] When comparing the samples to statically produced systems, it was apparent that by applying shear more nucleation was induced. However, no modification of the molecular packing of the crystals or the melting properties of the organogels was seen. When shearing was applied until reaching metastable conditions, the highest G' and yield stress were obtained. The study concluded that the application of shear determines both crystal size and the proportion of transient to junction zones, hence determining the hybrid organogel's rheological properties.

A number of researchers have reported that the application of continuous and constant shear during the cooling stage of an oleogel results in the formation of a weak gel structure. Often, this is attributed to an inappropriate proportion of permanent and transient junction zones required to develop a gel network.[68,72] In the case of CW oleogels, adding 1% of tripalmitin caused the stacking organization of the crystals to change, resulting in an increase in the solid fat content and the elasticity of the organogels.[68]

Doan *et al.* also produced hybrid shortenings by cooling under shear at 350 rpm to the pre-nucleation point and subsequently cooling further without shear. In their work, SW (1%) was combined with INT (25–14% in RBO).[70] Despite a small increase in solid fat content, a considerable increase in hardness values of the hybrid shortenings was found (Figure 9.5). They concluded that the addition of 1.0% SW enhanced the network strength and they could replace up to 40% of SFAs in the hybrid system (25% to 15% hardstock) to achieve the same hardness.

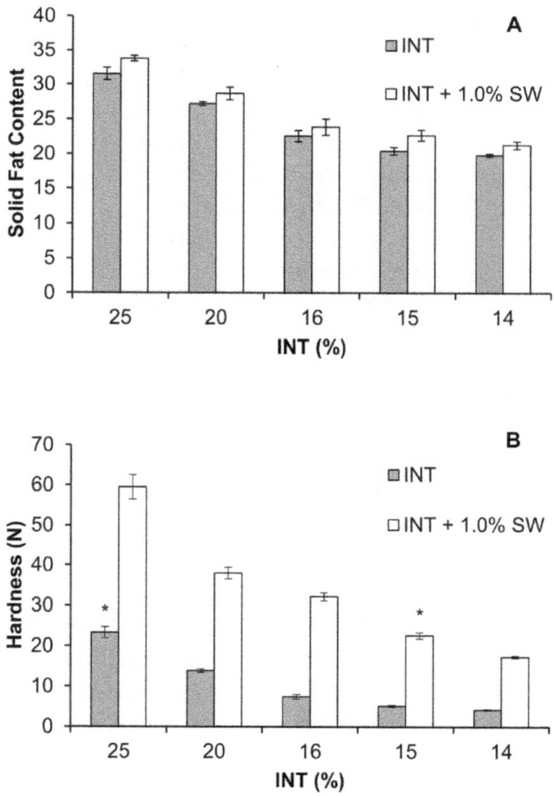

Figure 9.5 Solid fat content (A) and hardness (B) of the INT, RBO, and SW systems. On the *x* axis, different concentrations of INT are presented. Gray bars denote shortenings with 0% SW (reference) and white bars denote hybrid systems blended with 1% of SW. The asterisks indicate comparable hardness between the hybrid system with 15% INT + 1% SW and the 25% INT reference.
Adapted from ref. 70 with permission from John Wiley & Sons, Copyright © 2017 Society of Chemical Industry.

In addition, X-ray diffraction was used to further explore the appearance of the crystalline order and crystal morphology of this hybrid system.[70] The hybrid blend (INT–SW) crystallized in the β′ polymorph at 30 °C, probably due to the earlier crystallization of SW (which has been demonstrated to crystallize from the melt in orthorhombic symmetry).[73] When the temperature decreased, the INT α polymorph transformed into β′ morphology, and the SW β′ crystals transitioned into β morphology. Therefore, at 5 °C, indications were found for the structural arrangement and co-existence of different crystal morphologies (α, β′, and β) in the hybrid shortening.

9.3.1.2 Hybrid Emulsions

According to the EU-legislation, a margarine is a spreadable fat product in the form of a solid, malleable emulsion, principally of the water-in-oil (W/O) type, derived from vegetable and/or animal fats suitable for human consumption (milk fat content <3% of fat content) with a fat content of not less than 80% but less than 90%. When less fat is added, the terminology 'fat spread' is typically used. The water droplets are stabilized by an emulsifier and immobilized by the fat crystal network. Typically, a scraped surface heat exchanger is used to obtain a fine and uniform dispersion of the water droplets throughout the crystallizing fat phase.[74,75]

Waxes alone have been applied to stabilize W/O emulsions both in the presence and in the absence of surface-active emulsifiers.[45,53,76,77] The stability of the resulting W/O emulsions is strongly linked to the fat crystal spatial distribution.[78] If the wax has surface-active properties, it will behave as a Pickering particle and absorb in the water–oil interface (Figure 9.6A), allowing the formation of a physical barrier and preventing emulsion destabilization. High-melting wax particles are thought to crystallize and/or adsorb onto the oil–water interface, creating a steric barrier between dispersed water droplets and, thus, preventing the water collisions and subsequent coalescence.[70] Conversely, less surface-active waxes will interact with the water droplets to a lesser extent. If present in sufficiently large concentrations, they will form a three-dimensional network through the continuous phase, increasing the viscosity and enclosing the water droplets (Figure 9.6B). The synergy between Pickering stabilization and network stabilization (Figure 9.6C) is often seen in wax hybrid W/O emulsions.

Findings suggest that processing conditions, storage temperature, and the overall temperature history are key for the stability of the resulting emulsion, as they will determine the absorption of wax particles to the drop interface.[76] Binks and Rocher identified that emulsions with increased stability were formed by fast cooling of a pre-emulsion made with molten wax, oil, and water.[76] Wax type, wax–oil–water ratio, and shearing rate are other factors that are reported to be essential for the final stability of the system.

Wax-oleogels have been implemented as a fat phase to formulate margarines or fat spreads. SW, RBW, and CW oleogels prepared in soybean oil were used as the solely continuous fat phase without SFAs for the production

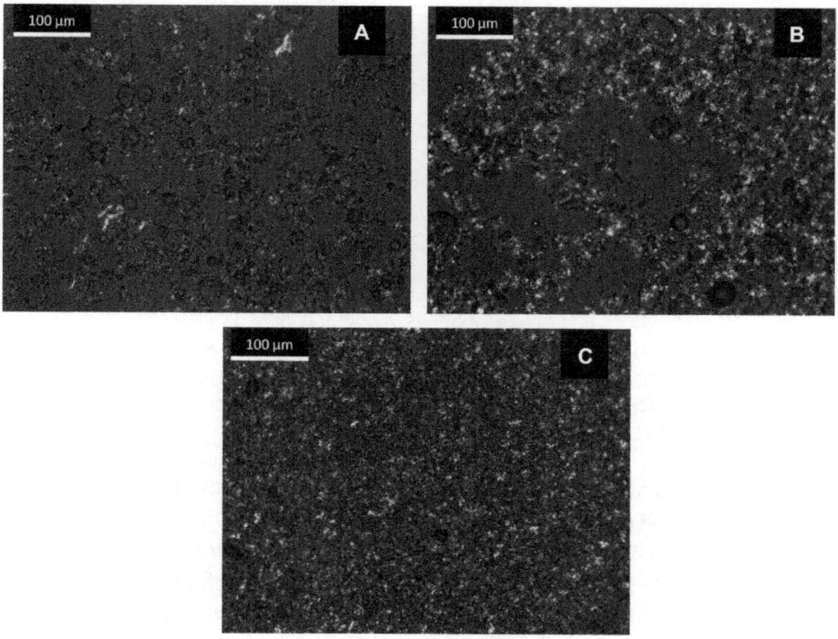

Figure 9.6 Polarized light micrographs of W/O emulsions prepared with CRW, sunflower oil, and 20% water. Images differ on wax concentrations and crystal spacial distribution: (A) 2.5%, (B) 5%, and (C) 7.5%.[79]

of a 20% W/O margarine.[53] SW showed the most potential as the products containing 2–6% wt SW in soybean oil had the same firmness as the margarine containing 18–30% wt hydrogenated soybean oil. RBW could not provide the desired firmness of the margarine although it formed a relatively firm organogel. Finally, CW showed phase separation after making the emulsion. Samples were made on a lab scale, and further upscaling needs to be investigated.

The application of wax-triglyceride hardstock fat blends in margarines or spreads has been scarcely investigated. da Silva *et al.* studied the effect of water addition on the physical properties of an emulsion gel system consisting of a blend of 2% CW, 2% MAG, and 2% HF in sunflower oil.[80] The samples were prepared on a lab scale with an Ultraturrax. Metastable emulsions were obtained with up to 20% water. Overall, the addition of water improved the physical properties of the system, such as hardness, viscosity, oil binding capacity, elasticity, and melting profile.

In a different study, Doan and co-workers investigated the combination of SW (1–2.5%) with INT for the production of 20% W/O emulsions.[70] It was found that the addition of 1.0% wt SW enhanced the solid content and network strength of the hybrid palm-based shortenings. SW helped to stabilize the water droplets inside the wax-based crystalline network without flocculation during shear-cooling. The impact of SW on the microstructure of the hybrid emulsions was observed using cryo-SEM (Figure 9.7).

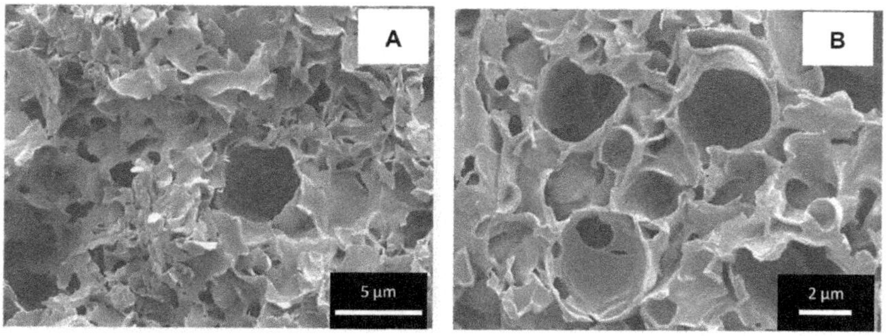

Figure 9.7 Crystal microstructures of 20% W/O emulsions. The fat phase consists of (A) a shortening (25% INT) or (B) a hybrid system (25% INT + 1% SW). Reproduced from ref. 70 with permission from John Wiley & Sons, Copyright © 2017 Society of Chemical Industry.

The water droplets were surrounded by platelet crystals of emulsifier and hardstock. The crystal shell surrounding the droplet was thin (≈8–10 nm) and not fully spherical. The addition of 1.0% wt SW increased the thickness of the fat crystal shell located at the droplet interface (approximately 200 nm), which helped to stabilize the droplets during shear. The maximum amount of SFAs being replaced in the hybrid emulsions by 1.0% wt and 2.5% wt SW in margarine were 25% and 40%, respectively.[70]

Altogether, this section highlighted the possibility of adding natural waxes to W/O emulsion systems. Many success factors make these systems of great interest, both with and without the addition of hardstock. Despite current efforts, the effect of several formulation parameters has not yet been addressed, and upscaling needs to be investigated.

9.3.1.3 Hybrid systems in bakery applications

A limited number of papers are available describing further application of hybrid fat systems in food applications. Some publications describe the use of hybrid systems in bakery applications. In short dough products, a plastic shortening plays a crucial role in determining the rheological properties of the dough and the quality attributes of the cookie. During kneading, fat crystals form a film around the flour particles composed of mainly starch and protein. Therefore proteins cannot interact with each other and starch molecules have limited access to water, so a weak gluten network is obtained.[69] An optimal amount of solid fat content (SFC) is required to obtain a good distribution of the fat in the short dough.[81] Next to the shortening effect, the fat crystals have a structural role regarding aeration of the dough and retaining air during the early stages of baking.[81] Finally, the fat phase impacts the dimensions and overall texture of the final product in which solid fat content is of importance.[81,82]

Wax oleogels have been implemented in cookie making but the functionality was not sufficient as short dough formation was not optimal and both spread ratio and cookie hardness increased.[58] When applying a hybrid system [CW (3/6%), canola oil, and shortening (30/60%)], more acceptable dough and cookie properties were achieved.[69] Blending oleogels with shortening allowed the reduction of SFA content. The obtained cookie dough properties, dimensions and hardness were closer to those of the reference shortening product.

Another study of the same group describes the use of BW oleogel–shortening hybrid systems in cake production.[67] The role of shortening in the cake baking process has been described as forming a stable emulsion which is able to endure the heat of baking and hence to facilitate aeration by the incorporation of small air cells into the plastic shortening during batter-mixing.[83] When using BW oleogels as a fat replacer in cakes, a reduced air-holding ability, porosity and specific volume was noticed compared to cakes made with a commercial cake shortening. However, comparable products were obtained when partially replacing the shortening with the oleogel in a hybrid system. Approximately 35% reduction in SFA content could be achieved with still acceptable gluten-free cake properties.[67]

In a similar study, CW oleogels were combined with shortening in muffin production.[84] Although a hybrid fat system was not really produced, the shortening could be successfully replaced with oleogel up to 25% by weight with comparable quality attributes to the control. Unfortunately, none of the above-mentioned studies included a sensory evaluation or consumer acceptance study, which is seen as a prerequisite for commercial applications.

Hybrid fat systems have also been applied in confectionary fillings.[85] BW oleogels in RBO were combined with palm oil in three replacement levels (17–33–50% wt) to form a hazelnut filling reduced in SFA. The maximum amount of palm oil that could be replaced by the wax-based oleogels was 17% while maintaining a similar gel strength and good oil binding property compared to the ref. (100% palm oil). A dilution effect was observed, and the melting enthalpy decreased with the increasing oleogel fraction. The solid fat content of the final filling products at body temperature was lower than 2.0% wt, suggesting a non-waxy mouthfeel.

9.4 Conclusion

Due to their excellent oil structuring ability, natural waxes are components which lately have received a lot of attention to serve as potential replacers of *trans* and SFAs. In contrast to the number of publications related to wax oleogel production and characterization, the interaction between waxes and hardstock fats has been less investigated. This chapter provided an overview on the formation of hybrid fat systems using waxes and hardstock fats. Until now, mainly CW and SW have been combined with different types of hardstock fats (including hydrogenated oils, interesterified fats, palm oil, milk fat). Overall, the composition and concentration of each structuring component seem to affect the crystal morphology and other physicochemical properties of the hybrid system. Only if enough wax is added, the heterogeneous crystallization of the hardstock is induced and co-crystallization obtained. The addition of more wax results in smaller crystals and has a positive influence on the texture. Altogether, hybrid systems have the possibility to achieve better elasticity, recovery, and oil loss value compared to pure wax oleogels.

However, a more in-depth characterization of the fat crystal network (polymorphism, domain size, storage stability, *etc.*) is required to further elucidate wax–triglyceride interactions. Furthermore, it needs to be noted that most studies have been carried out on a lab scale under static crystallization conditions. Future studies should also emphasize dynamic crystallization to make the link towards industrial applications, such as industrial shortenings and emulsions (margarines and fat spreads). In addition, far too little attention has been paid to the use of hybrid systems in other food applications. Described applications include cookies, cakes, and confectionary fillings, and these indicate that a partial replacement of SFA content could be achieved by hybrid systems while maintaining the desired product quality. However, there is still a lack of better understanding of the functionality of waxes in the final application and more data are required on the sensory properties of the foods which incorporate hybrid fat systems.

Acknowledgements

Vandemoortele Lipids NV is acknowledged for financially supporting the UGent Vandemoortele Centre on 'Lipid Science and Technology'. The Hercules foundation is recognized for its financial support in the acquisition of the scanning electron microscope JEOL JSM-7100F equipped with cryo-transfer system Quorum PP3000T (grant number AUGE-09-029). The Research Foundation – Flanders (FWO) is acknowledged for providing the SB grant 1SA5321N in support of the PhD candidate Ivana A. Penagos.

References

1. A. G. Marangoni and N. Garti, *Edible Oleogels: Structure and Health Implications*, Elsevier Inc., 2011.
2. A. G. Marangoni, J. P. M. van Duynhoven, N. C. Acevedo, R. A. Nicholson and A. R. Patel, *Soft Matter*, 2020, **16**, 289–306.
3. EC Eurostat, Causes of Death – Standarised Death Rate, *EU-27*, 2020.
4. CDC/National Center for Health Statistics, *Leading Causes of Death*, 2019.
5. EC, *Commission Regulation (EU) 2019/649 of 24 April 2019 amending Annex III to Regulation (EC) No 1925/2006 of the European Parliament and of the Council as regards trans fat, other than trans fat naturally occurring in fat of animal origin*, 2019.

6. World Health Organization, *Global Strategy on Diet, Physical Activity and Health*, Food and Nutrition Bulletin, 2004.
7. J. A. Nettleton, I. A. Brouwer, J. M. Geleijnse and G. Hornstra, *Ann. Nutr. Metab.*, 2017, **70**, 26–33.
8. FAO, *Fats and fatty acids in human nutrition: Report of an expert consultation* 978-92-5-106733-8, 2010.
9. World Health Organization, *Healthy Diet*, Regional Office for the Eastern Mediterranean, 2018.
10. D. J. Abdallah, L. Lu and R. G. Weiss, *Chemistry of Materials*, 1999, **11**, 2907–2911.
11. J. F. Toro-Vazquez, J. Morales-Rueda, V. A. Mallia and R. G. Weiss, *Food Biophys.*, 2010, **5**, 193–202.
12. A. R. Patel and K. Dewettinck, *Food Funct.*, 2016, 7, 20–29.
13. C. D. Doan, D. Van de Walle, K. Dewettinck and A. R. Patel, *J. Am. Oil Chem. Soc.*, 2015, **92**, 1739.
14. C. D. Doan, D. Van de Walle, K. Dewettinck and A. R. Patel, *J. Am. Oil Chem. Soc.*, 2015, **92**, 801–811.
15. H. S. Hwang, S. Kim, M. Singh, J. K. Winkler-Moser and S. X. Liu, *J. Am. Oil Chem. Soc.*, 2012, **89**, 639–647.
16. H. S. Hwang, S. Kim, K. O. Evans, C. Koga and Y. Lee, *Food Struct.*, 2015, **5**, 10–20.
17. M. Öğütcü, N. Arifoğlu and E. Yilmaz, *J. Am. Oil Chem. Soc.*, 2015, **92**, 459–471.
18. A. I. Blake, E. D. Co and A. G. Marangoni, *J. Am. Oil Chem. Soc.*, 2014, **91**, 885–903.
19. S. Martini, A. A. Carelli and J. Lee, *J. Am. Oil Chem. Soc.*, 2008, **85**, 1097–1104.
20. P. E. Kolattukudy, *Chemistry and Biochemistry of Natural Waxes*, Elsevier Scientific Pub. Co, 1976.
21. D. Chi Diem, C. M. To, M. De Vrieze, F. Lynen, S. Danthine, A. Brown, K. Dewettinck and A. R. Patel, *Food Chem.*, 2017, **214**, 717–725.
22. A. R. Patel, M. Babaahmadi, A. Lesaffer and K. Dewettinck, *J. Agric. Food Chem.*, 2015, **63**, 4862–4869.
23. JECFA, Evaluation of Certain Food Additives, 9241209348, FAO/WHO Expert Committee on Food Additives Geneva, 2006.
24. A. R. Patel, M. Babaahmadi, A. Lesaffer and K. Dewettinck, *J. Agric. Food Chem.*, 2015, **63**, 4862–4869.
25. J. F. Toro-Vazquez, J. A. Morales-Rueda, E. Dibildox-Alvarado, M. Charó-Alonso, M. Alonzo-Macias and M. M. González-Chávez, *J. Am. Oil Chem. Soc.*, 2007, **84**, 989–1000.
26. L. E. Vandenburg and E. A. Wilder, *J. Am. Oil Chem. Soc.*, 1970, **47**, 514–518.
27. L. S. K. Dassanayake, D. R. Kodali, S. Ueno and K. Sato, *J. Am. Oil Chem. Soc.*, 2009, **86**, 1163–1173.
28. S. R. Vali, Y.-H. Ju, T. Narayana, B. Kaimal and Y.-T. Chern, *J. Am. Oil Chem. Soc.*, 2005, **82**, 57–64.
29. T. C. S. Kanya, L. J. Rao and M. C. S. Sastry, *Food Chem.*, 2007, **101**, 1552–1557.
30. R. Aichholz and E. Lorbeer, *J. Chromatogr. A*, 1999, **855**, 601–615.

31. S. Bogdanov, *Beeswax: Production, Properties, Composition, Control*, 2016.
32. D. L. Grant, *Beeswax*, WHO/IPCS INCHEM Database, 2005.
33. A. P. Tulloch, *Lipids*, 1970, **5**, 247–258.
34. A. R. Patel and K. Dewettinck, *Eur. J. Lipid Sci. Technol.*, 2015, **117**, 1772–1781.
35. M. Romanus, F. Van Bockstaele and K. Dewettinck, Ghent Univeristy, Unpublished work, 2020.
36. E.-M. Schromm, *Oulu University of Applied Sciences*, 2018.
37. L. S. K. Dassanayake, D. R. Kodali and S. Ueno, *Curr. Opin. Colloid Interface Sci.*, 2011, **16**, 432–439.
38. A. I. Blake, E. D. Co and A. G. Marangoni, *J. Am. Oil Chem. Soc.*, 2014, **91**, 885–903.
39. NIIR Board of Consultants & Engineers, *The Complete Technology Book on Wax and Polishes*, Asia Pacific Business Press, 2011.
40. P. K. Okuro, I. Tavernier, M. D. Bin Sintang, A. G. Skirtach, A. A. Vicente, K. Dewettinck and R. L. Cunha, *Food Funct.*, 2018, **9**, 1755–1767.
41. C. D. Doan, PhD thesis, Ghent University, 2017.
42. EFSA, *Beeswax (E 901) as a Glazing Agent and as Carrier for Flavours – Scientific Opinion of the Panel on Food Additives, Flavourings, Processing Aids and Materials in Contact with Food (AFC)*, Wiley, 2007.
43. E. Yilmaz and M. Öğütcü, *J. Food Sci.*, 2014, **79**, E1732–E1738.
44. A. R. Patel, D. Schatteman, W. H. De Vos and K. Dewettinck, *RSC Adv.*, 2013, **3**, 5324–5327.
45. A. R. Patel, D. Schatteman, W. H. De Vos, A. Lesaffer and K. Dewettinck, *J. Colloid Interface Sci.*, 2013, **411**, 114–121.
46. Publications Office of the European Union, in *2008R1333*, EUR-Lex, 2016, 342.
47. EFSA, *Scientific Opinion on the Use of High Viscosity White Mineral Oils as a Food Additive*, Wiley-Blackwell Publishing Ltd, 2009.
48. EFSA, *Scientific Opinion on the Re-evaluation of Candelilla Wax (E 902) as a Food Additive*, Wiley-Blackwell Publishing Ltd, 2012.
49. EFSA, *Scientific Opinion on the Re-evaluation of Carnauba Wax (E 903) as a Food Additive*, Wiley-Blackwell Publishing Ltd, 2012.
50. I. A. Hansen and J. F. Mead, *Exp. Biol. Med.*, 1965, **120**, 527–532.
51. JECFA, *Toxicological evaluation of certain food additives and naturally occurring toxicants*, 9241660309, FAO/WHO Expert Committee on Food Additives, Rome, 1993.
52. T. C. S. Kanya, B. R. Lokesh and M. C. S. Sastry, *J. Food Biochem.*, 2008, **32**, 60–71.
53. H. S. Hwang, M. Singh, E. L. Bakota, J. K. Winkler-Moser, S. Kim and S. X. Liu, *J. Am. Oil Chem. Soc.*, 2013, **90**, 1705–1712.
54. C. D. Doan, I. Tavernier, P. K. Okuro and K. Dewettinck, *Innovative Food Sci. Emerging Technol.*, 2018, **45**, 42–52.
55. E. Yilmaz and M. Öğütcü, *RSC Adv.*, 2015, **5**, 50259–50267.
56. T. L. T. da Silva, K. F. Chaves, G. D. Fernandes, J. B. Rodrigues, H. M. A. Bolini and D. B. Arellano, *J. Am. Oil Chem. Soc.*, 2018, **95**, 673–685.

57. D. Chi Diem, I. Tavernier, M. D. Bin Sintang, S. Danthine, D. Van de Walle, T. Rimaux and K. Dewettinck, *Food Biophys*, 2017, **12**, 97–108.
58. B. Mert and I. Demirkesen, *LWT – Food Sci. Technol.*, 2016, **68**, 477–484.
59. C. Liu, Z. Zheng, Z. Meng, X. Chai, C. Cao and Y. Liu, *LWT*, 2019, **115**, 108446.
60. G. Fayaz, S. A. H. Goli, M. Kadivar, F. Valoppi, L. Barba, S. Calligaris and M. C. Nicoli, *LWT – Food Sci. Technol.*, 2017, **86**, 523–529.
61. J. F. Toro-Vazquez, M. Alonzo-Macias, E. Dibildox-Alvarado and M. A. Charó-Alonso, *Food Biophys.*, 2009, **4**, 199–212.
62. N. O. Ramírez-Gómez, N. C. Acevedo, J. F. Toro-Vázquez, J. J. Ornelas-Paz, E. Dibildox-Alvarado and J. D. Pérez-Martínez, *Food Res. Int.*, 2016, **89**, 828–837.
63. T. L. T. da Silva, D. B. Arellano and S. Martini, *J. Am. Oil Chem. Soc.*, 2018, **95**, 797–811.
64. T. L. T. da Silva, D. B. Arellano and S. Martini, *Food Res. Int.*, 2019, **121**, 900–909.
65. K. R. R. de Godoi, R. C. Basso, C. C. Ming, A. A. da Selva, L. P. Cardoso and A. P. B. Ribeiro, *Food Res. Int.*, 2020, 137.
66. R. M. Kerr, X. Tombokan, S. Ghosh and S. Martini, *J. Agric. Food Chem.*, 2011, **59**, 2689–2695.
67. I. Demirkesen and B. Mert, *J. Am. Oil Chem. Soc.*, 2019, **96**, 545–554.
68. M. Chopin-Doroteo, J. A. Morales-Rueda, E. Dibildox-Alvarado, M. A. Charó-Alonso, A. de la Peña-Gil and J. F. Toro-Vazquez, *Food Biophys.*, 2011, **6**, 359–376.
69. B. Mert and I. Demirkesen, *Food Chem.*, 2016, **199**, 809–816.
70. D. Chi Diem, I. Tavernier, S. Danthine, T. Rimaux and K. Dewettinck, *J. Sci. Food Agric.*, 2018, **98**, 1042–1051.
71. L. H. Wiedermann, *J. Am. Oil Chem. Soc.*, 1978, **55**, 823–829.
72. S. Da Pieve, S. Calligaris, E. Co, M. C. Nicoli and A. G. Marangoni, *Food Biophys.*, 2010, **5**, 211–217.
73. I. Tavernier, C. D. Doan, D. Van De Walle, S. Danthine, T. Rimaux and K. Dewettinck, *RSC Adv.*, 2017, **7**, 12113–12125.
74. I. P. Freeman and S. M. Melnikov, *Margarines*, Wiley-VCH Verlag GmbH & Co. KGaA, Weinheim, Germany, 2015.
75. A. J. Haighton, *J. Am. Oil Chem. Soc.*, 1976, **53**, 397–399.
76. B. P. Binks and A. Rocher, *J. Colloid Interface Sci.*, 2009, **335**, 94–104.
77. S. M. Hodge and D. Rousseau, *Food Res. Int.*, 2003, **36**, 695–702.
78. S. Ghosh and D. Rousseau, *Curr. Opin. Colloid Interface Sci.*, 2011, **16**, 421–431.
79. I. A. Penagos, F. Van Bockstaele and K. Dewettinck, Ghent University, Unpublished work, 2020.
80. T. L. T. da Silva, D. B. Arellano and S. Martini, *Food Biophys.*, 2019, **14**, 30–40.
81. L. S. Sciarini, F. Van Bockstaele, B. Nusantoro, G. T. Perez and K. Dewettinck, *J. Cereal Sci.*, 2013, **58**, 234–240.

82. Y. N. Manaf, J. M. N. Marikkar, S. Mustafa, F. Van Bockstaele and B. P. Nusantoro, *Int. Food Res. J.*, 2019, **26**, 1795–1802.
83. I. Demirkesen and B. Mert, *Crit. Rev. Food Sci. Nutr.*, 2020, **60**, 2460–2479.
84. J. Lim, S. Jeong, J. Lee, S. Park, J. Lee and S. Lee, *J. Sci. Food Agric.*, 2017, **97**, 3727–3732.
85. C. D. Doan, A. R. Patel, I. Tavernier, N. De Clercq, K. Van Raemdonck, D. Van de Walle, C. Delbaere and K. Dewettinck, *Eur. J. Lipid Sci. Technol.*, 2016, **118**, 1903–1914.

CHAPTER 10

Glycerolysis-structured Lipid Systems

R. A. NICHOLSON AND A. G. MARANGONI*

Food, Health, and Aging Laboratory, Department of Food Science, University of Guelph, 50 Stone Rd. E, Guelph, ON N1G 2W1, Canada
*Email: amarango@uoguelph.ca

10.1 Introduction

For the past two decades, many methods of structuring liquid oils to produce fat mimetics have been explored. While this area of study began with the purpose of replacing *trans* fats in foods, it remains an important topic as it is necessary for reducing the saturated fat contents of foods as well as for improving the sustainability of the fat source given the current usage of palm oil. The numerous fat mimetics produced thus far can be categorized as either mono and mixed component oleogels, structured biphasic systems, and polymer oleogels.[1] Monoglycerides (MGs) have proven to be effective in the field of fat mimetics. Previously, MGs were used to create mono-component gel systems.[2–5] To form these gel systems, saturated MGs (monopalmitin and monostearin) are mixed with a liquid oil at low concentrations (≤10% w/w), heated to ensure that the MGs are completely melted, and subsequently cooled. During cooling, the MGs form an inverse lamellar phase, before transforming to the sub-α crystalline phase below the crystallization point of the MGs. Later, the MG crystals transform into the β-crystalline phase.[4] Another MG-based strategy exploits the use of structured emulsions, where water swollen, crystallized MG layers stabilize the interface of dispersed oil droplets.[6,7]

Food Chemistry, Function and Analysis No. 31
Development of *Trans*-free Lipid Systems and their Use in Food Products
Edited by Jorge F. Toro-Vazquez

Published by the Royal Society of Chemistry, www.rsc.org

These fat mimetic applications employ MGs at low concentrations. But greater MG concentrations can further take advantage of the higher crystallization and melting temperatures of MGs (both saturated and unsaturated). This was shown previously when Vereecken and co-workers characterized the effects that MG addition, at higher concentrations (25%, 33%, 50%), had on the solid fat content (SFC) of triglyceride (TG) oils.[8] MGs, however, are not the only partial glyceride with higher crystallization and melting temperatures compared to TGs. In edible oils, diglycerides (DGs) exist in two isomeric forms, 1,3-DGs and 1,2-DGs, at a molar ratio of approximately 7 : 3. 1,2-DGs have a melting temperature similar to that of TGs, while 1,3-DGs, have a melting temperature that is approximately 10 °C higher than TGs of the same fatty acid composition.[9,10] This means that a lipid system containing substantial amounts of mixed glyceride species can form a structural fat. Another interesting property of DGs is that consumption of DGs instead of TGs has been shown to reduce serum TG levels and body fat storage along with low-density lipoprotein and total cholesterol levels.[11–14] This is a result of the different metabolic pathways for 1,3-DGs and TGs.

Even though partial glycerides are natural components of fats and oils that have a longstanding use in the food industry as emulsifiers, the ongoing "clean label" movement makes their addition to an ingredients list undesirable. Additionally, supplementing an oil with MGs and DGs would alter the fatty acid composition from that of the native oil. Furthermore, production of MGs and DGs can be performed through several processes. One of which is glycerolysis. Glycerolysis is traditionally used to produce partial glycerides for use as emulsifiers and other food ingredients. For this reason, the glycerolysis process has been well researched.[15–19] However, when the partial glycerides are not extracted following glycerolysis, their higher crystallization and melting temperatures can be exploited to directly structure the liquid oil into a solid fat. This is beneficial to consumers as there is no change in the fatty acid composition of the lipid system (*i.e.*, no increase in *trans* or saturated fatty acids). Our group has begun to characterize the effectiveness of enzymatic glycerolysis as a method of producing structural fats from plant-based oils for food applications.[20,21]

10.2 Effects of Glycerolysis Reaction Conditions

In the first of the two studies by our group, the effects of reaction conditions on the production of partial glycerides and subsequent changes in the physical properties of the system were explored.[20] These glycerolysis reactions were performed with cottonseed oil as the reaction substrate. All reactions were performed at 65 °C using the non-specific lipase *Candida antarctica* lipase B as the reaction catalyst, at 2% (w/w) in cottonseed oil. We showed that the production of partial glycerides was dependent on the reaction time (when a 2 : 1 glycerol : TG molar ratio was present). MG and DG concentrations increased rapidly throughout the first 18 h, with MGs having increased from less than 1% to 20%, while DGs increased from 7% to 43%. As the reaction continued, the MG content rose to 31.3% after 48 h, while the DG content rose to 49.1%. DGs had reached a plateau by this time, as the concentration was unchanged after 72 h (49.6%). During this time, the MG content increased only slightly (34.3%). Production of MGs and DGs was also shown to be influenced by the initial amount of glycerol. When glycerol was present at various glycerol : TG molar ratios (0.10 : 1, 0.25 : 1, 0.50 : 1, 1 : 1, 2 : 1, 4 : 1) for 24 h reactions, it was determined that the DG content was maximized (46.1%) with a glycerol : TG molar ratio of 0.5 : 1, but decreased only slightly to 42.3% as the molar ratio was raised to 4 : 1. Meanwhile, the MG concentration was maximized (27.5%) with a glycerol : TG molar ratio of 1 : 1.

Interestingly, it was shown that individual MG species were affected differently by the reaction time and glycerol content. A monolinolein concentration of around 6% was reached after 6 h and this concentration remained unchanged thereafter. However, monoolein and monopalmitin production continued throughout the entire reaction time range studied, increasing to 15.5% and 12.8%, respectively, after 72 h. Similarly, the glycerol : TG molar ratio had a limited effect on monolinolein production. Approximately 6% monolinolein was produced with a glycerol : TG molar ratio of 0.5 : 1 and only underwent a minor reduction as the ratio was increased to 4 : 1. Meanwhile, maximum concentrations of 11.0% and 10.4% were achieved for monoolein and monopalmitin, respectively, at a 1 : 1 glycerol : TG molar ratio.

We showed through differential scanning calorimetry (DSC), that the partial glycerides caused a dramatic change in the crystallization behavior of the cottonseed system. The native oil initially had a crystallization onset temperature of approximately −5 °C, however, following glycerolysis, a new crystallization peak was formed with a crystallization onset upwards of 15 °C. The reaction conditions were shown to change the crystallization behavior of the system. The most noticeable alteration being that the crystallization temperature of the high-temperature peak was maximized at a 1 : 1 glycerol : TG molar ratio. This coincided with the production of monoolein and monopalmitin which were maximized within this reaction formulation. Interestingly, the increased production of MGs and DGs after longer reaction lengths increased the height of this peak but did not change the temperature of crystallization. This demonstrated the importance of the glycerol content in tailoring the physical properties of the glycerolysis-structured system.

Our group explored the effects of the glycerol content and the resulting compositional differences on the physical properties of glycerolysis-structured systems further, with a particular focus on systems prepared with glycerol : TG ratios of 0.5 : 1, 1 : 1, and 2 : 1. The 1 : 1 molar ratio product resulted in faster isothermal crystallization kinetics at 5, 10, 15, and 20 °C and a lower half-time of crystallization compared to the other two glycerolysis products. Microstructural differences arose due to these variations. The fat crystal network of the 1 : 1 glycerol : TG molar ratio glycerolysis product contained smaller, more numerous fat crystals. This network also showed a higher degree of homogeneity in the spatial distribution of the crystalline material as was characterized by a greater box-counting fractal dimension (1.77 *vs.* 1.54 and 1.57, for 0.5 : 1 and 2 : 1 glycerolysis products, respectively).

Tailoring the crystallization properties with glycerol content had further implications on the oil binding properties of the systems. The oil binding capacity of a fat network is a key functionality parameter with implications pertaining to the stability and shelf-life of both single and multi-component fat-containing food products. The oil binding capabilities of the glycerolysis-structured systems were evidently improved when more partial glycerides were present, as was demonstrated by the inverse relationship between oil loss and reaction time. However, the composition of the MGs and DGs appeared to be more important than the overall partial glyceride content to the oil binding capabilities of the system. Oil loss was lowest for the 1 : 1 glycerol : TG molar ratio system following a 24 h reaction, even though it contained lower levels of both MGs and DGs than the 48 h and 72 h systems, which were reacted in the presence of a 2 : 1 glycerol : TG molar ratio. Evidently, the improved network homogeneity and smaller crystal size resulting from a higher concentration of MGs (particularly monopalmitin and monoolein), in relation to the DG content, benefitted the oil binding capabilities.

Perhaps the most important physical property factoring into the functionality of an edible fat system is the SFC. SFC has major implications on the structuring capabilities of the fat within a food product. The overall level of solids, at a given temperature, factors into whether a fat system is brittle or ductile and the reduction in SFC with respect to temperature affects the mouthfeel. Our group showed that glycerolysis is an effective means of increasing the SFC. Performing the reaction for just 6 h increased the SFC of the cottonseed oil system from 8.2% to 14.6% at 5 °C and the SFC continued to increase as the reaction length was extended to 72 h. Albeit, beyond 48 h the increases were very minor (26.2% to 27.7%) given the large increase in reaction time. Glycerol content affected the SFC at 5 °C in the sense that enough glycerol was required for production of substantial quantities of partial glycerides. Once glycerol was present at a molar ratio of 0.5 : 1, the SFC of the system reached 21.9% after 24 h, and although the maximum SFC was reached at a 1 : 1 glycerol molar ratio, only minor changes in the SFC took place between ratios of 0.5 : 1 and 4 : 1.

Furthermore, the importance of the glycerol content and resulting partial glyceride composition became evident at higher temperatures due to its effect on the SFC with respect to temperature. For the glycerolysis reaction time experiments, each of the structured systems displayed similar decreases in SFC as a function of increasing temperature. For this reason, the pattern for the SFC increase with respect to reaction time length was the same at 5 °C as it was at 20 °C. This was evidently because the proportions of monoolein and monopalmitin to DGs within each of these reaction products remained constant with increasing reaction time. These proportions changed depending on the glycerol content present during the reaction. As a result, the 1 : 1 ratio reaction product underwent a more gradual decrease in SFC as the temperature was increased and had a higher level of solids remaining at 20 °C compared to the glycerolysis-structured systems produced at the other glycerol contents. The ability to alter the SFC with respect to the temperature of these structural fats, even when the same base oil is employed, simply by changing the reaction conditions has much potential. A gradual reduction in SFC with respect to temperature resulting in a more structured lipid at room temperature may prove beneficial for some food applications, particularly non-refrigerated products. Whereas a more substantial reduction in SFC at lower temperatures would be more suited to other applications, such as refrigerated soft-tub margarines and spreads.

The potential of this reaction was also affected by enzyme characteristics. The source of lipase used as the catalyst can have an impact on the structuring capabilities of the reaction. Whether this is due to differing stereo specificities or simply the overall ability of the enzyme to produce partial glycerides remains to be seen. In previously unpublished work, glycerolysis reactions were performed with *Rhizomucor miehei* lipase to compare the structuring power of a 1,3-specific lipase with that of the non-specific *Candida antarctica* lipase B. Due to the lower enzyme activity (w/w) for the *R. miehei* lipase, the quantity of enzyme present was increased to achieve an overall activity matching that of the *C. antarctica* lipase B. To achieve an equilibrium SFC for the cottonseed system, the *R. miehei* lipase catalyzed reactions needed to be performed for upwards of 72 h. Furthermore, 72 h reactions were performed in the presence of various glycerol : TG ratios (1 : 1, 2 : 1, 4 : 1, 6 : 1, and 10 : 1). It should be noted that all *R. miehei* lipase catalyzed reactions were performed at 50 °C as this is the optimum temperature for this enzyme according to suppliers. The SFC at 5 °C was the highest for reactions containing a 6 : 1 glycerol : TG molar ratio. However, the maximum SFC achieved by *R. miehei* lipase catalyzed reactions was only 14.9%. This reaction product contained 10.9% MGs and 50.0% DGs. Evidently, the *R. miehei* lipase had a lower structuring capability due to a substantially reduced capacity for producing MGs. Again, whether this is enzyme specific or the result of inactivity of the 1,3-specific lipase at the *sn*-2 position is uncertain.

Furthermore, lipase concentration has a major influence on the kinetics of the glycerolysis reaction. Our group performed reactions with *C. antarctica* lipase B at 2% (w/w) relative to the cottonseed oil.[20] For this reason, the time it took to maximize the structuring potential and reach reaction equilibrium was upwards of 48 h. Reaction time could be substantially reduced if the quantity of enzyme was to be increased, making it much more feasible for industrial usage. Additionally, since immobilized lipases are reusable, it is foreseeable that the lipase could be separated upon completion of an industrial scale glycerolysis reaction and reused for additional reactions. One could also run the reaction in a continuous mode through a packed bed reactor, thus maximizing conversion.

10.3 Effects of Oil Composition

In our follow-up research, glycerolysis reactions were performed on a variety of oils using the same set of conditions (*C. antarctica* lipase B; 65 °C; 48 h; 1 : 1 glycerol : TG molar ratio) to characterize the effects of oil composition on the physical properties of glycerolysis-structured systems.[21] Glycerolysis was performed on tigernut, peanut, cottonseed, rice bran, olive, high oleic canola (HOCO), soybean, sesame, canola, and high oleic algal (HOAO) oils. Following this glycerolysis reaction, the oils each contained upwards of 30% MGs and 40–50% DGs, with around 20% TGs remaining. These oils offered a range of fatty acid compositions which were shown to affect the physical properties of the glycerolysis-structured systems.[21]

From Section 10.2 it is evident that the partial glycerides produced within various cottonseed glycerolysis systems caused a new high-temperature crystallization peak to appear. As a result, the glycerolysis-structured cottonseed systems had three crystallization regions. This was also the case with many of the other glycerolysis-structured systems.[21] Crystallization profiles are depicted in Figure 10.1 and subsequent melting profiles are shown in Figure 10.2. The first crystallization peak was due in large part to the saturated fatty acid (SFA) containing partial glycerides as they have the highest crystallization temperatures. The second crystallization peak region was composed mainly of the oleic acid containing species. This information could be deduced from the crystallization behavior of the HOAO system. HOAO contained 92.0% oleic acid, meaning it was composed mostly of triolein in its native state. The crystallization peak for native HOAO was located several degrees below −40 °C and production of MGs and DGs shifted this peak to just below −20 °C. The fact that this system contained one distinctive peak means that monoolein, diolein, and triolein were capable of co-crystallizing at the proportions present following glycerolysis. This peak corresponded with what was usually the second crystallization peak for the glycerolysis-structured systems. It was typical that the oleic acid species peak was the largest of the three for the glycerolysis-structured systems, which makes sense given that oleic acid is commonly present in high quantities in plant-based oils. The position of the oleic species crystallization region was influenced by the high-temperature crystallizing SFA species, which shifted it to a higher temperature. While the low-temperature crystallizing polyunsaturated fatty acid species shift the oleic species peak to a lower temperature due to dilution. Crystallization of the polyunsaturated fatty acid species corresponded with the third crystallization region.

In the cottonseed systems examined in Section 10.2, the high-temperature peak shifted to a higher temperature when the reaction was performed at a 1 : 1 glycerol : TG molar ratio, due to changes in the partial glyceride species (*i.e.*, monopalmitin). Maintaining consistent reaction conditions while changing the oil substrate allowed our group to observe the effects of differing fatty acid compositions on the crystallization behavior.[21] To begin with, the position of the SFA species crystallization peak shifted depending

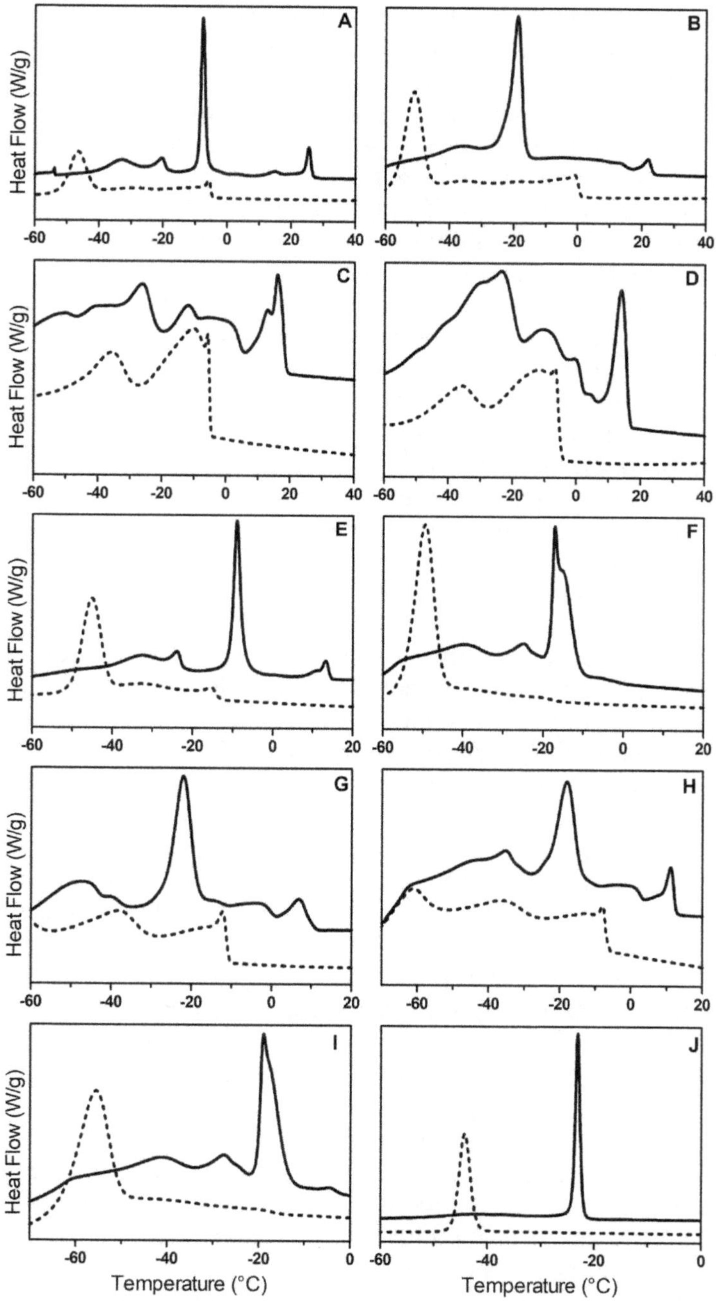

Figure 10.1 Differential scanning calorimetry crystallization (exothermic) profiles for glycerolysis reaction products (solid lines) and their respective oils (dotted lines). (A) tigernut; (B) peanut; (C) cottonseed; (D) rice bran; (E) olive; (F) high oleic canola oil (HOCO); (G) soybean; (H) sesame; (I) canola; (J) high oleic algal oil (HOAO).
Adapted from ref. 21 with permission from Elsevier, Copyright 2021.

on the type of SFA present. The tigernut oil system contained the highest concentration of stearic acid (6.6%) along with the highest onset of crystallization. Additionally, high levels of SFA with >18 carbons were present in peanut oil, and this system happened to be the only other system besides tigernut with a crystallization onset above 20 °C. Furthermore, the magnitude of changes in the three peak regions depended on the relative proportions of SFAs, oleic acid, and polyunsaturated fatty acids. For example, high SFA contents (>20%) combined with high linoleic acid contents (>30%) for both the cottonseed and rice bran oil systems, increased the size of the high-temperature and low-temperature peaks. When low SFA contents were present, the high-temperature crystallization peak was non-existent. HOCO and canola oil systems contained 6.9% and 7.1% SFAs, respectively, and showed no distinctive high-temperature peak. The only indication of these low concentrations of SFAs was a shoulder at the onset of the oleic species crystallization peak. Further to this point, the 2.8% SFAs within the HOAO system were not enough to provide any indication of SFA crystallization. Interestingly, even though a high-temperature crystallization peak for the SFA species was not produced within all oil systems, the crystallization onset increased by around 20 °C following glycerolysis, regardless of fatty acid composition.

The fat crystal network within the glycerolysis-structured systems was also affected by the fatty acid composition. XRD analysis showed that MG crystals were present in the β polymorphic form in every crystalline lipid system tested regardless of the fatty acid composition. When the oleic acid content of an oil was high enough, diolein co-crystallized with the MGs. This was the case for the tigernut, olive, HOCO, sesame, and canola oil systems.[21] Each of these systems, except for sesame oil (which did not behave in a manner that would be expected based on its fatty acid composition), contained >60% oleic acid. Furthermore, the diolein crystals were typically secondary crystals within the network, unless the oleic acid content was high enough, as was the case with the olive system (75.1%). With high oleic acid concentrations, it was possible for the diolein crystals to form the primary crystals within the structural network. However, in the event that an oil system was composed almost entirely of oleic acid (HOAO), no crystalline material was present at 5 °C within the glycerolysis product.

Analysis of the XRD spectra, particularly the 001 diffraction peak representing the long-spacing of the MG crystals, was used to identify changes in the molecular species present within the MG crystals. The fatty acid composition was shown to factor into the molecular composition of the MG crystals. It may simply have been the overall concentration of oleic acid present but given the interactions that arise between lipid molecules it was likely that the oleic–SFA balance and the oleic-linoleic acid (and sometimes other polyunsaturated fatty acids) balance both played roles in determining the subsequent balance of monoolein and monopalmitin within the MG crystals. In the cottonseed, soybean, and sesame oil systems, high SFA and linoleic acid contents resulted in MG crystals containing mostly monopalmitin,

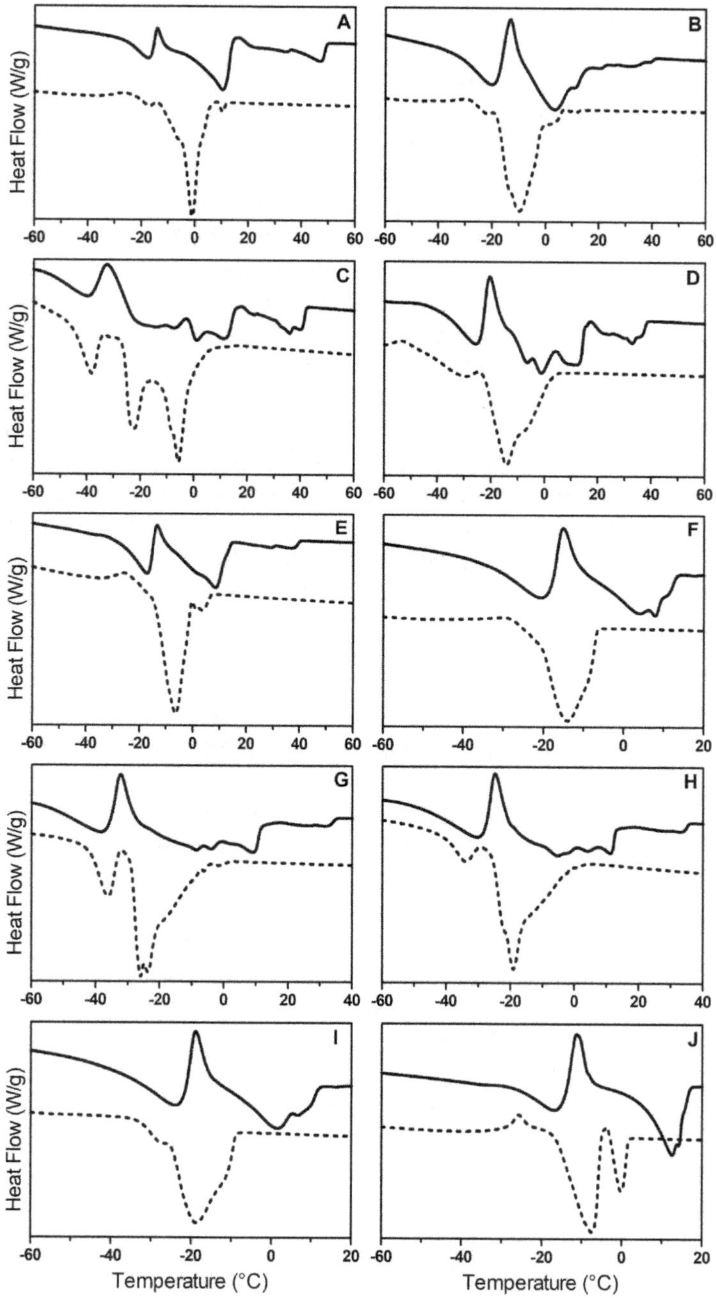

Figure 10.2 Differential scanning calorimetry melting (endothermic) profiles for glycerolysis reaction products (solid lines) and their respective oils (dotted lines). (A) Tigernut; (B) peanut; (C) cottonseed; (D) rice bran; (E) olive (F) high oleic canola oil (HOCO); (G) soybean; (H) sesame; (I) canola; (J) high oleic algal oil (HOAO).

with very little monoolein.[21] Furthermore, in the glycerolysis-structured rice bran oil system, monopalmitin was still the primary crystallizing molecule, however, a higher oleic acid content increased both the oleic-SFA balance and the oleic-linoleic balance allowing for more monoolein to be incorporated into the MG network. This was also the case with the tigernut and olive oil systems, with higher oleic acid contents resulting in still more monoolein within the MG crystals. For the olive oil system, the oleic acid content was high enough that the long-spacing of the MG crystals was closer to that of pure monoolein than it was pure monopalmitin. The peanut oil system was interesting because peanut oil contains a relatively high amount of oleic acid (58.4%) along with a high amount of SFAs (18.8%), with SFAs containing ≥20 carbons present in 8.0%. This resulted in a MG crystal long-spacing that was almost identical to that of pure monoolein, when in fact the MG network must have contained monopalmitin along with MGs containing the long-chain SFAs, in addition to a high amount of monoolein. When the SFA content of an oil was low (<10%) and the oleic acid content was high, the MG crystals within the glycerolysis-structured system appeared to be composed mostly of monoolein. This was the case for both the HOCO (6.9% SFA and 72.5% oleic acid) and canola oil (7.1% SFA and 63.1% oleic acid) systems.

Alterations in the molecular composition of the crystals coincided with microstructural differences between glycerolysis-structured systems (Figure 10.3). The majority of monopalmitin MG crystals (cottonseed, Figure 10.3C; soybean, Figure 10.3G; rice bran, Figure 10.3D) in the absence of diolein crystals formed crystals with a needle-like morphology. As was mentioned, based on the fatty acid composition of sesame oil (Figure 10.3H) the formation of diolein crystals was unexpected. In any event, the presence of diolein crystals with a majority of monopalmitin MG crystals disrupted the network. As a result, the morphology of these crystals was completely different from that of the MG crystals alone. Furthermore, MG crystals containing high amounts of long-chain SFAs, as was the case with the peanut oil system (Figure 10.3B), were very small and numerous with a platelet-like morphology. This was attributed to the occurrence of numerous nucleation sites due to a high crystallization onset temperature. Mixed monoolein–monopalmitin MG crystals in the presence of diolein crystals (tigernut, Figure 10.3A; olive, Figure 10.3E) formed small platelet-like crystals that aggregated to form larger structures. Diolein crystallization influenced the aggregation process, as the olive oil system with diolein as the primary crystalline material was less aggregated compared to the tigernut oil system. Finally, monoolein MG crystals co-crystallized with diolein crystals (HOCO, Figure 10.3F; canola, Figure 10.3I) formed large coral-like crystals extending throughout the oil phase.

In each of the systems studied, except for HOAO, the SFC increased following the production of partial glycerides through glycerolysis.[21] The changes that occurred following glycerolysis were more pronounced for some oil systems compared to others (Figure 10.4). When crystallized and

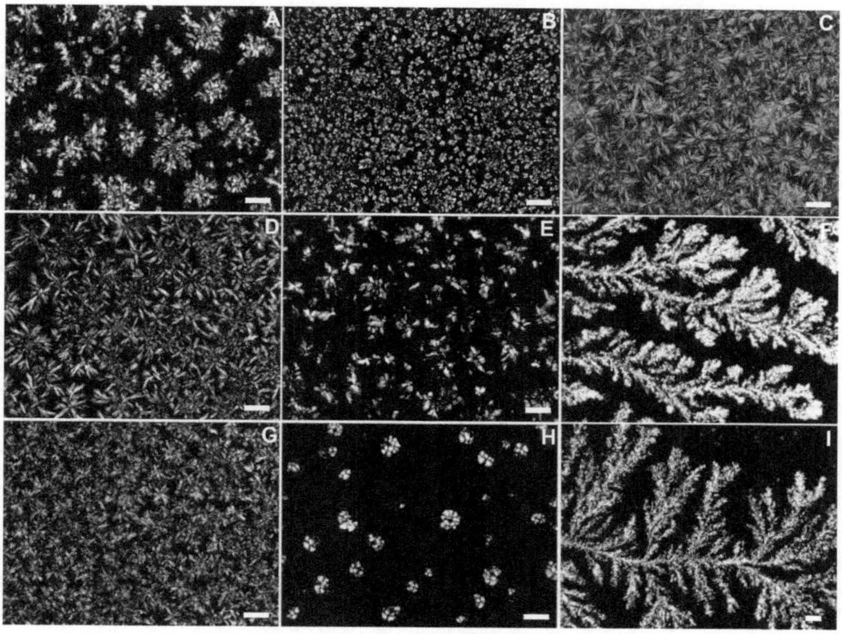

Figure 10.3 Micrographs for glycerolysis-structured systems of (A) tigernut oil; (B) peanut oil; (C) cottonseed oil; (D) rice bran oil; (E) olive oil; (F) high oleic canola oil (HOCO); (G) soybean oil; (H) sesame oil; (I) canola oil. Scale bar represents 100 μm. Imaging of the glycerolysis product samples was performed after crystallization and storage for one week at 5 °C. Adapted from ref. 21 with permission from Elsevier, Copyright 2021.

stored at 5 °C for one week, the glycerolysis-structured tigernut oil system had an SFC of 34.3%, compared to 7.7% solids in native tigernut oil. The SFC of the peanut oil system increased from 8.9% to 30.4%, while cottonseed oil increased from 8.2% to 29.1%. Rice bran oil underwent an increase from 0.3% solids to 26.7%. Each of the remaining native oils contained no solids at 5 °C, while SFC values of 24.1%, 18.7%, 16.3%, 10.7%, and 8.3% were observed for the olive, HOCO, soybean, sesame, and canola oil, respectively, following glycerolysis.

The SFC achievable at 5 °C was evidently affected by the fatty acids present in the native oil and the subsequent molecular composition of the crystals that formed within the glycerolysis-structured systems. As would be expected, a high SFA content produced a crystalline material in the lipid systems. In addition, the XRD results obtained by our group highlighted that oleic acid, both as monoolein and diolein can contribute to the solids at 5 °C. Cottonseed and soybean oil are examples of systems structured almost entirely by MGs containing SFAs. The SFC achieved at 5 °C for these glycerolysis-structured systems was only slightly greater than the SFA content (cottonseed: 29.1% SFC, 27.5% SFA; soybean: 16.3% SFC, 15.9% SFA). Although it is highly unlikely that the SFAs were entirely present as a

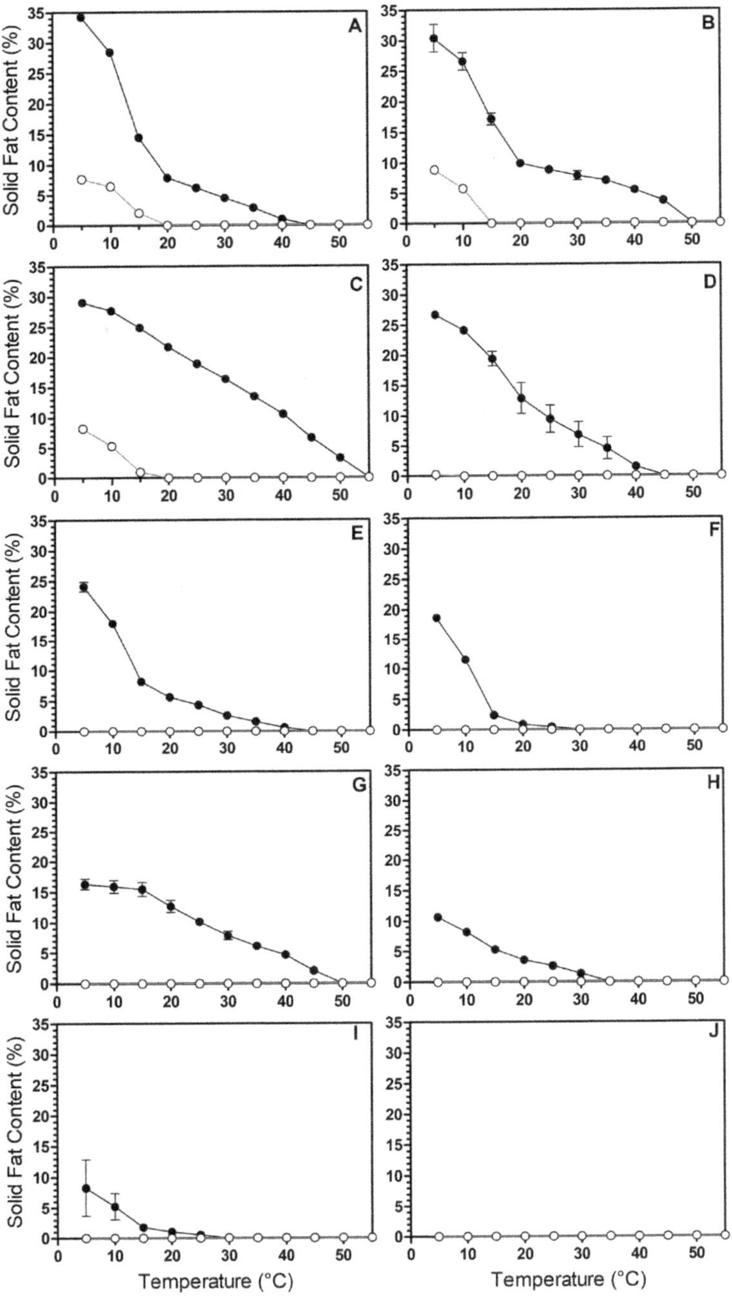

Figure 10.4 Solid fat content temperature profiles for glycerolysis-structured systems (closed circles; black lines) and their respective oils (open circles; gray lines). (A) Tigernut; (B) peanut; (C) cottonseed; (D) rice bran; (E) olive; (F) high oleic canola oil (HOCO); (G) soybean; (H) sesame; (I) canola; (J) high oleic algal oil (HOAO). The measurements were done after crystallization and storage at 5 °C for one week.
Adapted from ref. 21 with permission from Elsevier, Copyright 2021.

crystalline material within these systems and no unsaturated fatty acids became incorporated into the solids, this is still an interesting occurrence. Furthermore, the rice bran oil system was structured mostly by SFA containing MGs but contained some monoolein. Similar to cottonseed and soybean oils, the SFC (26.7%) was only slightly higher than the SFA content (23.3%). The structuring power of partial glycerides containing oleic acid was evident with the glycerolysis-structured tigernut oil system. This system achieved the highest SFC (34.3%) even though it contained less SFAs than either the cottonseed or rice bran oil systems. This was further exemplified by the peanut, olive, and HOCO systems. A comparison of HOCO and canola oil demonstrated that oleic acid needed to be present at a very high concentration (HOCO: 72.5%; canola: 63.1%) to achieve substantial increases in the SFC when low SFA levels were present. In this case, the higher concentration of polyunsaturated fatty acids in the canola oil was likely also a factor in the reduced structuring capabilities of the oleic acid containing species. Furthermore, even when very high levels of oleic acid were present (HOAO), SFAs needed to be present in appreciable quantities to form solid material at 5 °C through glycerolysis.

In addition to affecting the overall SFC achieved, the fatty acid profile and crystal composition influenced the SFC–temperature profile of glycerolysis-structured systems. In general, the glycerolysis-structured systems produced two distinctive SFC-temperature profiles. A two-segment profile, where the SFC decreased sharply with respect to temperature from 5–20 °C, before gradually decreasing as the temperature was increased further. For the second type of SFC–temperature profile, the SFC decreased gradually with respect to temperature throughout the entire temperature range. The occurrence of one SFC–temperature profile *versus* the other for glycerolysis-structured systems was heavily influenced by the amount of oleic acid within the crystalline material either as monoolein or diolein. Our research demonstrated that tigernut, peanut, olive, HOCO, and canola oil systems all displayed two regimes in the SFC–temperature profiles. Most of these systems contained more than 60% oleic acid, with peanut oil (58.4%) being the only system that contained less. Further to this point, XRD analysis showed that each of these glycerolysis-structured systems, besides peanut oil, demonstrated the presence of diolein crystals, and monoolein appeared to represent a substantial portion of the composition for all MG crystals. Meanwhile, cottonseed, soybean, and sesame glycerolysis-structured systems displayed gradual reductions in SFC across all temperatures. These systems all contained <40% oleic acid and XRD analysis appeared to show that the MG crystals of these systems contained mostly monopalmitin with only low amounts of monoolein.

Vereecken *et al.*[8] demonstrated that the oleic acid–linoleic acid ratio affected the SFC of MG systems at lower temperatures. It is likely that linoleic acid also influenced the SFC-temperature profiles of the glycerolysis-structured systems, as cottonseed, soybean, and sesame oil systems contained the highest levels of linoleic acid (53.4%, 54.4%, 43.7%, respectively).

However, it is difficult to accurately determine the full effect of linoleic acid because when the oleic acid content increases, without changing the SFA content, the linoleic acid content must decrease, and *vice versa*. Future studies in which the contents of various fatty acids are altered prior to glycerolysis would need to be performed to study this relationship further. This may clarify whether low levels of monoolein in the MG crystals come as a result simply of low overall oleic acid contents or due to a dilution effect from the high level of linoleic acid. Furthermore, due to either the slightly greater oleic acid content or the higher oleic acid–linoleic acid ratio, the glycerolysis-structured rice bran oil system contained more monoolein within its MG crystals compared to the cottonseed, soybean, and sesame oil systems. This could have caused the rice bran oil system to display an SFC–temperature profile which was an intermediate between the two-regimes and gradual curves. Finally, the presence of large amounts of long-chain SFAs caused a delayed reduction in SFC at higher temperatures, following the initial sharp reduction, leading to a higher end of melt temperature. This melting behavior was observed for the glycerolysis-structured peanut oil system.

When selecting a plant-based oil for structuring by glycerolysis for food applications, the achievable SFC and the melting characteristics are both of importance. These properties contribute to the maintenance of structure in a food product and the eating characteristics. From our investigation into glycerolysis-structured systems, several important points are evident. SFA content and oleic acid content are important factors contributing to the achievable SFC. The linoleic acid content may also be of importance, but when selecting an oil high in both SFAs and oleic acid, the linoleic acid content would naturally be low. Additionally, a glycerolysis-structured system in which partial glycerides containing oleic acid make a substantial contribution to the crystalline material will undergo a sharper reduction in SFC with respect to temperature. This leads to less solids at higher temperatures unless high-melting SFAs are present at high concentrations. Lower solids at high temperatures would be beneficial to the eating properties, reducing the waxy mouthfeel of the lipid, while higher solids would certainly be beneficial for temperature stability. Due to the negative health benefits and connotations associated with saturated fat, a low SFA content coupled with a high oleic acid content would certainly be welcome. However, the HOAO system showed that SFAs must be present for the partial glycerides containing oleic acid to crystallize at 5 °C. Furthermore, although the HOCO system achieved a relatively high SFC, low amounts of solids in the canola oil system demonstrated that SFAs should be present at appreciable quantities to maximize the structuring capabilities of the oleic species. A SFA content of ≥10% is likely necessary to achieve an adequate amount of crystalline material in high oleic systems, while a SFA content of 14%, as was present in the olive system, is better yet. More work still needs to be completed if we are to see just how high a SFC can be achieved in oils that would otherwise be liquid at room temperature. For example, the SFC achievable in a lipid system containing 27.5% SFAs (cottonseed oil) has the potential to be increased further if oleic acid is substituted for linoleic acid. But at the present time, based on what is known, a high SFA content within the lipid system may be detrimental to in-mouth sensory properties. For this reason, we suggest that 14–25% SFAs along with a high oleic acid content (>60%) may be optimal. This does, however, depend on the food application in question.

The potential of structuring oil systems containing more saturated fat than cottonseed oil is unknown at this point. However, at SFA contents much higher than that of cottonseed oil, it is likely that the TG system would already have some structure at room temperature. In this case, performing glycerolysis becomes superfluous. Furthermore, our group showed briefly that the structuring potential of glycerolysis was diminished when the starting material was already structured in its native TG state.[20] This was shown with mixtures of cottonseed oil and palm stearin. In addition, the melting point of a palm stearin–olein blend was unchanged following glycerolysis.[22] Regardless, glycerolysis has been shown to be effective at structuring liquid oils which would be the most suitable replacement for *trans* fat anyway, given the negative health effects of saturated fat.

10.4 Food Applications

Trans-free lipid systems are highly relevant to shortening as well as margarine and low-fat spreads. There is great potential to use the high-solid glycerolysis-structured oils in shortening applications and the systems that achieved lower levels of solids are a viable option for liquid shortenings, however, the following section will discuss the work that has been completed regarding margarines and spreads. Margarines and spreads, as well as other emulsified products, would be an excellent use for the glycerolysis-structured systems because in addition to the structuring potential of the crystalline material, the MGs and DGs present provide valuable emulsification properties.

Our group prepared an experimental margarine using the tigernut glycerolysis product as the lipid phase and no additional emulsifiers or stabilizers.[21] This experimental margarine was shown to be stable from oil separation for at least 10 months which is more than adequate, given that commercial margarines typically have a shelf life of 3–4 months. Furthermore, for margarine and spreads, the spreadability is crucial. Therefore, it is important to evaluate whether the final product will undergo brittle fracture or plastic flow during deformation. When a material is spread it undergoes a large deformation, making large deformation testing procedures ideal for examining the performance of a margarine or spread. Back extrusion testing offers a simple method of determining whether a material undergoes brittle fracture or plastic flow during deformation, based on whether the force-deformation curve of the material is jagged or smooth, respectively.[23] We performed back extrusion on the experimental tigernut margarine at a temperature typical of soft-tub margarine applications (5 °C). The force-deformation curve (Figure 10.5A) obtained for the tigernut margarine had a very similar appearance to that of commercially available margarine and butter tested at their respective application temperatures, 5 °C and 22 °C. Each of the samples produced smooth curves and demonstrated an absence of stress accumulation, indicating plastic flow behavior. In addition, the force necessary to deform the experimental tigernut margarine during the back extrusion procedure fell in between the forces required for the butter and margarine samples. These results indicate that margarine prepared with the glycerolysis-structured tigernut oil has a spreadability comparable to room-temperature butter and refrigerated soft-tub margarine commercial samples.

Another important characteristic of margarine is the melting behavior. This remains a negative aspect of using waxes for this application, which have otherwise shown promise as a *trans* fat replacement strategy.[24] Waxes have a high melting point and as a result a relatively high level of solid fat often remains in the mouth causing a residual waxy mouthfeel. This is therefore a very important aspect of margarine to examine when attempting to introduce a new fat source. We used both DSC and SFC melting profiles to examine the melting characteristics of the experimental tigernut margarine. The DSC melting profile (Figure 10.5B) demonstrated that the experimental tigernut margarine gradually melted from 5 °C to approximately 26 °C, before displaying a melting peak centered at around 32 °C. This translated to a gradual reduction in SFC as the temperature of the margarine was increased, with 3.2% solids remaining at 30 °C, before melting entirely by 35 °C (Figure 10.5C). Since no solids remained at in-mouth temperatures, the glycerolysis-structured tigernut oil demonstrated that it can successfully be used in margarine formulations to provide appropriate eating characteristics. Interestingly, the SFC of the commercial margarine tested was shown to decrease much more slowly with respect to temperature and at 35 °C over 4% solids remained. The corresponding DSC melting profile showed a melting peak just below 40 °C. The commercially available margarine examined by our group was formulated with palm oil as the structuring material, as is quite common for products of this sort. This indicated that glycerolysis-structured edible oils have the potential to perform better than the lipid sources currently being used (*i.e.*, palm oil) as *trans*-free lipid systems for margarine production.

The potential of glycerolysis-structured oils to be used in margarine is not limited to tigernut oil. Our group also demonstrated that margarine formulated with a cottonseed glycerolysis product demonstrated plastic flow behavior similar to that of a commercially available margarine.[20] However, the melting behavior of the glycerolysis-structured cottonseed margarine may not be adequate given that this system underwent a very gradual reduction in SFC with respect to temperature and still contained a substantial amount of crystalline fat at in-mouth temperatures. It should be noted that cottonseed oil systems structured under other reaction conditions had

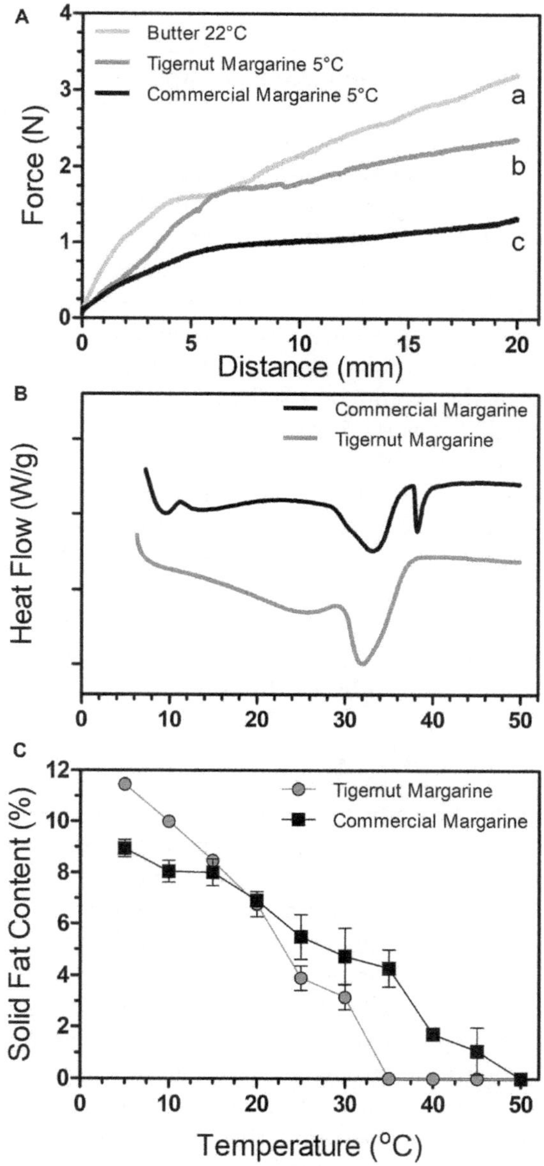

Figure 10.5 (A) Back extrusion force deformation profiles for margarine produced with glycerolysis-structured tigernut oil along with commercially available soft-tub margarine and butter. (a) Butter, (b) glycerolysis-structured tigernut margarine, and (c) commercial margarine. Experimental and commercially available margarines were measured at 5 °C, while butter was measured at room temperature (22 °C). (B) Differential scanning calorimetry melting (endothermic) profiles and (C) solid fat content temperature profiles for margarine produced with tigernut glycerolysis products (gray line) and a commercially available soft-tub margarine (black line). Adapted from ref. 21 with permission from Elsevier, Copyright 2021.

melting properties that differed from the particular glycerolysis-structured system used for this margarine. For a glycerolysis-structured lipid system to perform well in margarine formulations, as was the case with the tigernut system, a high level of solids is required at 5 °C for structuring and stability purposes. This also contributes to the textural properties of the margarine. Additionally, the glycerolysis-structured lipid system should demonstrate a segmented SFC–temperature profile, as described earlier, since this will result in a lower level of solids remaining at temperatures close to that of the mouth. Of the systems studied by our group, in addition to the tigernut oil system, rice bran, olive, and HOCO systems all shared these characteristics and therefore represent just some of the viable options for *trans*-free margarine and spread applications. Furthermore, there are potentially many more oil systems which could be used for this purpose, so long as they contain an intermediate level of SFAs, which contribute to a high SFC at 5 °C, and a high oleic acid content, contributing further to the SFC at 5 °C and to the sharp reduction in SFC as temperatures are increased. In addition, the reaction conditions, which were shown to affect the properties of glycerolysis-structured cottonseed systems, will also affect the properties of other oil systems. It is certainly possible then that reaction conditions could be employed as a further means of tailoring the physical properties of these systems. Perhaps this may prove beneficial in the case of some oils depending on the food application being investigated.

10.5 Future Directions

Glycerolysis-structured systems have shown their potential in margarine and spread formulations. However, these systems may be further reaching in the food industry as *trans*-free fat systems if they were to be used in shortening applications. While the SFC values attainable through glycerolysis are certainly very promising, the full potential of the glycerolysis-structured systems in shortening applications has not been explored at the present time. For starters, the performance of these glycerolysis-structured systems needs to be explored in baking applications. In addition, rheological characterization is necessary to compare the properties of glycerolysis-structured systems to shortenings which are currently available. Furthermore, when crystallized under shear on a stainless-steel tempering table, the crystal network of the cottonseed glycerolysis product changed dramatically. Figure 10.6 demonstrates that under shear a continuous structural network is formed containing very small fat crystals. Given the immense changes in the crystal network resulting from shear, the effects of controlled crystallization conditions (*i.e.*, shear rate, crystallization temperature) should be investigated further in each of the glycerolysis-structured systems. Altering the crystallization conditions is likely to cause changes in the molecular composition and physical properties of the fat crystals. This in turn will affect the physical properties of the glycerolysis-structured systems, which may improve their suitability for shortening applications.

Figure 10.6 Glycerolysis-structured cottonseed oil crystallized under shear (by hand) on a stainless-steel tempering table with a surface temperature of 5 °C. Scale bar represents 100 μm. Imaging of the glycerolysis product samples was done after storage for one week at 5 °C.

References

1. A. G. Marangoni, J. P. M. van Duynhoven, N. C. Acevedo, R. A. Nicholson and A. R. Patel, Advances in our Understanding of the Structure and Functionality of Edible Fats and Fat Mimetics, *Soft Matter*, 2020, **16**, 289.
2. N. K. O. Ojijo, I. Neeman, E. Eger and E. Shimoni, Effects of Monoglyceride Content, Cooling Rate and Shear on the Rheological Properties of Olive Oil/Monoglyceride Gel Networks, *J. Sci. Food Agric.*, 2004, **84**, 1585.
3. N. K. O. Ojijo, E. Kesselman, V. Shuster, S. Eichler, S. Eger, I. Neeman and E. Shimoni, Changes in Microstructural, Thermal, and Rheological Properties of Olive Oil/Monoglyceride Networks During Storage, *Food Res. J.*, 2004, **37**(4), 385.
4. C. H. Chen and E. M. Terentjev, Aging Metastab. Monoglycerides Hydrophobic Solutions, *Langmuir*, 2009, **25**(12), 6717.
5. C. H. Chen and E. M. Terentjev, in *Edible Oleogels*, ed. A. G. Marangoni and N. Garti, AOCS Press, Urbana, Illinois, 2nd edn, 2018, ch. 5, pp. 103–131.
6. H. D. Batte, A. J. Wright, J. W. Rush, S. H. J. Idziak and A. G. Marangoni, Effect of Processing Conditions on the Structure of Monostearin–Oil–Water Gels, *Food Res. J.*, 2007, **40**, 982.
7. A. G. Marangoni, S. H. J. Idziak, C. Vega, H. Batte, M. Ollivon, P. S. Jantzi and J. W. E. Rush, Encapsulation-Stucturing of Edible Oil Attenuates Acute Elevation of Blood Lipids and Insulin in Humans, *Soft Matter*, 2007, **3**, 183.
8. J. Vereecken, I. Foubert, W. Meeussen, A. Lesaffer and K. Dewettinck, Fat Structuring with Partial Acylglycerols: Effect on Solid Fat Profiles, *Eur. J. Lipid Sci. Technol.*, 2009, **111**, 259.

9. S. K. Lo, C. P. Tan, K. Long, M. S. A. Yusoff and O. M. Lai, Diacylglycerol Oil—Properties, Processes and Products: A Review, *Food and Bioprocess Technology*, 2008, **1**(3), 223.
10. H. D. Belitz, W. Grosch and P. Schieberle, in *Food Chemistry*, ed. H. D. Belitz, W. Grosch and P. Schieberle, Springer, New York City, New York, 4th edn, 2009, ch. 3, pp. 158–245.
11. B. D. Flickinger and N. Matsuo, Nutritional Characteristics of DAG Oil, *Lipids*, 2003, **38**(2), 129.
12. N. Matsuo, Nutritional Characteristics and Health Benefits of Diacylglycerol in Foods, *Food Sci. Technol. Res.*, 2004, **10**(2), 103.
13. T. Teramoto, H. Watanabe, K. Ito, Y. Omata, T. Furukawa, K. Shimoda, M. Hoshino, T. Nagao and S. Naito, Significant Effects of Diacylglycerol on Body Fat and Lipid Metabolism in Patients on Hemodialysis, *Clinical Nutrition*, 2004, **23**(5), 1122.
14. H. Yanai, Y. Tomono, K. Ito, N. Furutani, H. Yoshida and N. Tada, Diacylglycerol Oil for the Metabolic Syndrome, *Nutr. J.*, 2007, **6**(43), 1.
15. G. P. McNeill, S. Shimizu and T. Yamane, High-Yield Enzymatic Glycerolysis of Fats and Oils, *J. Am. Oil Chem. Soc.*, 1991, **68**(1), 1.
16. H. Noureddini and S. E. Harmeier, Enzymatic Glycerolysis of Soybean Oil, *J. Am. Oil Chem. Soc.*, 1998, **75**(10), 1359.
17. I. Elfman-Borjesson and M. Harrod, Synthesis of Monoglycerides by Glycerolysis of Rapeseed Oil Using Immobilized Lipase, *J. Am. Oil Chem. Soc.*, 1999, **76**(6), 701.
18. J. B. Kristensen, X. Xu and H. Mu, Process Optimization Using Response Surface Design and Pilot Plant Production of Dietary Diacylglycerols by Lipase-catalyzed Glycerolysis, *J. Agric. Food Chem.*, 2005, **53**, 7059.
19. P. B. L. Fregolente, L. V. Fregolente, G. M. F. Pinto, B. C. Batistella, M. R. Wolf-Maciel and R. M. Filho, *Monoglycerides and Diglycerides Synthesis in a Solvent-free System by Lipase-catalyzed Glycerolysis*, 2008, **146**, 165.
20. R. A. Nicholson and A. G. Marangoni, Enzymatic Glycerolysis Converts Vegetable Oils into Structural Fats with the Potential to Replace Palm Oil in Food Products, *Nature Food*, 2020, **1**(11), 684.
21. R. A. Nicholson and A. G. Marangoni, Lipase-catalyzed Glycerolysis Extended to the Conversion of a Variety Edible Oils into Structural Fats, *Curr. Res. Food Sci.*, 2021, **4**, 163.
22. E. Subroto, Supriyanto, T. Utami and C. Hidayat, Enzymatic Glycerolysis–interesterification of Palm Stearin–olein Blend for Synthesis Structured Lipid Containing High mono- and Diacylglycerol, *Food Sci. Biotechnol.*, 2019, **28**(2), 511.
23. A. J. Gravelle, M. Davidovich-Pinhas, S. Barbut and A. G. Marangoni, Influencing the Crystallization Behavior of Mixtures of Stearyl Alcohol and Stearic Acid (SOSA) Using Ethylcellulose, *Food Res. J.*, 2017, **91**, 1.
24. A. R. Patel, R. A. Nicholson and A. G. Marangoni, Applications of fat mimetics for the replacement of saturated and hydrogenated fat in food products, *Curr. Opin. Food Sci.*, 2020, **33**, 61.

CHAPTER 11

Modulating the Rheology of Oleogels to Mimic the Functionality of Commercial Fats

J. D. PÉREZ-MARTÍNEZ*[a] AND J. J. ORNELAS-PAZ[b]

[a] Universidad Autónoma de San Luis Potosí, Facultad de Ciencias Químicas-CIEP, Av. Dr. Manuel Nava 6, 78210, San Luis Potosí, México; [b] Centro de Investigación en Alimentación y Desarrollo A. C.-Unidad Cuauhtémoc, Av. Río Conchos S/N, 31570, Cd. Cuauhtémoc, Chihuahua, México

*Email: jdavidperez@uaslp.mx

11.1 Introduction

Plastic fats are partially crystallized systems that behave as solids able to resist small efforts, yielding once it exceeds a certain yield value (stress or deformation), flowing to generate a permanent deformation in its structure, and recovering the viscoelastic solid behavior and consistency when the stress is removed.[1,2] This rheological behavior enables plastic fats to perform several functions in foods, including the protection of a food's surface due to their barrier property, coating formation, modification of food texture, and acting as vehicles for water droplets, non-lipid solids, and gas bubbles. These functions are associated with sensory attributes in foods like lubrication and spreadability in margarine and butter, lightness in whipped

Food Chemistry, Function and Analysis No. 31
Development of *Trans*-free Lipid Systems and their Use in Food Products
Edited by Jorge F. Toro-Vazquez

Published by the Royal Society of Chemistry, www.rsc.org

creams, release of flavor and surface coating in confectionary products, softness in baked goods, and juiciness and tenderness in meat products.[1] The deleterious effects of saturated and *trans* fats on human health have driven the development of *trans* fat-free foods, reduced in saturated fat but maintaining their sensory attributes. Edible oleogels, with up to 98% of vegetable oil physically trapped within their structure, break and flow under stress, but most of them do not meet the mechanical strength and recovery after shear as saturated and *trans* fats do. Although the mechanical strength of oleogels increases with the gelator concentration, this strategy does not necessarily improve their mechanical recovery. Additionally, the development of off-flavors or unpleasant mouthfeel in food systems could arise in the product as the gelator concentration increases. Nonetheless, some changes in compositional factors (gelator refining, gelator mixing, use of additives, *etc.*), and processing conditions (cooling rate, setting temperature, shearing, *etc.*) have proven to be effective in modulating the oleogels' rheology and, tentatively, improving the functionality of plastic fats in a better way.

In this chapter, selected groups of gelators are discussed, describing the effects of their chemical composition and processing conditions on the corresponding oleogel's rheological properties and stability. The performance of gelators as plastic fat replacements in foods is also analyzed. Wax esters (WE), hydrocarbons (HC), free fatty acids (FFA), and free fatty alcohols (FFAl) impact the rheological properties and minimal gelling concentration of natural waxes. Wax components commonly overlooked (*e.g.*, hydroxylated esters, hydroxycinnamic esters, and resinous matters) also have relevant effects on an oleogel's rheological properties. The molecular weight of ethylcellulose (EC) and polar molecules in the oil phase of EC oleogels largely modifies the oleogel elasticity and mechanical recovery after shear. The addition of selected additives to oleogels structured with monoacylglycerols (MGs) delays the deleterious phenomena induced by polymorphic changes occurring in MG crystals within a week after the oleogel production. FFA and FFAl modify the rheological properties of oleogels as a function of the hydrocarbon chain length, but mixtures of FFA/FFAl (3 : 7 ratio) develop crystal compounds that outperform the effect of their individual components. Gelator mixtures closely mimicking plastic fats are analyzed in this chapter to establish the individual contributions of the components to the hybrid oleogel rheology. The replacement of plastic fats with oleogels is analyzed for each gelator group comparing the rheological parameters and sensory scores of oleogel-based foods with reference to fat-based foods. Processing conditions such as setting temperatures, cooling rate, and shearing play a significant role in the rheological properties of edible oleogels. Higher cooling rates improve the rheological properties of oleogels structured with crystallites, but not for those structured with EC. Shearing has a deleterious effect on the strength and oil binding capacity of wax and MGS oleogels. However, hybrid oleogels structured with two or three gelators have shown to be effective in overcoming these limitations.

In foods, the total or partial replacement of plastic fats with oleogels has been mainly focused on margarine-type products, baked goods, and meat products. Oleogels and W/O emulsions produced with <5% natural wax in the oily phase can match the consistency of reference products at refrigeration temperature, but they become too soft at room temperature. In many baked goods, oleogels are more effective than liquid oil in preventing the gluten-network development required to produce the crumb shortening. Nonetheless, oleogels with poor mechanical recovery after shear are not as efficient as commercial shortenings in avoiding the gluten network formation resulting in softening of baked products. Thus, oleogels must match the rheological behavior of the reference product during and after applying a level of shear that is characteristic of the production process to provide the required functionality. Instrumental measurements of texture show that full or partial replacement of animal fat with oleogels is more functional than replacing with liquid oil in finely-comminuted meat products, but it is not satisfactory in coarse-ground meat products. Nonetheless, the low correlation between instrumental texture measurements and sensory scores shows that there is pending work on the objective assessment of the rheological attributes of food products.

11.2 Rheological Methods for Oleogel Characterization

The rheological methods used for oleogel evaluation are commonly based on those used to determine the resistance to deformation and flow and the viscoelastic behavior of plastic fats.[3,4] Nonetheless, some methods have been specifically implemented or developed to assess the brittleness and mechanical reversibility (*i.e.*, partial thixotropy) of oleogels.[5–7] Overall, these methods can be classified as fundamental and empirical. Empirical methods (*e.g.*, texture profile analysis) are widely used to measure the resistance to deformation, fracture, and flow of plastic fats. In these methods, the nature and level of strain and stress are not well defined. As measuring conditions are arbitrary and the stresses and strains involved are not defined, empirical methods do not allow the measurement of fundamental rheological parameters.[8] Measurements obtained with these methods provide some indication of the textural characteristics but, on an arbitrary scale. Thus, results obtained from empirical methods are useful to compare one sample with another only if both have been analyzed under the same measurement conditions.[8] Comparisons of rheological measurements obtained in different studies using empirical methods can only be achieved through standardized methods. Nonetheless, non-standardized empirical methods are widely used due to their low cost, simplicity of implementation, easy pre-test sample handling, and direct correlation with sensory attributes.[8,9] Empirical methods, as texture profile analysis, currently used for the analysis of plastic fats and oleogels involve penetration, compression, or extrusion of the samples. In the penetration-based methods, the force–time curve is recorded by measuring the force applied by a probe (*i.e.*, cylinder, cone, ball, *etc.*) to achieve a specific value of penetration or deformation of the sample. The force–time curve of the sample provides the hardness (the maximum peak on the force curve), fracture (the force at the first major drop on the force curve), penetration work (the area under the force–distance curve), stickiness (the maximum force required to detach the probe from the sample), and adhesiveness (the area under the force–time curve required to pull the probe away from the sample) of the sample. Alternatively, the penetration depth of a probe can be measured under a constant load for a specified time. Additionally, it is possible to calculate the yield strength and hardness values from the penetration depth, load, and probe geometry. However, the penetration depth is the most reported.[2,9,10]

In the case of the compression-based methods, the force–time curve is recorded while the sample (*e.g.*, cylinder, cube, *etc.*) is compressed between two parallel plates using a constant crosshead speed (*e.g.*, 1–10 mm s^{-1}) until achieving a particular percentage (*e.g.*, 5–80%) of its original height. The sample is subjected to one or two compression cycles although for plastic fats the analysis is generally restricted to one compression cycle. As with the penetration-based methods, the hardness, fracture, penetration work, stickiness, and adhesiveness are calculated from the force–time curve.[2,9,10] In contrast, in the extrusion-based methods, the equipment records the force–distance curve generated to achieve a constant extrusion speed of a sample through one or more slots or holes located in the test cell. The maximum force and total work (area under the force–distance curve) required to achieve extrusion are usually the recorded parameters.[9]

On the other hand, fundamental rheological methods measure well-defined rheological properties (*e.g.*, viscosity, elastic modulus, shear modulus, *etc.*). The levels of deformation, shear and stress are well defined during the measurements by force and energy balances that consider the system geometry. Thus, the fundamental rheological properties measured with these methods will be independent of the measuring probe and the instrument, and therefore comparable between different studies. The most common fundamental rheological methods for the characterization of plastic fats and oleogels are small amplitude oscillatory shear (SAOS) and thixotropy tests.

The measurement of rheological properties in SAOS methods implies applying an oscillatory shear strain (γ) to the sample and recording the resulting stress (σ) or *vice versa*. The applied deformation must be within the limits of the linear viscoelastic region (LVR), which implies that the deformation does not generate breakage of structural elements of the sample. Therefore, the strain and stress generate sinusoidal functions in time (t), $\gamma = \gamma_0 \sin(\omega t)$ and $\sigma = \sigma_0 \sin(\omega t + \delta)$, where σ_0 is the maximum stress, ω is the strain frequency, γ_0 is the maximum applied strain, and δ is the phase shift angle. If the material is an ideal elastic solid, the stress is directly proportional to the deformation; that is, the functions $\sigma(t)$ and $\gamma(t)$ will be in phase (*i.e.*, synchronized), and $\delta = 0°$. In the case of an ideal viscous liquid, the function of $\sigma(t)$ is out-of-phase from the function $\gamma(t)$ and $\delta = 90°$. Viscoelastic materials, as plastic fats and oleogels, contain liquid and elastic contributions that are in-phase and out-of-phase. Therefore, in viscoelastic materials the function $\sigma(t)$ has values of δ that fall between the values of the ideal elastic solid (0°) and the ideal viscous liquid (90°). Thus, δ indicates the predominant character in the viscoelastic material. The liquid character prevails at δ values between 45° and 90° while the elastic character predominates at δ values between 0° and 45°. If the elastic and viscous behaviors are exactly balanced, as in the sol–gel transition, then $\delta = 45°$.[2] On the other hand, the overall resistance of the sample to deformation is determined by the complex shear modulus $G^* = \gamma(t)/\sigma(t)$. Although the results of an SAOS test can be analyzed from the values of G^* and δ as a function of ω, it is common to perform the analysis in terms of other rheological properties derived from them. The experimental stress curve is separated into two components; the stress that is in-phase with the deformation, which is defined as elastic or storage modulus (G') and calculated as $G' = \gamma_0/\sigma_0 \cos(\delta)$, and; the stress that is 90° out-of-phase with the strain, defined as the viscous modulus (G'') and calculated as $G'' = \gamma_0/\sigma_0 \sin(\delta)$. Both are related by $\tan \delta = G''/G'$.[2,9]

There are different methods to assess the mechanical recovery of materials. For oleogel's rheological characterization the three-interval thixotropy tests (3iTT) have shown to be particularly practical.[6] The 3iTT measure the material response (*e.g.*, G') resulting from stepwise changes in shear or strain rate. The viscometrical 3iTT test subjects the sample to three shear intervals; interval 1: low shear level (*e.g.*, 1 s^{-1}), interval 2: high shear level (*e.g.*, 100 s^{-1}) and, interval 3: the same shear level as in interval 1. The mechanical recovery is measured by comparing the steady-state viscosity at the end of the interval 1 against the viscosity at different times of the third interval.[2] Limitations of the viscometrical 3iTT method in oleogel characterization are addressed in Section 11.3.1.2. The oscillatory 3iTT test consists of stepwise changes (intervals) in oscillatory stress or strain; interval 1: stress or strain within the LVR, interval 2: stress or strain outside the LVR (*e.g.*, above yield point) and, interval 3: the same conditions as in interval 1. The G' measured within the LVR of interval 1 is compared with the G' values measured within the LVR of interval 3.[2] The mixed oscillatory-rotational-oscillatory 3iTT tests consist of interval 1: oscillatory strain within the LVR, interval 2: application of a rotational shear rate (*e.g.*, 10 s^{-1}) to perturb the internal structure of the material and, interval 3: oscillatory deformation under the same conditions as interval 1.[2] The mechanical recovery is established as the proportion of viscoelastic properties (*e.g.*, G') measured in the first and third intervals. The shearing levels applied during interval 2 are established according to the processing or final application conditions to which the material is subjected.

11.3 Edible Oleogels Mimicking the Properties of Plastic Fats

The rheological properties and textural attributes of oleogels can be modified by several strategies to approximate (*i.e.*, mimicking) the functionality of plastic fats. The main determining factors are changes in the composition and processing conditions during gel formation. The system composition has been, by far, the most studied factor in the quest for oleogels and gelled emulsions mimicking the functionality of plastic fats. Compositional changes to modify the rheology are driven by gelator type, concentration, and purity, as well as the addition of polar compounds to the gelled oil. The great variety of chemical species with the ability to gel vegetable oils can be grouped into natural waxes, EC, MGS, FFA, FFAl, among others.

11.3.1 Natural Wax Oleogels

11.3.1.1 Compositional Effect

Natural waxes from vegetables and insects are able to produce thermoreversible oleogels exhibiting moderate rheological changes between the body and refrigerator temperatures. Sunflower (SFW), candelilla (CW), rice bran (RBW), bees (BW), carnauba (CRW), and sugarcane (SCW) waxes are commonly used to produce edible oleogels.[11–20] These waxes are mainly composed of four groups of chemical species, which are WE, HC, FFA, and FFAl (Table 11.1). Each of these groups contains several compounds that contain fully saturated aliphatic hydrocarbon chains. In addition, some waxes contain specific groups such as hydroxyesters (28% in the BW and up to 14% in the CRW), hydroxycinnamic acid diesters (up to 23% of the CRW), free fatty aldehydes (FFAld; up to 60% of the SCW), and triterpenic alcohols (up to 24% of the CW).[15,16,19,21]

Wax composition is a determinant for crystal network structuration, wax solubility, interactions with polar compounds in the oil, and, consequently,

Table 11.1 Main components of natural waxes.[a]

	SFW[11–13] (%)	CW[13–15] (%)	RBW[13,17] (%)	BW[13,16] (%)	CRW[13,15,18] (%)	SCW[19] (%)
Wax esters	66–96	6–16	66–93	58–75[b]	36–62[c]	2
Hydrocarbons	0–7	41–73	0–7	15–27	0–1	6
Free fatty acids	3–16	8–17	6–16	8–9	7–11	3
Free fatty alcohols	0–13	4–6	0–13	0–6	0–12	30
Fatty aldehydes	0	0	0	0	0	60
Others	0	0–41	0	0–6	0–53	0

[a]SFW, Sunflower wax; CW, candelilla wax; RBW, rice bran wax; BW, bees wax; CRW, carnauba wax; SCW, sugarcane wax.
[b]35% wax esters (C40–C50), 28% hydroxy esters (C40–C50), 12% diesters (C56–C64).[16]
[c]38–40% wax esters, 20–23% *p*-hydroxycinnamic diesters, 12–14% ω-hydroxy esters 5–7%, *p*-methoxycinnamic diesters.[18]

the rheological properties of oleogels. The minimum amount of gelator able to form a self-supporting gel is denoted as the minimum gelling concentration (MGC). The MGC simultaneously depends on the solubility of the gelator in the solvent, the morphology and size of primary crystals as well as their subsequent aggregation in a crystal network.[22] Overall, the SFW has the lowest MGC, followed by CW and BW, RBW, CRW, and SCW (Table 11.2). Interestingly, RBW from different suppliers showed MGC values in soybean oil between 0.5 and 5.0%, indicating that the extent of refining and composition are of the outmost relevance for wax gelation.[23] The differences in MGC for RBW have been associated with the amount of resinous matter composed of a mixture of FFAld, FFAl, and FFA.[17] Similarly, Wild and Brazilian CRW showed different MGC (2 and 4%, respectively) and crystallization profiles in spite of their similar content of WE, HC, FFA, and FFAl.[13]

Table 11.2 Minimal gelling concentration of natural waxes.[a]

Gelator	MGC (%)	T_{set}	Oil	Method
SFW	0.5,[24] 1.0[13]	5 °C	Rice bran oil	SAFS
	0.5[7]	5 °C	High oleic sunflower oil	SAFS
	0.5[23]	RT	Soybean oil	Inverted tube
CW	0.5,[24] 1.0[13]	5 °C	Rice bran oil	SAFS
	0.5[25]	5 °C	High oleic safflower oil	Tilted tube
	0.5[26]	5 °C	Soybean oil	Tilted tube
	0.75[7]	5 °C	High oleic sunflower oil	SAFS
	1.0,[25] 2.0[14]	25 °C	High oleic safflower oil	Tilted tube
	1.0,[23] 2.0[23]	RT	Soybean oil	Inverted tube
	2.0[20]	20 °C	Olive oil	Tilted tube
	2.0[26]	25 °C	Soybean oil	Tilted tube
RBW	0.5[20]	20 °C	Olive oil	Tilted tube
	0.5,[23] 1.0,[23] 5.0[23]	RT	Soybean oil	Inverted tube
	4.5,[24] 5.0[13]	5 °C	Rice bran oil	SAFS
BW	1.0,[24] 1.5[13]	5 °C	Rice bran oil	SAFS
	1.0[7]	5 °C	High oleic sunflower oil	SAFS
	1.0[27]	RT	Sunflower oil	Tilted tube
	2.0,[23] 3.0[23]	RT	Soybean oil	Inverted tube
	3.0[28]	25 °C	Soybean oil	Inverted tube
	3.0[28]	25 °C	Camellia oil	Inverted tube
	3.0[28]	25 °C	Sunflower oil	Inverted tube
	4.0[28]	25 °C	Flaxseed oil	Inverted tube
CRW	4.0[7]	5 °C	High oleic sunflower oil	SAFS
	1.5,[24] 2.0[13b]	5 °C	Rice bran oil	SAFS
	3.5,[24] 4.0[13c]	5 °C	Rice bran oil	SAFS
	4.0[20]	20 °C	Olive oil	Tilted tube
	3–5[23]	RT	Soybean oil	Inverted tube
SCW	2.0[26]	5 °C	Soybean oil	Tilted tube
	4.0[26]	25 °C	Soybean oil	Tilted tube

[a]SFW, sunflower wax; CW, candelilla wax; RBW, rice bran wax; BW, beeswax; CRW, carnauba wax.

[b]Wild carnauba wax.

[c]Brazilian carnauba wax; SCW, sugarcane wax; RT, room temperature; MGC, minimal gelling concentration; T_{set}, setting temperature; SAFS, small amplitude frequency sweep.

This might be attributed to unquantified/unidentified components of the resinous component of waxes.

The relative proportion of chemical species in waxes produces important changes in the microstructure and physical properties of oleogels.[13,29,30] Doan *et al.*[13] observed that the content of WE, HC, FFA, and FFAl in several waxes (SFW, CW, BW, RBW, CRW, and Berry wax) determined the mechanical properties of the corresponding oleogels prepared at the MGC (Table 11.2). Within this context, G' and the oscillatory elastic limit of oleogels showed a negative correlation with the FFAl fraction and a positive correlation with the HC, FFA, and WE fractions, in decreasing order of correlation strength.[13] Scharfe and Flöter[31] observed that despite the compositional similarity and the difference in wax concentration, SFW (MGC = 1%) showed a G' eight times higher than RBX (MGC = 5%). The differences in G' and MGC for these waxes were associated with the morphological features of the crystals developed by the RBX and the SFW, and finally with the differences in the hydrocarbon chain length of the WE. Nonetheless, the melting thermograms of RBW showed, in addition to the melting peak of WE detected in both waxes at ~75 °C, another endothermic peak at 60 °C. The additional endothermic peak suggested the melting of crystals of compounds (*e.g.*, FFA, FFAl, or FFAld) that were not completely removed during the refining process.[17] In the same manner, Palla *et al.*[32] reported that the enthalpy of commercial SFW (214.20 J g^{-1}) was substantially greater than that of a laboratory-refined SFW (175.64 J g^{-1}), in spite of both SFWs showing similar WE content (0.732 and 0.739 kg of WE per kg of wax, respectively). The content of minor components in each wax had a different effect on the crystallization of major components, affecting the rheology of oleogels.[32] In the quest for the optimal composition for oleogelation with BW, Sarkisyan *et al.*[29] studied the interactions among six fractions of BW with a high proportion of HC (F1), monosters (F2), mono-, di- and triesters (F3), mono-, di- and triesters and FFA (F4), hydroxymonoesters, hydroxy-polyesters, FFA and FFAl (F5), and FFA and FFAl (F6). They observed by binary phase diagrams of these fractions, that F1 and the other fractions formed eutectic systems, but mixtures of other fractions produced solid solutions.[29] This implies that BW oleogels could be structured by more than one type of crystal. BW fractions and their combinations produced oleogels with significant differences in microstructure and textural properties. The oleogels structured with the fractions rich in HC (F1) and mono-, di- and triesters (F2, F3, and F4) showed branched microplates of larger sizes than the BW crystals, resulting in a yield value greater than that of BW (Figure 11.1). In contrast, the fractions rich in FFA and FFAl (F5 and F6) showed crystal sizes and yield values lower than those of BW (Figure 11.1). The comparative analysis of oleogels obtained with the individual fractions of BW and their mixtures showed antagonistic or synergistic interactions on the rheology of the system. Notably, the mixture of F1, F3, and F5 showed a synergistic effect on the mechanical properties of the system, reaching a Young's modulus greater than that of the individual

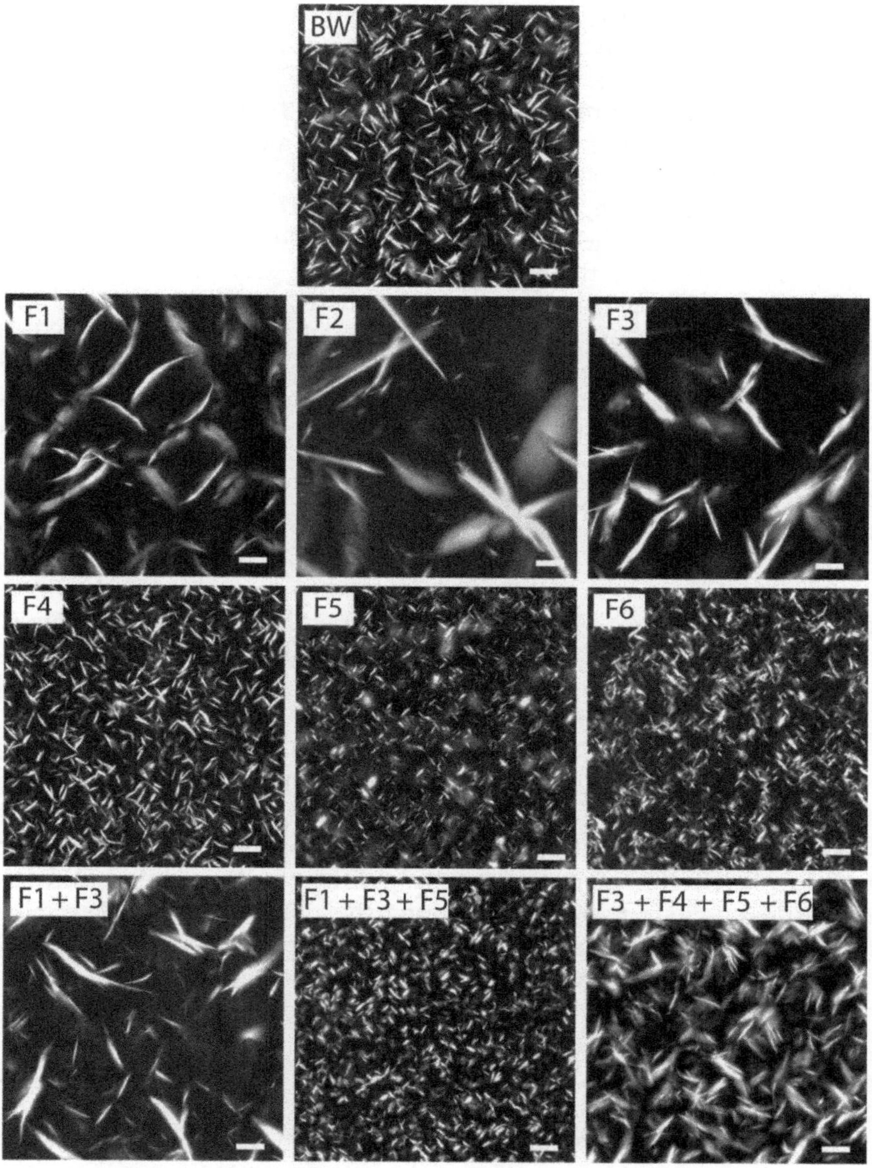

Figure 11.1 Polarized light micrographs of crystals obtained for oleogels formulated with bees wax (BW) and bees wax fractions with a high proportion of HC (F1), monoesters (F2), mono-, di- and triesters (F3), mono-, di- and triesters and FFA (F4), hydroxy-monoesters, hydroxy-polyesters, FFA and FFAl (F5), FFA and FFAl (F6), and their combinations (horizontal scale equals 100 μm).
Adapted from ref. 29 with permission from John Wiley & Sons, Copyright © 2020 AOCS.

fractions and BW.[29] The increase in the concentration of FFAl led to a decrease in the elasticity of oleogels.[29] Serrato-Palacios *et al.*[30] observed a remarkable effect of the HC/FFA ratio on the melting profile, crystalline habit, and rheological properties of CW oleogels. These authors tested the effect of several ratios of *n*-hentriacontane (C31) to melissic acid (C30-A), the most abundant HC and FFA in CW, on the oleogel properties and observed that the melting profile, crystalline habit, and rheological properties of the gels strongly depended on the molar fraction of C31 (x_{C31}). The phase diagrams showed that the crystallization of the C31/C30-A mixtures at $x_{C31} \leq 0.95$ produced at least two crystalline phases, one rich in C31 and the other in C30-A. At room temperature, both C31 and C30-A crystals had orthorhombic subcell packing and, therefore, they were undistinguishable by wide angle X-ray diffraction. In the study of complex mixtures, such as CW, this can be misinterpreted as the crystallization of all compounds into a single crystalline form (*i.e.*, formation of a solid solution). A model mixture of CW (C_{31}/C_{30}-A with $x_{C31} = 0.8$) showed a granular microstructure with a crystal size extremely different from that of the pure compounds and significantly close to that of CW crystals (Figure 11.2a).[30] The melting profiles of oleogels of the model mixture and CW showed melting endotherms at very close temperatures, although the melting enthalpy for CW was significantly lower than that of the model mixture (Figure 11.2b). The lower enthalpy for CW could be associated with its high content of resinous material (*e.g.*, triterpenic alcohols) that did not crystallize in the system. Interestingly, the CW oleogels showed higher hardness than those produced exclusively with C31, C30-A, and the model mixture,[30] suggesting that the non-crystallizing species cause significant structural changes and promote inter-crystal interactions leading to a higher mechanical strength. These studies collectively evidenced that the native wax composition and the compositional changes originated during the refining process determine the gelling capacity and rheological properties of oleogels. Unfortunately, the wax resinous matter and other compounds are not usually determined in the studies aiming to correlate composition and oleogel properties.

The effect of gelator concentration on the rheology of wax oleogels has been extensively assessed by non-standardized empirical methods. The mechanical resistance to deformation, breakage, and flow of wax oleogels are commonly reported as hardness (*e.g.*, firmness), stickiness, penetration work, and penetration depth. The effect of wax concentration has been generally evaluated at two or three concentration levels established *a priori* to produce oleogels with hardness or melting points close to those of a reference food product (*e.g.*, margarine, shortening, topping, *etc.*). Independently of the type of wax and vegetable oil, the mechanical resistance of wax oleogels increases proportionally with wax concentration.[14,20,23,25,26,32–34] In measurements carried out by SAOS, the G' (Pa) of oleogels with 1–3% CW developed under static conditions increased in the form of a power law equation with the mass percentage of CW ($G' = 968.55 \times CW^{5.23}$) in a similar way as in plastic fats structured with

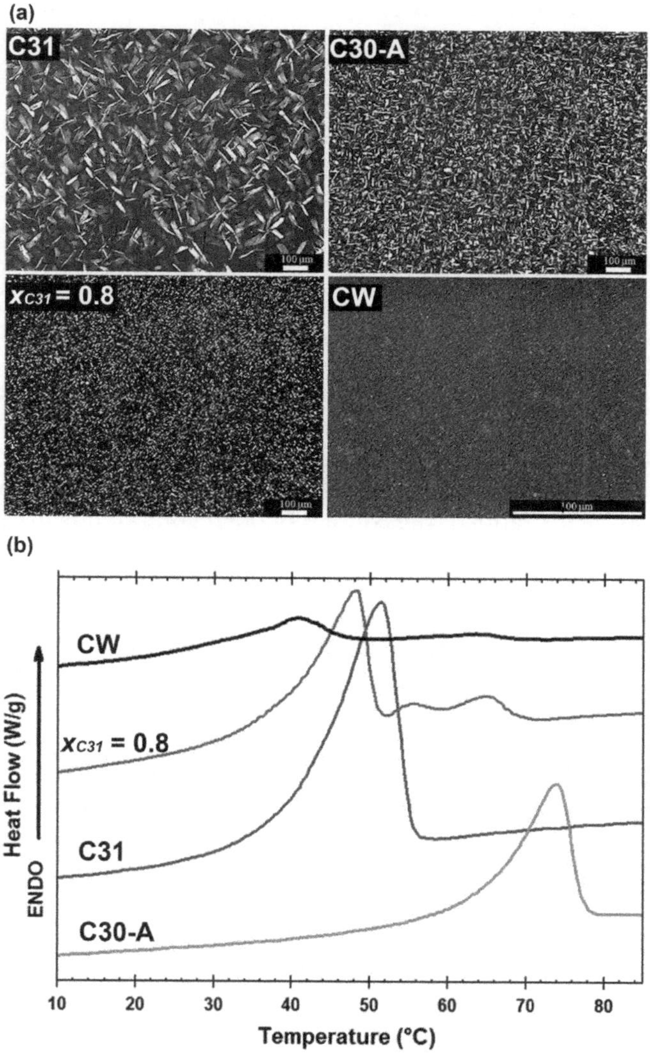

Figure 11.2 Polarized light micrographs (a) and melting thermograms (b) for oleogels with 4% (wt/wt) *n*-hentriacontane (C31), melissic acid (C30-A), *n*-hentriacontane/melissic acid mixture with $x_{31} = 0.8$ (C31/C30-A), and candelilla wax (CW).
Adapted from ref. 30 with permission from John Wiley & Sons, Copyright © 2015 American Oil Chemists' Society (AOCS).

5–15% fully hydrogenated soybean oil (FHSBO) ($G' = 2.30 \times \text{FHSBO}^{4.20}$).[5] A similar trend was observed for oleogels with 2–6% shellac wax.[35] This rheological behavior suggested that wax oleogels, as plastic fats, are structured as viscoelastic gels of aggregated colloidal particles.[7] The elastic property scaling of CW oleogels over a wide concentration range (1–8%) fitted to the fractal model for colloidal gels in the weak link regime.[36] The fractal dimension (D) of the CW oleogels at 5 and 25 °C were within the range observed for milk fat and palm oil structured through reaction-limited aggregation. It is worth noting that the scaling of the elastic properties has not been evaluated for other wax oleogels and does not seem to follow an exponential growth for oleogels formulated with 5% to 15% of natural wax.[37,38] The scaling behavior of the elastic properties of CW oleogels indicates that the rheological properties of natural wax-oleogels are affected by the size of the primary particles (wax crystals) and the aggregation process necessary to structure the spanning crystal network. Thus, compositional changes or processing conditions affecting the size of primary crystals and the aggregation process are key factors for engineering the rheology of these gels.

Overall, the effect of oil composition on the rheological properties of wax oleogels is limited. Thus, the hardness of SFW oleogels showed slight changes with fatty acid compositions of the vegetable oil.[39] In particular, the changes in hardness did not correlate with major fatty acids, unsaturated to saturated fatty acids ratio, or the content of polar compounds in oils. A small but significant inverse correlation between the content of polar compounds and the hardness of margarine produced with these SFW oleogels was the only compositional effect associated with the vegetable oil composition.[39] Similarly, BW oleogels produced with oils rich in medium-chain or long-chain fatty acids showed similar microstructural features and thermal parameters and both oleogels showed the same G^* value under isothermal conditions (25 °C).[40] RBW oleogels produced with commercial corn oil exhibited higher G' and G'' than unrefined corn germ oil oleogels. The low elasticity shown by oleogels prepared with unrefined oil was attributed to the development of a smaller amount of crystalline phase, resulting from interactions between polar lipids of the oil (*e.g.*, endogenous phospholipids, tocopherol, and certain phenolics, *etc.*) and hydroxylated compounds of BW.[41] These studies indicate that the rheological properties of wax oleogels are more affected by the polar lipid concentration than by the TAG composition of the vegetable oil.

11.3.1.2 Processing Conditions

The cooling rate modifies the mechanical properties of oleogels. Fast cooling during gelling significantly increases the hardness of SFW and sorghum wax oleogels.[23,42] The hardness of 3% SFW soybean oil oleogels increases up to 4 times by increasing the cooling speed from 1 to 4 °C min^{-1} (Figure 11.3).[23] Higher cooling rates induce higher nucleation rates, reducing the crystal size and forming more extended networks with a greater number of bonding zones between wax crystals (Figure 11.4).[43] Increasing the cooling rate from 1.5 to 5 °C min^{-1} causes a decrease in crystal length to more than half, decreasing the porous area fraction of the lattice, modifying the fractal dimension of the crystal network, and increasing the oil binding capacity in CW, RBW, and SFW oleogels.[44] The microstructural changes suggest that the

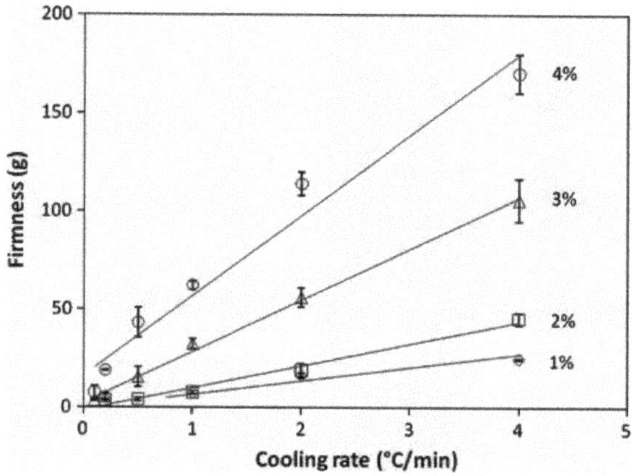

Figure 11.3 Effect of cooling rate on the firmness of 1%, 2%, 3%, and 4% (wt/wt) sunflower wax (SFW) in soybean oil. Error bars show the standard deviation of three replicates. The mean values were significantly different as a function of the cooling rate for the same concentration, except the values obtained from assays conducted at 0.2 °C min^{-1} and 0.5 °C min^{-1} with 2% SFW and at 0.1 and 0.2 °C min^{-1} with 3 and 4% (wt/wt) SFW. Mean comparisons were performed with a Tukey–Kramer HSD test ($P < 0.05$).
Reproduced from ref. 23 with permission from John Wiley & Sons, Copyright © 2012 American Oil Chemists' Society (AOCS).

elasticity of oleogels produced with these waxes could also increase if the cooling rate is increased during the gel formation stage.

Shearing effects on wax oleogels have been studied during gelling process and on oleogels structured under static conditions. Toro-Vazquez *et al.*[45] studied the effect of shearing during the gelling process of wax oleogels and observed that the application of constant shear (30 to 600 s^{-1}) to CW solutions during cooling (90 °C to 5 °C) produced poorly structured oleogels, which were weaker than those formed under static conditions (CS-30 to CS-600 in Figure 11.5). The G' of CW oleogels decreased as the shear rate increased. The oleogels produced under constant shear increased their elasticity under quiescent conditions for 60 min at 5 °C, however, their G' values remained up to three orders of magnitude below those of oleogels structured under static conditions. Shear application during cooling up to 10 °C above the dynamic crystallization temperature onset ($T_C = 52$ °C) and subsequent cooling under static conditions (S52 in Figure 11.5) generated oleogels with larger microplatelets and G' slightly higher than those gelled under static conditions. Oleogels S52 structured at 300 s^{-1} showed the highest G', which was associated with the higher solid phase content (SPC), and the development of mesophases that crystallized and developed a three-dimensional network with a higher degree of interaction between the microplates and greater elasticity.[45]

The mechanical recovery or partial thixotropy of wax oleogels is low compared to those of plastic fats. Structural perturbation occurs when the

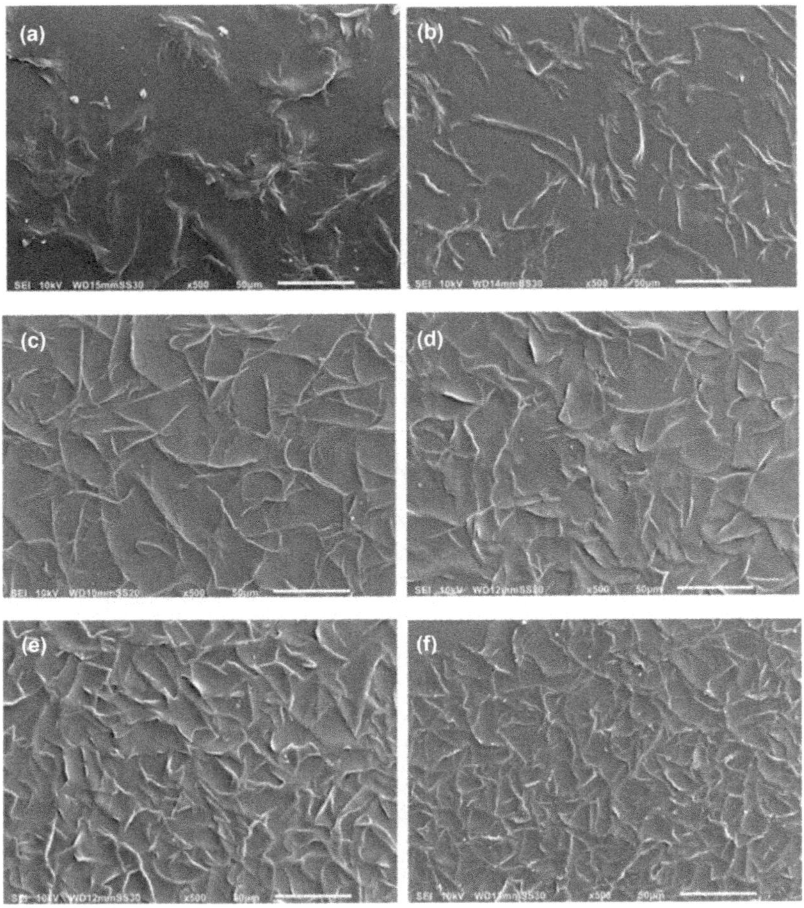

Figure 11.4 Scanning electron micrographs of 5% SFW oleogels in soybean oil obtained by cooling at (a) 0.5 °C min^{-1}, (b) 1 °C min^{-1}, (c) 3 °C min^{-1}, (d) 5 °C min^{-1}, (e) 10 °C min^{-1}, and (f) 20 °C min^{-1}.
Reproduced from ref. 43 with permission from Elsevier, Copyright 2015.

stress exceeds the yield point, causing a drop in the mechanical resistance (*e.g.*, shear viscosity) (Figure 11.6a) that is proportional to the shear rate, as shown by the drop in apparent viscosity of oleogels (Figure 11.6b). The recovery of viscosity (16–81% of viscosity) of wax oleogels as determined by viscometrical 3iTT is generally higher than the mechanical recovery assessed by oscillatory-flow-oscillatory 3iTT, where G' decreases up to 100 times after the shear interval. The rationale is that in viscometrical 3iTT the reference viscosity is taken at the end of the first test interval, once the oleogel structure has been perturbed at a low shear level (0.1 s^{-1}) during ~10 min, causing a viscosity drop of approximately one order of magnitude (Figure 11.6).[7,24] Moreover, the structural loss during the shearing in the first and second intervals of the 3iTT can cause the loss of the solid-like behavior

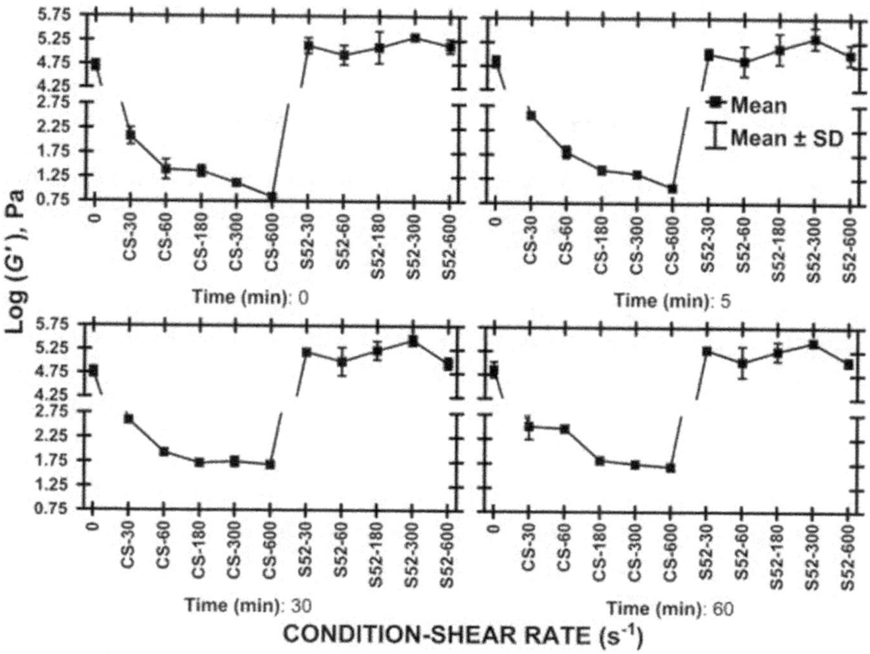

Figure 11.5 Logarithm of the elastic modulus [Log(G')] as a function of shearing conditions (CS: constant shearing; S52: shearing up to 52 °C), shear rate, and time under storage at 5 °C.
Reproduced from ref. 45 with permission from Elsevier, Copyright 2012.

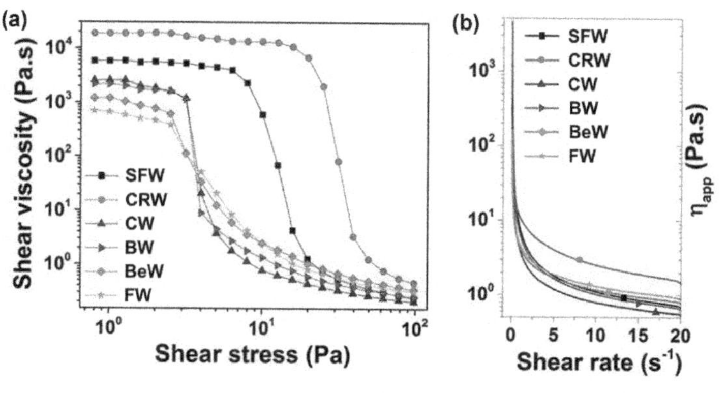

Figure 11.6 (a) Data from the stress ramp shown as shear viscosity *versus* shear stress and (b) apparent viscosity (η_{app}) *versus* shear rate for natural wax oleogels produced at the minimal gelling concentration. Sunflower wax (SFW), carnauba wax (CRW), candelilla wax (CW), beeswax (BW), berry wax (BeW), and fruit wax (FW).
Adapted from ref. 7, https://doi.org/10.1021/acs.jafc.5b01548, with permission from American Chemical Society, Copyright 2015.

of wax oleogels,[7] and so, the capacity to mimic the rheological behavior of plastic fats. Considering the incapability to determine whether the system behaves as a gel or as a crystal suspension after shearing and the structural perturbation inflicted to the gel during the first interval to set the reference viscosity, the viscometrical 3iTT method alone is insufficient to test the rheological similarity of oleogels and TAG-based plastic fats after shearing.

The structural loss in wax oleogels is evident when analyzed by oscillatory-flow-oscillatory 3iTT. CW oleogels showed G' recoveries after shearing lower than 10% of the reference value, which corresponded to the G' of the oleogel developed under static conditions.[5] Applying shearing levels used for dough mixing to oleogels of 2.5–5% CW, 2.5–5% CRW, and a commercial shortening caused decreases in G' and significant increases in tan δ values (a more viscous character) in all systems, but the commercial shortening showed the highest G' recovery (~50%).[46,47] Moreover, the increase in wax concentration reduced the G' recovery, independently of wax type.[5,46,47]

Shearing also produced structural changes that modified the scaling behavior of the elastic properties of CW oleogels. The G' scaling behavior of sheared CW oleogels did not fit the fractal model of colloidal gels and showed a semi-logarithmic correlation with the SPC.[36] The oleogels of particles subjected to stresses undergo an elastic deformation of their network structure and a gradual breakage of junction zones.[48,49] Crystal–crystal junction zones can be reversible or irreversible. Reversible junction zones occur between neighboring crystals attracted by van der Waals forces. In contrast, irreversible junction zones are caused by interpenetration of microplates or growth of crystalline dislocations. Reversible junction zones broken under shear reshape, allowing macroscopic recovery of elasticity. On the contrary, breakage of irreversible junction zones produces permanent softening.[5,36] Thus, the poor mechanical recovery of wax oleogels can be attributed to a high proportion of irreversible junction zones in the crystal network. To increase the structural reversibility, the ratio of irreversible/reversible junction zones should increase. Waxes and saturated TAGs develop mixed crystal networks with enhanced elasticity and improved mechanical recovery.[50,51] CW and FHSBO crystals in hybrid oleogels showed a synergistic effect on G', which was attributed to the large number of grainy FHSBO crystals nucleated on the surface of CW microplatelets during cooling. Mixed oscillatory-flow-oscillatory 3iTT showed that the mechanical recovery of hybrid oleogels after shearing was larger than that of oleogels structured solely with CW. For example, adding 10% FHSBO to 3% CW oleogels increased the G' recovery from 9% to 60%.[5]

11.3.1.3 Wax Oleogels in Foods

Replacement of plastic fats with oleogels will be feasible only if oleogels mimic the rheology of the reference fat without affecting the mouthfeel, flavor, and appearance of foods. The issues associated with flavor and the general appearance of foods will be left from this chapter, although they are of utmost importance. Strategies such as wax fractionation and use of flavors (*e.g.*, diacetyl) have been used to improve these attributes.[29,52,53] Increasing the wax concentration in oleogels increases their resistance to deformation, breaking and flow, but also increases the oleogel's melting temperature, the development of brittle consistency, and increases the possibility of leaving residual waxy flavors. Therefore, the wax concentration in oleogels is usually ≤10%. Öğütcü and Yilmaz observed that wax oleogels (*e.g.*, SFW, BW, and CRW) can achieve levels of hardness and stickiness similar to those of commercial margarines at room temperature.[33,54] Under refrigeration conditions, however, the hardness of margarine can be up to 5 times greater than that of oleogels with 10% CRW or SFW.[55] The hardness variation as a function of storage temperature is minimal for wax oleogels as compared to commercial margarine. This attribute of wax oleogels can be favorable to develop refrigerated ready-to-use spreads. Oleogelled margarines with stable consistency during storage and high resistance to heating/cooling cycles have been produced using 3.5% CW and 2–5% SFW oleogels.[32,52] However, the hardness of commercial margarine is more than double that of oleogelled margarine with ∼3.5% CW.[52] SFW margarines showed $G' \gg G''$ and their properties were independent of the deformation frequency, corresponding to the strong gel behavior. Nonetheless, the G' of commercial margarine was one to two orders of magnitude higher than those of emulsions gelled with 5% and 2% natural wax, respectively.[32] These findings suggest that oleogels with <5% wax could be suitable for developing ready-to-use refrigerated soft spreads instead of stick or bakery margarines.

Plastic fats can be totally or partially replaced with oleogels in margarine and bakery formulations. The replacement of 100% of butter in sweet bread with 10% CW oleogel increased the bread hardness (3.71 N), but the partial substitution (75% and 25–50%) of butter allowed bread to be produced (3.01 N) that matched the hardness of the reference product (2.86 N) or were softer (2.55–2.28 N).[56] The substitution of 0–100% butter with 5% CW oleogel in sponge cakes also caused increases in softness (from 29.61 N to 12.37 N) and specific volume.[57] However, the performance of oleogels and liquid oil was not compared in these studies; so, the effect of the oil structuration was unknown. Total or partial replacement of shortenings with CW and CRW oleogels in cookies causes increases of hardness in the baked product, with these increases being proportional to the percentages of oleogel substitution and inversely proportional to the wax content in the oleogels.[46,47] Furthermore, partial substitution of shortenings with oleogels increases the extensibility of doughs, with this increase being proportional to the percentage of oleogel, suggesting that the oleogel crystals are less effective in preventing the formation of a gluten network during kneading. However, the hardness of the cookies produced with oleogels is generally lower than that of cookies produced with liquid oil, suggesting that the structuring of the vegetable oil in the oleogel is more effective than liquid oil in shortening the gluten network.[47] Overall, wax oleogels have shown a better performance than liquid oils in dough shortening, a crucial aspect for developing smooth texture and flavor in many baked goods (*e.g.*, cookies, cakes, *etc.*).

Partial or total replacement of animal fat with wax oleogels has been tested in finely comminuted and coarse-ground meat products. Replacing pork backfat with RBW oleogels or liquid oil produced slight increases in the hardness and chewiness, small differences in the size dispersion of fat globules, and a flavor reduction in some meat products.[58,59] Frankfurter-type sausages produced with RBW oleogels or pork fat showed similar firmness, chewiness, and springiness. This effect was produced with either 2.5 or 10% RBW oleogels but not with the liquid oil. Nonetheless, frankfurters produced with oleogels and liquid oil did not match the hardness assessed by sensory evaluation and incisor probe analysis.[59] On the other hand, replacing 42% of pork backfat with liquid oil or oleogels with 2.5 or 10.0% RBW in bologna did not produce significant changes in the final texture profile analysis parameters, incisor peak force, and sensory texture. Bologna raw batters containing pork backfat or oleogels were more stable than those containing liquid soybean oil, but the yield [yield = (finished weight/raw weight) × 100%)] was unaffected by the type of fat.[58] Low-fat pork burgers containing 11% BW oleogel produced softer products than burgers containing pork backfat, although these differences in hardness were not detected in a hedonic sensory analysis. The panelists did not find differences in color, flavor, and acceptability between the control and oleogel-based burgers.[60] On the other hand, the partial replacement of 20–80% of pork backfat with 8–10% BW oleogels in fermented sausages produced important reductions in hardness, gumminess, and chewiness as compared to the original product. In addition, the sausages produced with the oleogels scored lower visual appearance, flavor, and aroma than the original product.[61,62] As this type of sausage is a coarse-ground product, the macroscopic fragments of fatty tissue contribute to the drying process, consistency, and sensory perception of the product. Wax oleogels not able to withstand the grinding were evenly distributed in the product affecting the overall perception even at 20% replacement of fat.[61] Overall, these studies showed that wax oleogels are suitable to replace, at least partially, the animal fat in cold-serve finely comminuted meat products.

11.3.2 Ethylcellulose Oleogels

11.3.2.1 Compositional Effects

EC oleogels are thermoreversible plastic materials showing significant changes in their viscoelastic properties as a function of EC molecular weight and concentration, and the quantity and type of polar molecules in the oil. Notably, EC oleogels have shown higher mechanical recovery after shearing than oleogels formulated with other gelators (*e.g.*, waxes, MGs).[6] EC is a linear polymer derived from cellulose, produced by the partial substitution of terminal hydroxyl groups of cellulose with ethoxy groups.[63] The degree of substitution (DS) or content of ethoxy groups determines the solubility of EC. Solubility of EC in water is reached at DS of 1.0–1.5, while the solubility of EC in organic solvents (*e.g.*, vegetable oil) is achieved at DS of 2.4–2.5 DS.[63] EC has average molecular weights of 19, 29, 52, 73, 71, and 144 kDa, which at 5% (w/v) concentration in a mixture of 80% toluene and 20% ethanol provide viscosities of 4, 10, 20, 45, 200, and 300 cP, respectively. EC is a semi-crystalline powder showing slight variations in the glass transition temperature ($T_g = 130$ °C) and melting peak ($T_m = 180$ °C) as a function of its molecular weight.[63] EC oleogels are produced by dispersing the polymer in liquid oil, which is generally heated for 30 min at temperatures between ~20 °C and 30 °C above the T_g to achieve EC solubilization. The gelation occurs as EC solutions at concentrations above the MGC (Table 11.3) are cooled below the gelling temperature.[64] ECs showing the highest gelling effect on vegetable oils provide viscosities between 10 cP and 100 cP. ECs providing viscosities lower than 10 cP show very little oil binding capacity at concentrations below 12%, while ECs that provide viscosities above 100 cP require temperatures above the EC melting temperature (180 °C) and very long heating times (> 30 min) to achieve solubilization in vegetable oils.[65]

The mechanical strength of EC oleogels, measured as hardness, increased up to 2000% as EC viscosity increased from 10 cP to 100 cP.[68,72] The partial polymer solubilization of ECs providing viscosities greater than 100 cP limited the formation and thickening of polymer strands in the three-dimensional network, producing weak oleogels with macroscopic aggregates of EC.[72] Increasing the mixing temperature above the melting temperature of EC 45 cP (180 °C) allowed the removal of the EC crystalline fraction, exposing an additional number of EC chains that facilitated the formation of additional EC–EC junction zones. Thus, gels mixed above T_m

Table 11.3 Minimal gelling concentration of common gelators.[a]

Gelator	MGC (%)	T_{set}	Oil	Method
EC 4 cP	<6.0[66]	15 °C	Canola oil	N. D.
EC 20 cP	8.0[67]	RT	Canola oil	Texture analysis
EC 45 cP	4.0[68]	3 °C	Soybean oil	Back extrusion
	6.0[68]	3 °C	Canola oil	Back extrusion
	6.0[69]	4 °C	Canola oil	Back extrusion
MGs	3.5[70]	20 °C	High oleic safflower oil	Inverted tube
Long-chain fatty acids (16–22 carbons)	≥2.0[71]	5 °C	Sunflower oil	N. D.
Long-chain fatty alcohols (16–22 carbons)	≥2.0[71]	5 °C	Sunflower oil	N. D.
Behenyl alcohol	3.0[100]	25 °C	Sunflower oil	Inverted tube
Behenic acid	2.0[100]	25 °C	Sunflower oil	Inverted tube
Stearic acid	2.0[100]	25 °C	Sunflower oil	Inverted tube
Palmitic acid	4.0[100]	25 °C	Sunflower oil	Inverted tube
Myristic acid	6.0[100]	25 °C	Sunflower oil	Inverted tube
LC	5.07[70]	20 °C	High oleic safflower oil	Inverted tube

[a]EC, ethylcellulose; MGs, monoacylglycerols; LC, fully hydrogenated soy lecithin; MGC, minimal gelling concentration; T_{set}, setting temperature; RT, room temperature; N. D., not declared.

(*i.e.*, 190 °C) were up to 6 times harder than those mixed just above the T_g (*i.e.*, 140 °C) of the EC.[64] The strength of oleogels increased exponentially with the concentration of EC at concentrations between 7% and 15% for EC of 10, 20 and 45 cP.[68,73] This exponential growth is maintained independently of the type of oil and the type and concentration of surfactant molecules added to the oil phase. However, the increase in hardness did not follow the same trend at concentrations above 15%.[64]

Surfactant molecules can form hydrogen bonds, increase the polarity of vegetable oil, and thus significantly affect the mechanical properties of EC oleogels. The re-evaluation of the effect of solvent on the mechanical strength of EC oleogels showed that rather than the fatty acid composition (*e.g.*, degree of unsaturation) of vegetable oils, the content of free fatty acids and other polar molecules were the main causes of increased hardness of EC oleogels.[68,74,75] The increase in polarity of refined vegetable oil has a direct effect on the mechanical behavior of EC oleogels due to an increase in polar functional groups in the lipid phase capable of interacting with the polymer chains.[74] The largest changes in gel strength are produced by adding small surfactant molecules with polar groups (*e.g.*, FFA, FFAl, MGs, sorbitan monosters).[67,73,74,76] The increase in the surfactant/EC ratio generally leads to a transition from a synergistic to an antagonistic effect on the mechanical properties of oleogels (Figure 11.7). The addition of small quantities of surfactants (*e.g.*, oleyl alcohol, oleic alcohol, *etc.*) produces an increase in the mechanical resistance (*i.e.*, Young's modulus) of these gels (ϕ_s <0.05 in Figure 11.7).[74] This effect is attributed to the interaction of polar heads of surfactants with polar regions of EC at the oil/polymer interface to minimize the contact with the hydrophobic environment, allowing the interaction among the polar groups of the EC chains. Above this concentration (ϕ_s> 0.05 in Figure 11.7), a decrease in gel

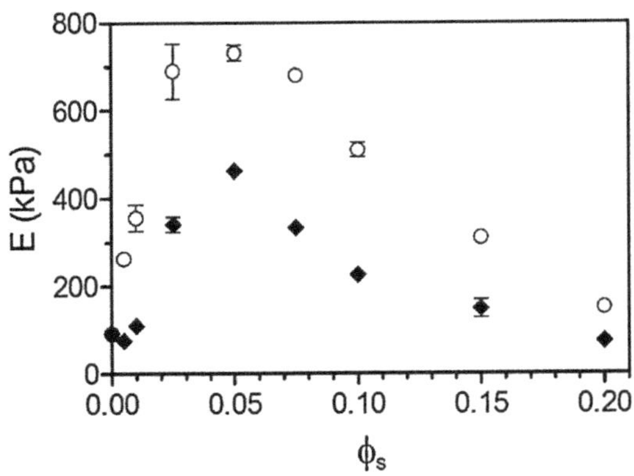

Figure 11.7 Effect of incorporating oleic acid (open circles) and oleyl alcohol (filled diamonds) on Young's modulus (E) of soybean oil-based oleogels prepared using 11% (wt/wt) ethylcellulose of 45 cP. ϕ_s indicates the mass fraction of the oil component which was substituted with oleic acid or oleyl alcohol.
Adapted from ref. 74 with permission from Elsevier, Copyright 2016.

strength is observed due to a plasticizing effect of surfactants on the EC network in the oleogel as well as an increase in EC solubility.[74,77] Surfactants (*e.g.*, FFA, MGs, SMS, *etc.*) also modify the rheology of EC solutions at elevated temperatures and the sol–gel transition temperature. The addition of surfactants to EC solutions increases their viscous character (increase of δ) and decreases their resistance to deformation (reduction of G^*, η^*) at temperatures above the sol–gel transition ($G' = G''$).[6,65,67,78,79] Additionally, the sol–gel transition temperature is displaced to lower temperatures by adding surfactants, both in the cooling and heating cycles.[67,74,76] These changes are driven by hydrogen bonding between hydroxyl groups of EC and polar heads of surfactants causing limited intra- and inter-chain interactions of EC polymeric chains, hindering the formation of supramolecular structures and junction zones.[75,76] Overall, the surfactant/EC polymer ratio and the induced changes in solvent polarity play a major role on oleogel formation, solution viscoelasticity, sol–gel temperature transition, and EC solubility. These effects can be used to mimic in a better way the thermo-mechanical properties of plastic fats.

11.3.2.2 Added Surfactants and Mechanical Reversibility

Adding surfactants to EC oleogels increases their mechanical recovery after shearing.[6,67,77] The thixotropy of EC oleogels has been modulated by changing the glycerol monooleate (GMO)/oil ratio. These oleogels achieved a viscosity recovery from 0% to 100% after shearing, depending on how close the Hansen solubility parameters of hydrogen bonding were for the liquid phase and the EC. The total viscosity recovery of EC 10 cP oleogels, intending to replace a petrolatum jelly, was achieved with concentrations of 40% to 50% GMO.[77] Adding 10% lauric acid to 8% EC 20 cP oleogels increased the G' recovery after large oscillatory deformation from 16% to 34%.[67] Using mixed oscillatory-flow-oscillatory 3iTT, it was shown that EC 45 cP oleogels added with saturated MGs have G' recoveries significantly higher than those of oleogels produced only with MGs or CW (Figure 11.8).[6] EC 45 cP oleogels with 5% MGs with or without CW recovered more than 100% of their elasticity after oleogel shearing at 10 s^{-1} for 1 min (Figure 11.8). The recovery of elasticity was associated with the reestablishment of the polymer–polymer interactions of EC by the hydroxyl groups of the MGs in the solvent. It was hypothesized that the recovery resulted from re-formation of junction zones of the type EC-[MGs]n-EC, with $n \geq 1$, that were broken under shearing and, then, quickly restructured by MG molecules in the oil phase.[6] Remarkably, the mechanical recovery of EC oleogels with MGs and CW were very close to those of edible shortenings solidified under similar conditions.[6]

11.3.2.3 Effects of Processing Conditions

The effects of processing conditions on the rheology of EC oleogels have been scarcely studied. Davidovich-Pinhas *et al.*[64] evaluated the effect of solubilization temperature above the EC glass transition (140 °C) or melting

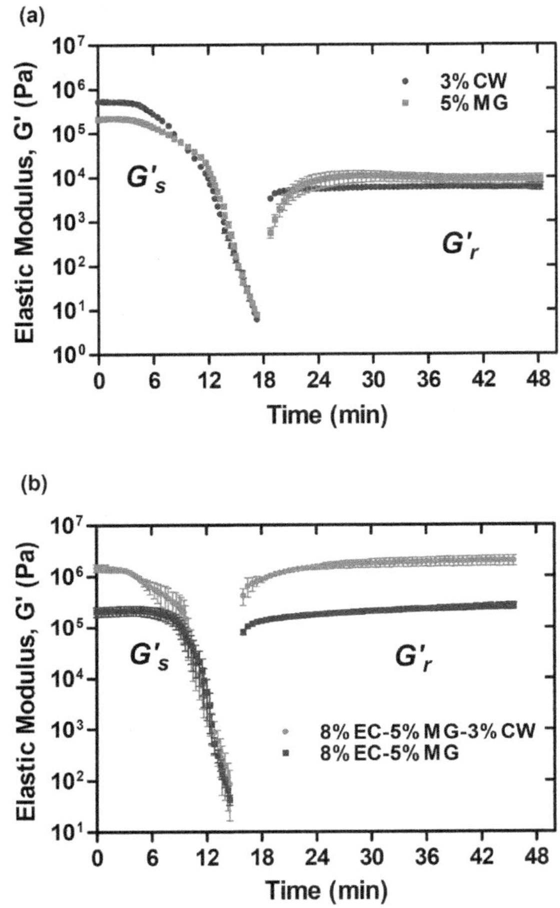

Figure 11.8 Oscillatory-flow-oscillatory 3 interval thixotropic test of oleogels produced with a single gelator (a) and gelator mixtures (b). The content of candelilla wax (CW), ethylcellulose (EC), and monoacylglycerols (MG) in oleogels is indicated in the legends. Elastic modulus (G') measured in the strain sweep (step 1) of oleogels structured under static conditions is denoted by G'_{s}. G'_{r} was G' measured during the recovery (step 3) after shearing oleogels for 1 min at 10 s^{-1}.
Reproduced from ref. 6 with permission from Elsevier, Copyright 2021.

temperature (190 °C), and annealing temperature (−20, 4, 20, 40, 50, 60, 80, and 100 °C for 24 h) before incubation and mechanical testing at 25 °C for the EC gels. They observed that EC solutions heated at 140 °C showed gel strengthening as the annealing temperature increased because of more efficient polymer–polymer hydrogen bonding in the polymeric network. Nonetheless, this effect was not observed for EC solutions heated above the polymer melting temperature (190 °C). The causes of this behavior were not fully clarified but, it was proposed that heating above the melting temperature increased the number of accessible sites for hydrogen bonding in the polymeric chain, facilitating the formation of additional EC–EC junction zones, creating a new gel network with enhanced physical properties. The cooling rate during the gelling process has an important role in the homogeneity of EC oleogels and polymer network structuration.[80] Although structural changes also impact the rheological properties of oleogels, these were not assessed. Overall, these studies offer a glimpse at how the processing conditions (*e.g.*, cooling rate, shearing, *etc.*) alter the rheological properties of EC oleogels and reveal that much research is still needed on the matter.

11.3.2.4 Ethylcellulose Oleogels in Foods

EC oleogels have been mainly incorporated into finely comminuted meat products (*e.g.*, frankfurters) to replace the animal fat fully or partially. Replacing beef fat with liquid canola oil increased the hardness of frankfurters, however, the incorporation of canola oil structured in a 10% EC oleogel reduced the hardness to similar values of the original product.[68] The same behavior has been observed for EC oleogels produced with a variety of vegetable oils.[81] As EC oleogels can be strengthened by adding surfactants like sorbitan monostearate (SMS) and glycerol monostearate (GMS), they have commonly been incorporated into EC oleogels intending to replace the animal fat in meat products.[82–86] The effect of EC 10 cP concentration (8–14%) on the hardness, Warner–Bratzler shear force, and the score of sensory texture of frankfurters was modified by the amount of SMS.[85] The use of more than 10% EC 10 cP did not produce additional softening and even increased the frankfurter's hardness, although there was not a straight correlation between instrumental and sensory scores. Frankfurters produced with 8% and 10% EC 10 cP oleogels and containing 1.5% or 3.0% SMS exhibited lower sensory hardness than those formulated without SMS.[85] On the other hand, adding 0%, 1.5%, or 3.0% GMS gradually increased the hardness of 12% EC 10 cP oleogels and caused softening of cooked meat batters to the levels of the original product, however, this softening was not directly proportional to the increase of oleogel hardness.[84] Besides this positive effect of EC oleogels on texture, oleogels do not significantly affect the water and fat loss during the cooking process.[82,84] The challenge ahead is to achieve a better approximation to sensory attributes such as the juiciness and oiliness of the cooked product.[86]

11.3.3 Monoacylglycerol Oleogels

11.3.3.1 Compositional Effects

MG oleogels are thermoreversible plastic materials with viscoelastic properties that are mainly affected by MG concentration, additives, and shearing during the gelling process. The oil binding capacity and rheological properties of MG oleogels can change during storage as polymorphic changes alter the crystal network topology.[87]

A detailed description of the mesophases, crystal phases, and polymorphic transitions that occur in MGs during the development of oleogels is given in Chapter 8. Here, we present a discussion on MG behavior in vegetable oil and the rheology of MG oleogels. MG molecules develop an inverted lamellar phase stabilized by hydrogen bonds between the polar heads as the MGs solution is cooled. MG aliphatic chains in the bilayer crystallize to the sub-α phase as temperature decreases. The inverted-lamellar phase and the sub-α crystalline phase are metastable, and they transform to the β crystalline polymorph during aging over a wide temperature range (5–45 °C).[87,88] The transformations from metastable phases to the β phase induce crystal aggregation and growth, and, eventually, the collapse of the three-dimensional crystal network.[87] The sub-α→β transformation starts soon after crystallization but, the complete transformation occurs within a week at room temperature. Therefore, MG oleogels suffer significant changes in rheology during this period.[87–89]

The sol–gel transition of MGs oleogels occurs at temperatures just below the isotropic liquid→inverted-lamellar phase transition.[87] The inverse-lamellar phase→sub-α transition produces an increase of G' in one order of magnitude in oleogels with 0.75–8% commercial MGs (93.51% GMS). Furthermore, sub-α→β transformation and the concomitant microstructural changes produce an additional increment of G' of up to one order of magnitude over 3 h at 5 °C.[88] However, over long-term storage at 4 °C and 20 °C, oleogels with 3–10% MGs based on hydrogenated palm oil showed gradual decreases in hardness and stickiness from day 30 to 90.[54] MG-based chocolate spreads have also showed slight drops in hardness after 30 days of storage, a behavior that contrasts with wax-based spreads that slightly increased their hardness under the same storage conditions.[90] The instability of the inverse-lamellar phase and sub-α crystal, and the rheological changes associated with polymorphic transitions, could limit the application of this type of oleogels in large scale production. This limitation has been overcome by adding EC or saturated soybean lecithin (LC) to MGs oleogels to slow down the sub-α→β transitions.[66,70] The addition of 6% EC 20 cP hindered the sub-α→β transition and microstructural changes in two commercial MGs in vegetable oleogels over 14 days at 15 °C. MG crystals and EC showed a synergistic interaction effect on oleogel strength (*i.e.*, G^*), with formulations containing 8% MGs and 6% EC showing mechanical strength close to that of commercial shortenings. Additionally, MG–EC oleogels showed lower oil loss during storage than MG oleogels developed without EC.[66] In subsequent research, hybrid oleogels were produced using 2% MGs and 0.25% to 2.5% LC, with these gelator concentrations being below the MGC of pure components (Table 11.3). The MG–LC mixtures developed well-structured oleogels with variations on G' as a function of LC content. Interestingly, the highest G' was obtained with the lowest LC content (*i.e.*, lowest SPC) and the addition of 2.5% to 10% water/total mass of the gelator resulted in oleogels with higher G'. Calorimetric and X-ray diffraction studies showed that LC incorporation favored the development of a mixture of MG–LC co-crystals and MG crystals. MG–LC co-crystals showed a limited inverse-lamellar phase to β polymorphic transition, favoring the stability of the gels and preventing the phase separation over 12 months at 15 °C. This effect was greater with the increase of LC content and with the addition of 10% of water to the systems.[70]

The G' values of MG oleogels increase exponentially with MG concentration. This behavior was observed for pure and commercial MGs at different aging times and temperatures.[88] G' increased exponentially with the volume fraction of network solids (φ) in MG oleogels produced under quiescent conditions. That is, $G' \sim \varphi^{4.74}$ in the range $0 < \varphi \leq 0.063$.[91] This behavior provides evidence that the structuration of MG oleogels follows the mechanisms of colloidal particle gels, where small particles (*e.g.*, inverted lamellas or crystals) aggregate to form flocs, which bundle in clusters to form a spanning network. As in wax-oleogels and plastic fats, the scaling behavior of the elastic properties of MG oleogels with $0 < \varphi \leq 0.063$ is determined by the aggregation process, and the primary particle's size and shape. At $\varphi > 0.063$, the G' scaling deviates from this behavior tentatively due to the formation of a secondary structure.[91] Under large deformation texture analysis, MG oleogels showed a linear increase in hardness as MG concentration increased. Adhesiveness and fracturability showed a similar trend, while cohesiveness decreased with an increase of MGs in oleogels.[92] Based on these trends, the gelator content and processing conditions (*e.g.*, setting temperature) were optimized to develop MG oleogels mimicking commercial margarine. The optimized oleogels and the reference product showed similar viscoelastic properties at small deformations and hardness, but their adhesiveness and cohesiveness were different.[92]

11.3.3.2 Effects of Processing Conditions

The application of shearing during crystallization and gelation of MG oleogels affects their nano- and micro-structure. MG gelling under static conditions allows the structuration of strong networks characterized by a high oil binding capacity and G' values over one magnitude order as compared to those gelled under constant laminar shear (100–2000 s^{-1}). The elasticity loss produced by shearing remained after 24 h of recovery under static conditions at 20 °C. Constant shear not only produced weak gels but also decreased their low oil binding capacity. Apparently, these levels of shearing did not affect the MG polymorphism.[93] The application of shearing over 60 s (75–600 s^{-1}) at the starting (47 °C) or preceding (55 °C) gel formation temperature showed a differential effect on the G' of MG oleogels equilibrated at 25 °C ($G'_{@25\ °C}$). Oleogels sheared at 47 °C exhibited $G'_{@25\ °C}$ about four times higher than those sheared at 55 °C. Additionally, the $G'_{@25\ °C}$ of samples sheared at 47 °C reached a maximum with 300 s^{-1}. Conversely, the elasticity of samples sheared at 55 °C was always lower than that of unsheared samples.[91] The application of shearing to the gelling solution at temperatures (70, 80, and 90 °C) well above the isotropic → lamellar phase transition temperature did not affect the rheological properties of the oleogels.[92] The rheological effect of the modification of the cooling rate during gelling has been evaluated only at very low cooling rates. Increasing the cooling rate from 0.42 °C to 1.09 °C min^{-1} led to a monotonic increase in hardness of the MG oleogels. Faster cooling rates shift the sol–gel transition to lower temperatures.[91] Both the increased hardness and the decline in the sol–gel transition temperature generated by fast cooling rates might be associated with the modifications of the MG non-isothermal crystallization kinetics. The increase of the cooling rate during non-isothermal crystallization leads to a decrease in the onset temperature of crystallization, which in turn causes increases in the degree of supercooling and nucleation rate.[94] Thus, the crystal network is strengthened by the greater number of smaller crystals and junction zones produced during the crystallization process.

11.3.3.3 Monoacylglycerol Oleogels in Foods

MG oleogels in foods have been tested in margarine-type products, baked goods, confections, and filling creams. Total replacement of plastic fats with MG oleogels in these products seems to be on the right way, as most rheological and functional properties have been mimicked. However, the instability of MG oleogels still remains the challenging issue. The concentrations of MGs (10.8% and 12.7%) and the processing conditions required to produce oleogels mimicking the rheological parameters of commercial margarines have been determined using multi-objective experimental designs. MG oleogels produced using this strategy showed the same hardness and G' values as margarines, but their adhesiveness and cohesion values were significantly lower.[92] Using the same optimization strategy, oleogels with 8% and 6% MGs were produced using different processing conditions to replace light margarine in muffins. The light margarine was replaced by MGs oleogel or vegetable oil, and a water amount proportional to that of the light margarine. MGs' oleogels and liquid oil produced the same hardness in muffins. Muffins containing oleogels and control fat showed similar cohesiveness, specific volume, and oil loss but the greatest similarity in these attributes was observed for muffins containing control fat and liquid oil. An advantageous aspect is that MG oleogels reduced the oil loss from muffins over 10 days of storage, which improved the appearance of the product.[95] MG oleogels have performed well as confectionery filling fats. 10% MG oleogels at concentrations of 26% in biscuit filling formulas caused viscoelasticity values that did not differ significantly from the commercial filling cream used as a reference. Increasing the oleogel content allowed the development of softer filling creams, which showed higher adhesiveness and increased ability to keep the cookies together with the filling.[96] The viscoelastic properties of a fat model of praline have been successfully imitated by a 6% MG oleogel. Unfortunately, this oleogel exhibited a melting point of 46 °C, which could cause a pasty mouthfeel in confectionery and toppings. Reducing to 3% MGs, the oleogel melted at 38 °C, but also reduced its G' and yield point to levels much lower than that of the reference product.[97]

11.3.4 Oleogels of Fatty Acids and Fatty Alcohols

FFA and FFAl produce thermoreversible oleogels exhibiting large rheological changes between the body and refrigeration temperatures. Thus, the gel–sol transition temperature can be adjusted according to the final product application by proper selection of type and concentration of FFA or FFAl. Advantageously, FFAs in vegetable oil can crystallize in the C-stable polymorph form, producing oleogels that are stable over one year at room temperature.[98] The thermodynamic properties and polymorphism of FFA and FFAl have been widely studied,[99] however, the gelation capacity of a homologous series of saturated aliphatic FFA and FFAl with 10 to 31 carbon atoms have been studied only by Daniel & Rajasekharan[100] and Gandolfo *et al.*[71] The MGC of the series does not follow a simple trend with the chain length. The MGC of FFA is reduced with the chain length only between 10–18 carbon atoms (Table 11.3) but, FFA with 18–31 carbon atoms showed a MGC$\leq$3%. FFAl with 22–30 carbon atoms showed MGC values (3–4%) slightly higher than those of FFA of the same chain length.[100]

The effect of chain length on oleogel hardness is markedly higher for FFA than for FFAl in the range of 16 to 22 carbon atoms.[71] No differences in hardness were produced as a function of FFAl chain length, but palmitic acid developed oleogels with higher hardness than the rest of the fatty acids at the same gelator concentration. However, the highest mechanical strength and oil binding capacity have been obtained with mixtures of FFAl/FFA in a mass fraction ratio of 7 : 3 for stearyl alcohol/stearic acid (SO/SA), palmitoyl alcohol/palmitic acid and behenyl alcohol/behenic acid.[101] The synergistic interaction of these mixtures on oleogel hardness originated from mixed crystals (*i.e.*, crystalline compound) of FFAl/FFA at a molar ratio of 3 : 1, achieving the quasi-complete crystallization of compounds in solution and producing favorable microstructural changes.[101,102] The elasticity changes as a function of temperature for pure or hybrid oleogels of FFA and FFAl resembles the behavior of structured plastic fats produced with saturated TAGs. Hybrid oleogels of 5% SO/SA in a 7 : 3 ratio showed G' values ($2–3\times10^6$ Pa) close to those reported for commercial shortenings in the range of 5 °C to 25 °C and fully melted at 35 °C.[102] Oleogels of pure stearic acid, pure stearic alcohol, and other mixtures of SO/SA showed similar changes as a function of temperature but with less consistency.[101,102]

A SO/SA mixture (7 : 3) and EC 45 cP can develop hybrid oleogels with flow curves comparable to those of margarine at refrigeration temperature. The rheological properties of these oleogels were measured by back extrusion. Flow curves of oleogels were analyzed quantitatively to evaluate the brittleness associated with the SO/SA crystal network, and the stress overshoot generated by some concentrations of EC (Figure 11.9). Flow curves of oleogels with 8% and 12% SO/SA exhibited a large number of small discontinuities associated with the formation of local fractures and regular oscillations produced by large-scale fractures (Figure 11.9a and c). Regular oscillations suggested that SO/SA crystals formed a brittle network. This was quantified using the Brittleness factor, which is equivalent to the root mean square deviation (RMSD) of a second-order polynomial equation fitted to the flow curve in the steady-state region (15–30 mm displacement). The stress overshoot was determined as the ratio of the peak force (Fmax) and the force recorded at the same penetration distance on the straight line fitted to the flow curve in the steady-state region (Flinear) (Figure 11.9b).[69] Using these metrics, it was established that SO/SA oleogels increased the Brittleness factor from 2.8 to 6.16 as the concentration increased from 8% to 12%. On the other hand, the 12% EC 45 cP oleogel showed a low Brittlenes factor (1.79) but its stress overshot was relatively large (1.37). The addition of 4% and 6% EC to 8% SO/SA oleogels did not modify the crystal polymorphism but produced microstructural changes in SO/SA crystals and reduced the Brittleness factor to 0.25 and 1.79, respectively. The stress overshoot and the

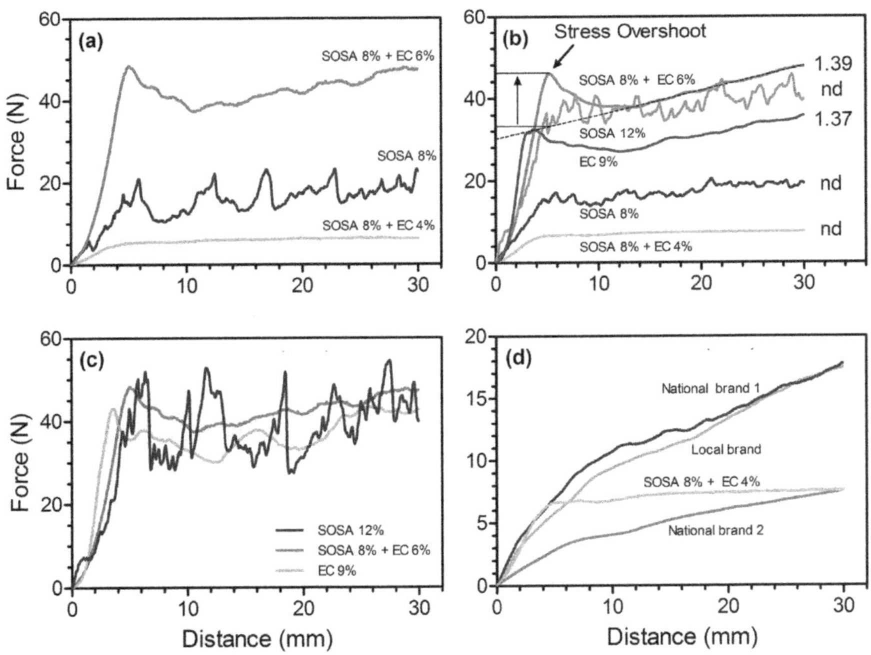

Figure 11.9 Representative (a, c) and average (b, d) back extrusion flow-curves of oleogels structured with 7 : 3 stearyl alcohol/stearic acid (SOSA), ethylcellulose (EC), or a mixture of SOSA and EC. Panel (b) indicates the stress overshoot (nd = not determined). Panel (d) compares the flow behavior of the 8% SOSA + 4% EC oleogel with that of commercial margarines.
Adapted from ref. 69 with permission from Elsevier, Copyright 2017.

Brittleness factor were avoided and reduced by adding EC at a concentration below the MGC (*i.e.*, 4% EC). Interestingly, this hybrid oleogel mimicked the smoothness of commercial margarines under large-deformations and flow conditions (Figure 11.9d). FFA oleogels could be functional in some foods, considering that ≤3% FFA do not develop unpleasant flavors in vegetable oils (*e.g.*, virgin olive oil).[1] Overall, these results encourage further investigations of FFA and FFAL oleogels as plastic fat mimetics.

11.4 Conclusions

Since early reports on the gelation of vegetable oils, several strategies have been implemented to produce oleogels mimicking closely the rheology of plastic fats. A strategy vastly investigated involves the use of different types and concentrations of single or multiple gelators to produce oleogels with increased elasticity, higher mechanical reversibility, reduced brittleness, and improved thermal stability. Gelator concentration significantly modifies the oleogel elasticity but the addition of surfactants and gelator mixtures seems to be more effective to modulate the mechanical reversibility and brittleness of oleogels. The effect of processing conditions on oleogel properties has been barely addressed. Shearing wax and MG oleogels during gelation has a deleterious effect on the strength and oil retention capacity of oleogels. This represents an important limitation for its handling, storage and use in many food production processes. However, this limitation could be overcome by hybrid oleogels (*e.g.*, CW-TAG, EC-MG, EC-MG-CW, *etc.*) showing high levels of mechanical recovery after shearing (partial or full thixotropy). Faster cooling rates increase the elasticity and hardness of oleogels structured by crystalline networks. Despite the scarce information on this matter, it can be inferred that a precise control of this variable could lead to a better design of this type of oleogels. In finely-comminuted meat products, cookies, bread, filling creams, and margarines the full or partial replacement of plastic fats with oleogels provides more functionality to the food products than the direct use of vegetable oil. The use of oleogels or liquid oil is indistinct in the case of other products (*e.g.*, muffins). In coarse-ground meat products, the replacement of adipose tissue by oleogels has not produced satisfactory results because oleogels are fragmented into smaller pieces than the adipose tissue. In general, there is a low correlation between instrumental and sensory measurements. Establishing a better correlation between such measurements will increase the effectiveness of the efforts to match the rheological properties of oleogels and plastic fats.

References

1. R. D. O'Brien, *Fat and Oils: Formulating and Processing for Applications*, CRC Press/Taylor & Francis, Boca Raton, 3rd edn, 2009.
2. T. G. Mezger, *The Rheology Handbook*, Vincentz Network, Hanover, 4th edn, 2014.

3. J. M. deMan and L. deMan, in *Physical Properties of Lipids*, ed. A. G. Marangoni and S. S. Narine, Marcel Dekker, New York-Basel, 1st edn, 2002. Texture of fats.
4. A. J. Wright, M. G. Scanlon, R. W. Hartel and A. G. Marangoni, *J. Food Sci.*, 2001, **66**, 1056.
5. N. O. Ramírez-Gómez, N. C. Acevedo, J. F. Toro-Vazquez, J. J. Ornelas-Paz, E. Dibildox-Alvarado and J. D. Pérez-Martínez, *Food Res. Int.*, 2016, **89**, 828.
6. A. K. Rodríguez-Hernández, J. D. Pérez-Martínez, J. A. Gallegos-Infante, J. F. Toro-Vazquez and J. J. Ornelas-Paz, *Carbohy. Polym.*, 2021, **252**, 1.
7. A. R. Patel, M. Babaahmadi, A. Lesaffer and K. Dewettinck, *J. Agric. Food Chem.*, 2015, **63**, 4862.
8. J. F. Steffe, *Rheological methods in food process engineering*, Freeman Press, East Lansing, 2nd edn, 1996.
9. M. C. Bourne, *Food Texture and Viscosity: Concept and Measurement*, Academic Press, San Diego, 2nd edn, 2002.
10. P. F. Fox, T. P. Guinee, T. M. Cogan and P. L. H. McSweeney, *Fundamentals of Cheese Science*, Springer, New York, 2nd edn, 2017.
11. T. C. Sindhu Kanya, L. Jaganmohan Rao and M. C. Shamanthaka Sastry, *Food Chem.*, 2007, **101**, 1552.
12. M. C. Chalapud, E. R. Baümlamer and A. A. Carelli, *Eur. J. Lipid Sci. Technol.*, 2016, **118**, 1.
13. C. D. Doan, C. M. To, M. De Vrieze, F. Lynen, S. Danthine, A. Brown, K. Dewettinck and A. Patel, *Food Chem.*, 2017, **214**, 717.
14. J. F. Toro-Vazquez, R. Mauricio-Pérez, M. M. González-Chávez, M. Sánchez-Becerril, J. J. Ornelas-Paz and J. D. Pérez-Martínez, *Food Res. Int.*, 2013, **54**, 1360.
15. A. P. Tulloch, *J. Am. Oil Chem. Soc.*, 1970, **50**, 367.
16. A. P. Tulloch, *Lipids*, 1970, **5**, 247.
17. S. R. Vali, Y. Ju, T. Kaimal Balagopala and Y. Chern, *J. Am. Oil Chem. Soc.*, 2005, **82**, 57.
18. L. E. Vandenburg and S. C. Wilder, *J. Am. Oil Chem. Soc.*, 1970, 514.
19. T. M. Attard, C. B. McElroy, C. A. Rezende, P. Polikarpov, J. H. Clark and A. J. Hunt, *Ind. Crops Prod.*, 2015, **76**, 95.
20. L. S. K. Dassanayake, D. R. Kodali, S. Ueno and K. Sato, *J. Am. Oil Chem. Soc.*, 2009, **86**, 1163.
21. J. F. Toro-Vazquez, J. A. Morales-Rueda, A. Torres-Martínez, M. A. Charó-Alonso, A. V. Mallia and R. G. Weiss, *Langmuir*, 2013, **29**, 7642.
22. J. F. Toro-Vazquez and J. D. Pérez-Martínez, in *Molecular gels: Structure and Dynamics*, ed. R. G. Weiss, Thermodynamic aspects of molecular gels, RCS Publishing, Croydon, 1st edn, 2018, pp. 57–87.
23. H. S. Hwang, S. Kim, M. Singh, J. K. Winkler-Moser and S. X. Liu, *J. Am. Oil Chem. Soc.*, 2012, **89**, 639.
24. D. C. Doan, D. Van de Walle, K. Dewettinck and A. R. Patel, *J. Am. Oil Chem. Soc.*, 2015, **92**, 801.

25. J. F. Toro-Vazquez, J. A. Morales-Rueda, E. Dibildox-Alvarado, M. A. Charó-Alonso, M. Alonzo-Macias and M. M. González-Chávez, *J. Am. Oil Chem. Soc.*, 2007, **84**, 189.
26. J. C. Barbosa Rocha, J. D. Lopes, M. C. N. Mascarenhas, D. B. Arellano, L. M. R. Guerreiro and R. L. da Cunha, *Food Res. Int.*, 2013, **50**, 318.
27. E. Yılmaz, E. K. Uslu and B. Toksöz, *J. Oleo Sci.*, 2020, **69**, 1317.
28. M. Pang, Z. Shi, Z. Lei, Y. Ge, S. Jiang and L. Cao, *Grasas Aceites*, 2020, **71**, 1.
29. V. Sarkisyan, R. Sobolev and Y. Frolova, *J. Am. Oil Chem. Soc.*, 2021, **98**, 281.
30. L. L. Serrato-Palacios, J. F. Toro-Vazquez, E. Dibildox-Alvarado, A. Aragón-Piña, M. D. R. Morales-Armenta, V. Ibarra-Junquera and J. D. Pérez-Martínez, *J. Am. Oil Chem. Soc.*, 2015, **92**, 533.
31. M. Scharfe and E. Flöter, *Eur. J. Lipid Sci. Technol.*, 2020, **122**, 1–24.
32. J. Merchán Sandoval, A. Carelli, C. Palla and E. Baümler, *J. Food Sci.*, 2020, **85**, 2866.
33. E. Yilmaz and M. Öğütcü, *J. Food Sci.*, 2014, **79**, E1732.
34. M. Öğütcü, N. Arifoğlu and E. Yilmaz, *J. Am. Oil Chem. Soc.*, 2015, **92**, 459.
35. A. R. Patel, D. Schatteman, W. H. De Vos, A. Lesaffer and K. Dewettinck, *J. Colloid Interface Sci.*, 2013, **411**, 114.
36. J. D. Pérez-Martínez, M. Sánchez-Becerril, A. G. Marangoni, J. F. Toro-Vazquez, J. J. Ornelas-Paz and V. Ibarra-Junquera, *Food Res. Int.*, 2019, **122**, 471.
37. A. I. Blake, E. D. Co and A. G. Marangoni, *J. Am. Oil Chem. Soc.*, 2014, **91**, 885.
38. G. Fayaz, O. Polenghi, A. Giardina, V. Cerne and S. Calligaris, *Int. J. Food Sci. Technol.*, 2021, **56**, 1040.
39. H. S. Hwang, M. Singh, J. K. Winkler-Moser, E. L. Bakota and S. X. Liu, *J. Food Sci.*, 2014, **79**, C1926.
40. A. J. Martins, M. A. Cerqueira, L. H. Fasolin, R. L. Cunha and A. A. Vicente, *Food Res. Int.*, 2016, **84**, 170.
41. M. Zhao, Y. Lan, L. Cui, E. Monono, J. Rao and B. Chen, *Food Chem.*, 2020, **309**, 125704.
42. L. Liu, S. A. Ramirez, J. Yang and O. N. Ciftci, *Food Chem.*, 2020, **309**, 125567.
43. H. S. Hwang, S. Kim, K. O. Evans, C. Koga and Y. Lee, *Food Struct.*, 2015, **5**, 10.
44. A. I. Blake and A. G. Marangoni, *Food Biophys.*, 2015, **10**, 403.
45. F. M. Alvarez-Mitre, J. A. Morales-Rueda, E. Dibildox-Alvarado, M. A. Charó-Alonso and J. F. Toro-Vazquez, *Food Res. Int.*, 2012, **49**, 580.
46. B. Mert and I. Demirkesen, *Food Chem.*, 2016, **199**, 809.
47. B. Mert and I. Demirkesen, *LWT - Food Sci. Technol.*, 2016, **68**, 477.
48. Y. Miyazaki and A. G. Marangoni, *Mater. Res. Express*, 2014, **1**, 025101.

49. M. Van den Tempel, *J. Colloid Sci.*, 1961, **16**, 284.
50. C. D. Doan, A. R. Patel, I. Tavernier, N. De Clercq, K. Van Raemdonck, D. Van de Walle, C. Delbaere and K. Dewettinck, *Eur. J. Lipid Sci. Technol.*, 2016, **118**, 1903.
51. S. L. da Silva, J. Tomazzetti Amaral, M. Ribeiro, E. Endler Sebastião, C. Vargas, F. de Lima Franzena, G. Schneidera, J. M. Lorenzo, L. L. Martins Friesa, A. J. Cichoskia and P. C. Bastianello Campagnol, *Meat Sci.*, 2019, **149**, 141.
52. T. L. T. da Silva, K. F. Chaves, G. D. Fernandes, J. B. Rodrigues, H. M. A. Bolini and D. B. Arellano, *J. Am. Oil Chem. Soc.*, 2018, **95**, 673.
53. E. Yilmaz and M. Öğütcü, *RSC Adv.*, 2015, **5**, 50259.
54. M. Öğütcü and E. Yilmaz, *Grasas Aceites*, 2014, **65**, 1.
55. M. Öğütcü and E. Yilmaz, *Int. J. Food Prop.*, 2015, **18**, 1741.
56. D. Jung, I. Oh, J. Lee and S. Lee, *LWT - Food Sci. Technol.*, 2020, **125**, 109194.
57. J. Alvarez-Ramirez, E. J. Vernon-Carter, Y. Carrera-Tarela, A. Garcia and C. Roldan-Cruz, *LWT - Food Sci. Technol.*, 2020, **130**, 109701.
58. R. Tarté, J. S. Paulus, N. C. Acevedo, K. J. Prusa and S. L. Lee, *LWT - Food Sci. Technol.*, 2020, **131**, 109659.
59. T. L. Wolfer, N. C. Acevedo, K. J. Prusa, J. G. Sebranek and R. Tarté, *Meat Sci.*, 2018, **145**, 352.
60. J. Gómez-Estaca, A. M. Herrero, B. Herranz, M. D. Álvarez, F. Jiménez-Colmenero and S. Cofrade, *Food Hydrocolloids*, 2019, **87**, 960.
61. D. Franco, A. J. Martins, M. López-Pedrouso, L. Purriños, M. A. Cerqueira, A. A. Vicente, L. M. Pastrana, C. Zapata and J. M. Lorenzo, *Foods*, 2019, **8**, 366.
62. T. Pintado and S. Cofrades, *Foods*, 2020, **9**, 830.
63. M. Davidovich-Pinhas, S. Barbut and A. G. Marangoni, *Cellulose*, 2014, **21**, 3243.
64. M. Davidovich-Pinhas, A. J. Gravelle, S. Barbut and A. G. Marangoni, *Food Hydrocolloids*, 2015, **46**, 76.
65. A. K. Zetzl, A. J. Gravelle, M. Kurylowicz, J. Dutcher, S. Barbut and A. G. Marangoni, *Food Struct.*, 2014, **2**, 27.
66. A. Lopez-Martínez, M. A. Charó-Alonso, A. G. Marangoni and J. F. Toro-Vazquez, *Food Res. Int.*, 2015, **72**, 37.
67. A. H. Eisa, S. Laufer, J. Rosen-Kligvasser and M. Davidovich-Pinhas, *Eur. J. Lipid Sci. Technol.*, 2020, **122**, 1900044.
68. A. K. Zetzl, A. G. Marangoni and S. Barbut, *Food Funct.*, 2012, **3**, 327.
69. A. J. Gravelle, M. Davidovich-Pinhas, S. Barbut and A. G. Marangoni, *Food Res. Int.*, 2017, **91**, 1–10.
70. M. Aguilar-Zárate, A. De la Peña-Gil, F. M. Álvarez-Mitre, M. A. Charó-Alonso and J. F. Toro-Vazquez, *Food Biophys.*, 2019, **14**, 326.
71. F. G. Gandolfo, A. Bot and E. Flöter, *J. Am. Oil Chem. Soc.*, 2004, **81**, 1.

72. X. Ye, P. Li, Y. M. M. Lo, H. Fu and Y. Cao, *J. Food Sci.*, 2019, **84**, 1456.
73. A. J. Gravelle, S. Barbut, M. Quinton and A. G. Marangoni, *J. Food Eng.*, 2014, **143**, 114.
74. A. J. Gravelle, M. Davidovich-Pinhas, A. K. Zetzl, S. Barbut and A. G. Marangoni, *Carbohydr. Polym.*, 2016, **135**, 169.
75. T. Laredo, S. Barbut and A. G. Marangoni, *Soft Matter*, 2011, 7, 2734.
76. M. Davidovich-Pinhas, S. Barbut and A. G. Marangoni, *Carbohydr. Polym.*, 2015, **127**, 355.
77. T. A. Stortz and A. G. Marangoni, *Green Chem.*, 2014, **16**, 3064.
78. P. Ahmadi, M. Tabibiazar, L. Roufegarinejad and A. Babazadeh, *Int. J. Biol. Macromol.*, 2020, **150**, 974.
79. J. Rosen-Kligvasser and M. Davidovich-Pinhas, *Food Chem.*, 2021, **334**, 127585.
80. A. J. Gravelle, S. Barbut and A. G. Marangoni, *Food Funct.*, 2013, **4**, 153.
81. S. Barbut and A. G. Marangoni, *Meat Sci.*, 2019, **149**, 9.
82. S. Barbut, J. Wood and A. G. Marangoni, *J. Food Sci.*, 2016, **81**, C2183.
83. M. López-Pedruso, J. M. Lorenzo, B. Gulló, P. C. B. Campagnol and D. Franco, *Curr. Opin. Food Sci.*, 2021, **40**, 40.
84. M. Alejandre, I. Astiasarán, D. Ansorena and S. Barbut, *Food Res. Int.*, 2019, **122**, 129.
85. S. Barbut, J. Wood and A. G. Marangoni, *Meat Sci.*, 2016, **122**, 155.
86. S. Barbut, J. Wood and A. G. Marangoni, *Meat Sci.*, 2016, **122**, 84.
87. C. H. Chen and E. M. Terentjev, *Langmuir*, 2009, **25**, 6717.
88. A. López-Martínez, J. A. Morales-Rueda, E. Dibildox-Alvarado, M. A. Charó-Alonso, A. G. Marangoni and J. F. Toro-Vazquez, *Food Res. Int.*, 2014, **64**, 946.
89. M. D. Bin Sintang, T. Rimaux, D. Van de Walle, K. Dewettinck and A. R. Patel, *Eur. J. Lipid Sci. Technol.*, 2017, **119**, 1500517.
90. G. Fayaz, S. Amir, H. Goli, M. Kadivar, F. Valoppi, L. Barba, S. Calligaris and M. C. Nicoli, *LWT - Food Sci. Technol.*, 2017, **86**, 523.
91. N. K. O. Ojijo, I. Neeman, S. Eger and E. Shimoni, *J. Sci. Food Agric.*, 2004, **84**, 1585.
92. C. Palla, A. Giacomozzi, D. B. Genovese and M. E. Carrín, *Food Struct.*, 2017, **12**, 1.
93. S. da Pieve, S. Calligaris, E. D. Co, M. C. Nicoli and A. G. Marangoni, *Food Biophys.*, 2010, **5**, 211.
94. A. G. Marangoni and L. H. Wesdorp, *Structure and Properties of Fat Crystal Networks*, CRC/Taylor & Francis, Boca Raton, 2nd edn, 2013.
95. A. S. Giacomozzi, M. E. Carrín and C. A. Palla, *J. Food Sci.*, 2018, **83**, 1505.
96. C. A. Palla, M. F. Wasinger and M. E. Carrín, *J. Food. Sci. Agric.*, 2021, **101**, 2398.
97. H. Si, L. Z. Cheong, J. Huang, X. Wang and H. Zhang, *J. Am. Oil Chem. Soc.*, 2016, **93**, 1075.

98. S. S. Sagiri, V. K. Singh, K. Pal, I. Banerjee and P. Basak, *Mater. Sci. Eng., C*, 2015, **48**, 688.
99. F. Kaneko, in *Crystallization Processes in Fats and Lipid Systems*, ed. N. Garti and K. Sato, Polymorphism and phase transitions of fatty acids and acylglycerols, CRC Press, New York, 1st edn, 2001, pp. 53–98.
100. J. Daniel and R. Rajasekharan, *J. Am. Oil Chem. Soc.*, 2003, **80**, 417.
101. M. Callau, K. Sow-Kébé, L. Nicolas-Morgantini and A. L. Fameau, *J. Colloid Interface Sci.*, 2020, **560**, 874.
102. H. M. Schaink, K. van Malssen, S. Morgado-Alves, D. Kalnin and E. van der Linden, *Food Res. Int.*, 2007, **40**, 1185.

CHAPTER 12

Sensory Properties and Aromatics Profile of Edible Oleogels

E. YILMAZ

Çanakkale Onsekiz Mart University, Faculty of Engineering, Department of Food Engineering, 17020, Çanakkale, Turkey
Email: eyilmaz@comu.edu.tr

12.1 Introduction

As a subclass of lipids, edible fats and oils comprise one major component of the human diet. Edible fats and oils provide energy, essential fatty acids, fat-soluble vitamins, and some minor components as valuable nutrients. In nature, edible fats and oils exist in liquid, semi-liquid, and solid (plastic consistency) phases. Although there are numerous plant and animal-based sources, just around 20–30 are economically important and supply almost all of the fats and oils for edible and non-edible (oleochemical, cosmetics, paints, energy, *etc.*) applications.[1,2] The physical state of a fat/oil is primarily dependent on its fatty acid composition. Thus, sources with more saturated and long-chain fatty acids exist as solids or semi-solids. Additionally, the concentration of unsaturated fatty acid isomers also affects the fat/oil physical state significantly. It is well known that fatty acids with the same chain length and the same unsaturation extent differ significantly in their melting temperature because of differences in the content of *cis* or *trans* isomers, with *trans* isomers showing around 10–20 °C higher melting temperatures than the *cis* counterparts. Although some *trans*-fatty acids

Food Chemistry, Function and Analysis No. 31
Development of *Trans*-free Lipid Systems and their Use in Food Products
Edited by Jorge F. Toro-Vazquez

Published by the Royal Society of Chemistry, www.rsc.org

exist as native components of fats and oils, most are formed during partial hydrogenation or heat processing of vegetable oil. Even though *trans* fats provide useful textural and melting properties to lipid-based food systems, the negative cardiovascular effects of them are well documented. Lastly, the triacylglyceride configuration could change the melting behavior of fats/oils. Natural plastic fats (butter, cacao butter, tallow, lard, chicken fat, coconut fat, palm, and palm kernel oil) are less abundant than edible liquid vegetable oils. Solid edible fats have some textural properties that cannot be provided by vegetable oils. The textural properties like plastic consistency, spreadability, hardness, and brittleness, and the sensory properties like appearance, palate, creaminess, slippiness, cooling effect, and mouthfeel are only possible with solid fats in food products like margarine and spreads, shortenings, bakery fats, chocolate and confectionery, meat products, some dairy products, and others. There is a constant demand for solid fats by the food industry, and due to the shortage of natural sources, some technological modifications have been in practice to convert natural vegetable oils into semi-solid or solid fats. The best-known technologies are hydrogenation/partial hydrogenation, interesterification, fractionation, and blending with saturated fats. These technologies have been around for more than a century but still have some limitations mainly associated with high costs (*i.e.*, interesterification), production of *trans* fatty acids, and use of saturated fatty acids. Hence, the new ways of oil structuring are an open, challenging and active research field.[1–4]

12.2 Physical Properties of Solid and Semi-solid Fats

The melting behavior of crystallized edible oils results from their triacylglyceride (TAGS) composition, configurations, and crystal polymorphic phase. The polymorphic type and phase or solution dynamics of the TAGS crystals define their melting interval and melting temperature. The ideal solubility behavior of TAGS is present in a liquid state, but ideal and non-ideal solubility behavior can be observed in solid fats. Even in the same oil source, solid solution, monotectic, eutectic, and compound behaviors could occur between different TAGS due to the molecular heterogeneity. Since edible fats/oils are complex heterogeneous mixtures of TAGS, they significantly differ in their melting behaviors. Furthermore, they could not have a definite melting or crystallization temperature but a temperature interval. Higher melting temperature TAGS in melted or edible oils form crystals upon cooling, while other TAGS remaining in the liquid state are immobilized within the crystal three-dimensional network TAGS resulting in a system with semi-solid or solid fat consistency. Because of the molecular heterogeneity among the constitutive TAGS of vegetable oils, the applied shear, the saturation extent, and the time–temperature conditions applied (*i.e.*, cooling rate), different types of fat crystals (*i.e.*, polymorphs) could be developed. Polymorphic forms are the different solid phases of the same chemical composition that differ in crystal organization but, upon melting, result in identical liquid phases. For long-chain fatty acids and TAGS, polymorphism is associated with different packing arrangements of the hydrocarbon chain or different tilt angles. The glass transition state (γ polymorph) forms as a transition state once the melted fat source rapidly cooled. This form is transparent and unstable. It immediately converts to a more stable α polymorph. One of the three main forms, the α form, is a fine, waxy, less stable, least dense, lowest melting point form with hexagonal chain packing, with 4.2 Å short spacing having a single band at 720 cm^{-1} characteristic infrared spectra. An intermediate density, medium melting point, orthorhombic chain packing β' type polymorph shows 3.8 and 4.2 Å short spacings and a double infrared spectrum at 727 cm^{-1} and 719 cm^{-1} bands. This form yields a very fine, creamy, soft texture, and it immobilizes the maximum amounts of liquid oil due to higher surface area. It is known that under certain conditions, a fat with a β' polymorph could transform to a β polymorph. The β polymorph form presents triclinic chain packing with 4.6 Å short spacing and a single infrared band at 717 cm^{-1} and yields a coarse and sandy texture. Among the three common polymorphs, the β form is thermodynamically the most stable and higher melting point form. The fat consistency, plasticity, sandiness, spreadability, melt cooling, creaminess, and other physical properties are mainly determined by the crystalline polymorph type and the proportion of solid and liquid TAGS. The solid fat index (SFI) was defined as the total amount of solid TAGS in a fat source at a given measurement temperature. It was well established that sensory properties, palate, and mouthfeel of solid fat products are quite positively related to crystal polymorph types and SFI values.[4–9]

The melting behavior of fat could be defined by its molecular composition and polymorphic type successfully, but its mechanical behavior (rheology) cannot be predicted only with molecular data. Oscillatory rheometers are usually used to assess the flow behavior of fats. Complex viscosity or its components, the storage modulus (G') and the loss modulus (G''), were usually the measured parameters. Furthermore, structural recovery, the effect of different temperatures on flow, and other properties are measured by rheometers. Most of these rheologic parameters were correlated with the sensory properties of fat samples. More details can be found in the cited references or in other literature.[9–13]

The main functional properties of solid fats are hardness, spreadability, air incorporation, brittleness, crispness, melting rate and range, melt cooling, liquid oil separation, re-crystallization ability, starch and gluten network protection, flavor release, controlled delivery, oil migration suppression, and others. There are some analytical or instrumental techniques used to predict product performance. All instrumental data must be correlated with sensory data and then with consumer tests to manage engineering to possibly develop successful food products for the rapidly changing and challenging markets.[1,6,9,14]

12.3 Oil Structuring Through Oleogelation

Edible liquid vegetable oils can be converted into hard fat stocks by some technologies. Beyond structuring, most of the time a significant increase in saturated and *trans* fatty acids also occur after the modifications. Since the negative health impacts of saturated and *trans* fatty acids have been established, the World Health Organization (WHO) has suggested certain limitations. There are regulatory issues limiting these fatty acids in different countries. Ideally, any technology converting liquid oil into hard fat without enhancement of saturated and/or *trans* fatty acids is demanded. The oleogelation concept was introduced to fulfill this need.[8,9,15] A detailed description of oleogelation technology could be found elsewhere, but for the main purpose of this chapter, the concept is briefly discussed below.

Oleogels are three-dimensional, self-standing, thermo-reversible, anhydrous, viscoelastic materials created by adding organogelators into liquid oil. Although during oleogelation, the liquid oil is converted to semi-solid or solid fats with various hardness and melting properties, the fatty acids and minor components composition of the starting liquid oil are not changed. This is the main advantage of the technology. If food grade and suitable organogelators are used, safe, edible, and sensorially acceptable oleogels could be prepared with technological feasibility.[8,9,16,17]

The success and properties of an oleogel are mostly governed by the organogelator molecule used. Food-grade and legally permitted organogelators are classified into two main classes: low molecular weight gelators (LMWG) and polymeric gelators (PG). LMWGs include monomeric subunits, which are able to self-assemble into fibrillar networks that entrap the liquid oil to immobilize it. The best-known LMWGs include saturated mono-, di-, and triacyl glycerides, various fatty acids and alcohols, hydroxylated fatty acids, lecithin, some plant waxes and wax esters, ceramides, sorbitan derivatives, phytosterols, γ-oryzanol + phytosterol, sucrose esters, some amphiphiles, and some amino acid derivatives. The polymeric gelators (PG) include some food hydrocolloids (xanthan gum, gum arabic, locust bean gum, *etc.*), some carbohydrate polymers (cellulose derivatives, modified chitin and chitosan, waxy starch, *etc.*), some food proteins (gelatin, zein, gluten, lactoglobulin, *etc.*), and their combinations. Basically, the polymer strand–strand and polymer strand–liquid oil interactions yield some junction zones, and in turn, provide a network to entrap the liquid oil into a gel state. Furthermore, emulsion templates are prepared to immobilize the liquid oil. Oleogels can also be classified according to the gelling mechanism (self-assembly, crystal particles, micelles, bilayers, tubules, agglomerates), type of the gelators (low molecular weight gelators, polymeric gelators, inorganic gelators), and the number of gelator components (mono-component gelators like waxes, multi-component gelators like sitosterol + oryzanol). New oleogels formed with different approaches were also named with 'oleocolloid, oleofoam, and oleofilm' terms.[3,8,9,11,15,17,18]

Clearly, the organogelator kind, its concentration, and the strategy implemented to prepare the oleogel defines the macroscopic features (rheology, texture), stability, melting behavior (melting and crystallization ranges and peak temperatures), and sensory properties of the final products. Although not completed in most studies, the sensory properties of the oleogels must be defined, and the acceptance of the products by the consumers must be evaluated before any possible successful commercialization.[4,8,9,19]

Most of the current oleogel literature deals with the microstructural, rheological, and physico-chemical characterizations of the oleogels. A less common portion includes various food product (margarine and spreads, bakery goods, confectionery and chocolates, dairy products, comminuted meat products, and others) applications of the oleogels. Discussion about the applications of oleogels in different food products are available in book chapters[19–22] and in recent reviews.[9,17,18,23–26] According to the purpose of this chapter, sensory properties and aromatics profiles of edible hard fats and oleogels will be discussed in the next sections.

12.4 Sensory Assessment of Edible Fats/Oils

Humans have been evaluating their consumed products for goodness and badness since the early centuries of history. After the rise of trade, sensory evaluation has become a more formal reality in human life, applied between the traders and consumers to value and price the goods. Over time, grading products like wine, tea, coffee, butter, and others with primitive sensory assessments became a reality. After 1965, sensory science became an analytical discipline. Scientists have developed sensory test methodology, formalized and structured new tests, and codified the methods. Today, sensory analysis is accepted as an interdisciplinary science.

Sensory evaluation is basically divided into seven consecutive tasks: determining the objective of the tests (the reason to take the test), determining the test objective (selecting the most suitable test technique for the particular reason determined previously), screening the test samples (could guide the best suitable test techniques among many others), designing the tests (selection and training of subjects, scales, test room, test conditions, *etc.*), conducting the test (ensure all requirements were met), data analysis (correct statistical analyses), and interpretation of results (make recommendations for any action to fulfill the product requirements). Clearly, the sensory evaluation of foods is not a simple task but a complex process that produces essential data for product development, product comparing, and trading purposes.[27–31]

12.4.1 Sensory Vocabulary Used for Fats/Oils and Oleogels

Sensory evaluation is the ultimate method to evaluate the quality and stability of fats, oils, and similar foods. There are no equal instrumental or chemical methods that can take the place of the human senses of nose, tongue, mouth, and touch so far. Selection processes and training facilities of human subjects for sensory evaluations can be found elsewhere. In this brief section, the focus will be given to the sensory attributes of fats, sensory vocabulary, test techniques, consumer tests, and their interrelationships.[27,29,31,32]

For any food, including fats/oils, the product sensory attributes are perceived by humans in the order of appearance, aroma/odor/smell, texture/consistency, and taste/flavor. The human senses of vision, olfaction, gustation, trigeminal nerves, touch, and hearing are used simultaneously within and between sensory modalities for various sensory evaluation tests. Factors affecting the sensory verdict, techniques to measure human responses, factors to control sensory test rooms, sample preparations, and the panel training and orientation are deeply discussed in many textbooks, and not included in this chapter.[27–32]

There are numerous terminology and sensory vocabulary that have been used for different food products. For edible fats and oils, the most common vocabulary and sensory terms will briefly be discussed. Bland is a common term in edible fats/oils and defines a total absence of flavor and odor. Hydrolytic rancidity and oxidative rancidity could produce both sour and metallic tastes, and greasy and popcorn-like odors in all edible fats/oils.[27–29,32] The American Society for Testing and Materials (ASTM) data series publication DS 66 was published in the 'Aroma and Flavor Lexicon for Sensory Evaluation: Terms, Definitions, References, and Examples' title.[33] In this publication, the sensory description terms for edible oils were listed as bacony, beany, bitter, bland, burnt, buttery, cardboardy, corny, deep-fried, fishy, fruity, grassy, green, hay, heated oil, hully, hydrogenated oil, lard oil, light-struck oil, melon, metallic, musty, nutty, overheated oil, oxidized, painty, paraffin oil, peanut oil, pecan oil, petroleum, phenolic, pine oil, plastic, pumpkin, rancid, raw, reverted oil, rubbery, soapy, sour, sulfur, sunflower seed shell, watermelon, waxy, weedy, and woody. Furthermore, specific definitions, references, and preparation of the references of the terms are given in that publication. Some of the most commonly used sensory descriptors and their reference standards for refined rapeseed oils are given in Table 12.1.[28] In that study, the sensory terms used for rapeseed oils include quite common definition terms, which could also be used for other edible fats/oils.

It has been well established that for every food sample, different sensory description terms could be defined by the particular panel evaluating the sample. The American Oil Chemists' Society (AOCS) has also published a method (Cg 2–83, flavor panel evaluation of vegetable oils),[34] and that method collaborated with different oil samples.[35] The AOCS method Cg 2–83 suggests descriptors similar to the DS 66 document.[33,34] Also, mouthfeel properties like cooling, hot, coating, pungent, velvet, and fatty/watery were listed as other attributes. For margarine, spreads, oleogels, and similar products, some of these kinesthetic or mouth feeling properties were shown to be essential for quality evaluation and consumer acceptance.[36–39]

Table 12.1 Sensory attributes, definitions, and reference products to evaluate refined rapeseed oil according to ASTM. Reproduced from ref. 28 with permission from John Wiley & Sons, Copyright © 2019 AOCS.

Sensory attributes	Definition	Reference product	Score
Yellow color	The intensity of yellow color	Yellow color intensity wheel	0; 5; 10
OOI	The intensity of all odor attributes taken together	No perception-mineral water, 1-freshly rapeseed oil, 10-aged rapeseed oil with high intensity of negative attributes	0; 1; 10
Seed odor	The odor of fresh seeds	Freshly cold pressed sunflower oil	8
Fishy odor	The odor reminiscent of cod liver oil	Linseed oil aged 3 months at room temperature	10
Nutty odor	The odor reminiscent of fresh, sweet nutmeat	Freshly cold-pressed peanut oil	8.5
Buttery odor	The odor reminiscent of fresh, sweet, unsalted butter	Freshly processed unsalted butter	7
Hay odor	The odor reminiscent of dried grass character of air-dried grain or vegetation	Crude soybean oil from heat-treated beans diluted in good-quality soybean oil (5 : 95)	10
Grassy odor	The odor reminiscent of the green character of mowed grass	Fresh cut grass	10
OFI	The overall flavor intensity	No perception-mineral water, 1-freshly rapeseed oil, 10-aged rapeseed oil with high intensity of negative attributes	0; 1; 10
Acid taste	A basic taste simulated by such substances as acid	Refined rapeseed oil with bleached rapeseed oil (99 : 1)	8.5
Sweet taste	A basic taste simulated by such substances as sugar	Fresh sesame oil	9.5
Bitter taste	A basic taste simulated by such substances as quinine and caffeine	0.2% of caffeine in water	7
Buttery flavor	The flavor reminiscent of fresh, sweet, and unsalted butter	Fresh, sweet, unsalted butter diluted in good quality soybean oil (1 : 99)	9
Seed flavor	The flavor reminiscent of fresh seed	Freshly cold pressed sunflower oil	9.5
Rancid flavor	The flavor reminiscent of highly oxidized oils such as sunflower, cottonseed or peanut oils containing high amount of linoleic acid	Good quality cottonseed oil aged 4 days at 60 °C or until PV ∼ 5.0 meq O_2 kg^{-1} is reached	10
Fishy flavor	The flavor reminiscent of cod liver oil	Linseed oil aged 3 months at room temperature	10
Grassy flavor	The flavor reminiscent of the green character of mowed grass	Rapeseed oil aged 12 months and diluted in linseed oil (98 : 2)	9
Hay flavor	The flavor reminiscent of dried grass character of air-dried grain or vegetation	Crude soybean oil from non-heat-treated soybeans diluted in good quality soybean oil (5 : 95)	8
Painty flavor	The flavor reminiscent of oils such as linseed and rapeseed oils containing linolenic acid; it cannot be noted in non-linolenic acid oils such as peanut oil	Good quality canola oil aged 4–8 days at 60 °C or Until PV = 10.0 meq O_2 kg^{-1} is reached	10
Burnt flavor	The flavor reminiscent of burnt popcorn or grains	Crude unprocessed corn oil	10

Texture properties of edible fats were usually evaluated, but for liquid oils, only viscosity or fluidity was evaluated. The most common descriptors used for fat texture were hardness, firmness, spreadability, melting rate, creaminess, sandiness, grittiness, flakiness, *etc.* These textural/structural evaluations were usually linked to overall product quality, product specifications set by codexes, and to some instrumental analysis data (texture measurements with TA-XT, *etc.*). Moreover, consumer expectations for product goodness, freshness, and wholesomeness could be linked to textural sensory evaluations.[29,30,35–37]

12.4.2 Sensory Techniques Used for Fats/Oils and Oleogels

Food hard fats and other similar products have been evaluated by different sensory analysis techniques. After a discussion of the common sensory vocabulary for edible fats/oils, it would be better to briefly familiarize with the commonly used sensory test techniques. The sensory techniques could be classified into three main categories: discrimination testing, descriptive analysis, and affective testing. Discriminating tests are used to find out whether a sensory difference exists between two samples or not. This group of tests can legitimately be set up in hundreds of different ways, but there are practically preferred techniques. The most common tests include paired-comparison, duo-trio test, triangle test, and others.[27,28,30,40]

Descriptive analysis methods involve the detection (discrimination) and description of the sensory properties of samples by trained assessors, also called panelists or judges. It includes the most sophisticated test methods available for professionals. These tests provide vocabulary (word) descriptors of a product with some actual numbers on selected scales to compare and prefer different samples. These tests could be used to detect consumer demands, make quality and/or legislative controls, determine changes during storage and shelf-life, relate with the findings of instrumental analysis results, and measure attribute changes over time.[27,28,30,40] The components of different descriptive profiles commonly used for many food products are presented in Table 12.2. The most commonly used descriptive techniques are Flavor Profile, Texture Profile, Quantitative Descriptive Analysis (QDA), Spectrum Descriptive Analysis, and Free-Choice Profiling.[27,28,30,40]

The third category of sensory evaluations is the 'Affective Testing' methods. The main purpose of these tests is to assess the personal response of a product, product idea or specific product character by a habitual or potential consumer of that product. The aim of these tests is to measure liking or preference for a product. Preference is measured among products to select the one getting more appeal. Acceptance measures the general level of acceptability of a sample by the expected perceptions of the consumers. Since affective testing measures the final point (consumers) of a food product for marketing success, they are essential for every sensory evaluation program to be integrated with other sensory tests and analytical data. Depending on the target demography, 9-point or 5-point hedonic scales were frequently used to

Table 12.2 Common sensory descriptive terms used for the evaluation of fats, oils, and similar products.

Appearance attributes	
a. Color	– hue, chroma, uniformity, depth
b. Surface structure	– shine, smoothness/roughness
c. Size and shape	– dimensions and geometry
d. Interactions of pieces/particles	– stickiness, agglomeration, loose particles
Aroma attributes	
a. Olfactory sensations	– vanilla, fruity, floral, green, *etc.*
b. Nasal feelings	– cool, pungent, astringent, *etc.*
Flavor attributes	
a. Olfactory sensations	– vanilla, fruity, floral, chocolate, skunky, rancid
b. Taste sensations	– salty, sweet, sour, bitter, umami
c. Oral feeling factors	– heat, cool, burn, astringent, metallic
Mouthfeel attributes	
a. Mechanical parameters	– hardness, viscosity, fracturability, elasticity, adhesiveness, cohesiveness, gumminess, *etc.*
b. Geometrical parameters	– gritty, grainy, flaky, stringy, sticky
c. Fat/Moisture parameters	– oily, greasy, juicy, moist, wet, dry, watery,
Skinfeel attributes	
a. Mechanical parameters	– thickness, ease to spread, slipperiness, denseness
b. Geometrical parameters	– gritty, foamy, flaky, sandy, creamy
c. Fat/Moisture parameters	– greasy, oily, dry, wet
d. Appearance parameters	– gloss, whitening, peaking

measure preference and acceptance. The scales ranged from 'like extremely' (9 or 5) to 'dislike extremely' (1). Also, the face or 'smiley' scale is used to measure product acceptance with basic pictures of facial emotions.[27,29,30,40]

Usually, consumer data collected with the affective testing methods are related to the descriptive sensory data, and when possible, with ingredient and processing variables, or with instrumental or chemical analyses to guarantee product success at the bottom line, consumer acceptance. Consumer inputs at the early stages of the product development process can help to identify the quality signals, and in turn, allow the product developer to remain on course with the development and take necessary actions when needed to ensure that the final product is acceptable. Hence, a consumer-guided quality system could be generated and governed if the food companies use and generate an integrated and interactive database. Especially for oleogels and oleogel-used food products, this comprehensive approach seems ultimate for the success of new products.[27,30,40,41]

The Flavor and Texture Profile Analyses, the QDA method, and the hedonic consumer affective tests were the most frequently used techniques among many other sensory tests for edible fats and oils. Besides, the data gained from instrumental texture measurements, which provide numerical results of hardness, spreadability, adhesiveness, cohesiveness, and other similar parameters, could be related to the sensory Texture Profile or the

QDA test results. Also, instrumental rheological analyses providing the modulus values, complex viscosity, coefficients of friction, thixotropic recovery, tribology data, and others could be related to the descriptive sensory data. Likewise, rheometry or differential scanning calorimetry (DSC) determined thermal behavior could be related to sensory attribute measurements. In addition, the sensory Flavor Profile or QDA results for aroma, flavor, and aftertaste attributes could be related to analytical findings of free fatty acidity and peroxide values, color values, phenolics composition, volatiles composition, and others. Through interrelationship with the consumer tests, the defects could be identified first in sensory terms and then in analytical data to make the corrective actions.

12.5 Sensory Properties of Oleogels and Foods Containing Oleogels

In this section, studies involving sensory evaluations for oleogels and foods containing oleogels will be discussed in different sub-sections.

12.5.1 Sensory Properties of Oleogel Products

There are quite a few studies reporting the sensory properties of prepared oleogels directly.[26] In fact, in most of the oleogel studies, sensory analyses were performed in the products where the prepared oleogel was used as an ingredient. In one of our studies, the oleogels of virgin olive oil (VOO) and hazelnut oil (HO) were prepared with beeswax and sunflower wax at 5 wt% concentration.[42] The hazelnut oil oleogels were also added with 0.5% of diacetyl for butter flavor. Commercial spreadable margarine and butter were the control samples. The VOO oleogels were compared with the commercial spreadable margarin, and the HO oleogels were compared to commercial butter. In addition to the compositional and physico-chemical measurements, a quantitative sensory descriptive analysis (QDA) with a trained panel composed of 10 individuals was completed with thirteen sensory definition terms (hardness, spreadability, liquefaction, grassy, milky, rancid, fatty, sweet, salty, waxy, grittiness, cooling, and mouth-coating). In addition, a consumer preference test with a 5-point hedonic scale was completed for the appearance, odor, flavor, and spreadability attributes with 120 volunteer consumers. The QDA test results of HO oleogels and a commercial butter sample is provided in Table 12.3.[42]

As could be observed from the table, there were significant differences among the samples for the sensory descriptors. The HO oleogels were less hard but more spreadable in comparison with commercial butter. The amount of melted sample once it was spread over a bread loaf was measured with a liquefaction term, and butter had quite low scores compared to the oleogels. Among the aroma and aftertaste attributes, the grassy one was significantly different and lower in the oleogel samples. Basically, the

Table 12.3 Quantitative Descriptive Analysis (QDA) results of the hazelnut oil oleogels prepared with beeswax and sunflower wax. The results are reported as mean ± standard error, followed by the median value. Reproduced from ref. 42 with permission from the Royal Society of Chemistry.[a]

Descriptor	HBW	HSW	CBT
Hardness	2.44 ± 0.29 2.00[c]	2.50 ± 0.21 2.75[c]	7.81 ± 0.50 8.50[b]
Spreadability	7.88 ± 0.35 8.00[b]	7.75 ± 0.57 8.00[b]	2.94 ± 0.76 2.75[c]
Liquefaction	5.50 ± 0.88 5.75[b]	5.38 ± 0.87 5.00[b]	1.69 ± 0.95 0.75[c]
Grassy	4.44 ± 0.79 4.50[c]	4.81 ± 1.00 5.75[c]	8.81 ± 0.31 9.00[b]
Milky	2.44 ± 0.80 2.00[b]	1.00 ± 0.28 1.00[bc]	0.50 ± 0.25 0.25[c]
Rancid	1.19 ± 0.30 1.25[b]	1.25 ± 0.44 0.75[b]	0.75 ± 0.35 0.50[b]
Fatty	6.75 ± 1.04 7.50[b]	8.44 ± 0.27 8.75[b]	9.00 ± 0.19 9.00[b]
Sweet	2.44 ± 0.56 1.75[b]	2.44 ± 0.68 2.25[b]	2.44 ± 0.48 3.00[b]
Salty	0.88 ± 0.26 0.50[c]	0.69 ± 0.19 0.75[bc]	1.75 ± 0.38 1.75[b]
Waxy	4.06 ± 0.98 3.50[b]	3.56 ± 0.79 3.25[bc]	1.19 ± 0.43 0.75[c]
Grittiness	1.75 ± 0.37 2.00[b]	1.44 ± 0.35 1.00[b]	0.88 ± 0.28 0.50[b]
Cooling	1.31 ± 0.56 1.00[b]	1.19 ± 0.44 1.00[b]	1.56 ± 0.38 1.50[b]
Mouth coating	3.06 ± 0.55 3.25[b]	2.69 ± 0.57 2.50[b]	2.94 ± 0.68 2.50[b]

[a]Lowercase letters compare the samples within each row ($p \leq 0.001$).
[b]HBW: hazelnut oil–beeswax oleogel; HSW: hazelnut oil–sunflower wax oleogel.
[c]CBT: commercial butter.

sensory evaluation results pointed out the most important similarities and differences between diacetyl aromatized HO oleogels and commercial butter to re-design the formulation to get better samples for higher consumer acceptability. In fact, in the same study, the consumer hedonic study showed that especially the appearance and odor attributes of the HO oleogels needed improvements to catch the scores of the commercial butter sample. Furthermore, consumer buying decision data revealed that more than 50% of the consumers would buy or try once and then buy the oleogel products. Overall, this study clearly showed the essentiality of sensory studies in the product development phases of oleogels. Even if all the physical and chemical properties of the prepared oleogels matched the desired qualities, they must have good enough sensory properties, which were liked or accepted by the consumers.

In another study, aromatized (strawberry, butter, and bananas), and vitamin-enriched (500 mg kg^{-1} Vit. E and D3, and 25 mg kg^{-1} β-carotene) hazelnut oil-beeswax and hazelnut oil-sunflower wax (5 wt% wax) oleogels were prepared and characterized.[43] In these samples, both descriptive sensory analyses (QDA) and volatile component measurements were done. Three appearance attributes (roughness, gloss, homogeneity), four texture attributes (firmness, knife stickiness, spreadability, liquefaction), four aroma attributes (fruity, added aromas – strawberry, diacetyl or bananas, sweet aromatic, rancid), and one flavor attribute (sweet) were quantified. The interrelationship between texture measurement and sensory texture attributes and volatile compound concentrations and sensory aroma attributes were discussed. The study successfully proved that instrumental data were very well correlated with the sensory results. Moreover, the added aromatics were quantified both by GC/MS volatile analysis and by sensory QDA, and well related. Also, after 3 months of storage, the presence of added aromas was verified, indicating the volatile retention potential of the oleogels. Another similar study,[44] reported VOO oleogels prepared with the same added aromas and vitamins, and their storage stabilities were determined. The same instrumental and sensory analyses were completed, and the results indicated that both oleogel structures and added aromas were fairly stable over 3 months of storage. In particular the rancid attribute of QDA analysis was very effective in determining the sensory oxidative stability of the oleogel samples.[44]

In order to mask the fishy odor, fish oil oleogels with beeswax (5 wt% concentration) and externally added commercial liquid strawberry and lemon flavors (0.5%) were prepared and characterized.[45] The 5-point hedonic scale was used, and 150 consumers participated in the tests to evaluate the consumer preferences for appearance, odor, flavor, and spreadability. The results indicated that aroma-enriched fish oil–beeswax oleogels were structurally very good and well-accepted, and the added aromas partly masked the fishy odor. Strawberry flavored samples had the highest scores compared with the control (no flavor added) and lemon-flavored samples. This study has pointed out the undeniable place of consumer tests for various purposes in oleogel development. Lately, we have developed some animal wax-based oleogels (beeswax, shellac wax, whale spermaceti wax, and lanolin wax), and 11 trained panelists described the oleogel samples with the QDA technique on a 10 cm line scale.[46] A distinct sheep odor was detected in the lanolin wax oleogel, and that sample was found unacceptable as a food, while a whale spermaceti wax oleogel was found to be quite similar to the beeswax oleogel and evaluated as a suitable candidate for food applications.

Corn oil–beeswax oleogels were used to carry β-cyclodextrin-fish oil complexes to protect the omega fatty acids.[47] In the meantime a panel of 14 adults was used to make a just about right sensory test to identify the odd sample (fish oil) among the oleogel and whey protein coated samples. The results indicated that the fish odor was decreased in the whey protein-coated oleogel samples.

Since most sensory studies were accomplished in oleogel-containing food samples, the next paragraphs will be on the discussion of different groups of food samples for various sensory findings.

12.5.2 Sensory Properties of Bakery Products Containing Oleogels

Bakery products have been one of the important sectors of oleogel applications. In one study, a shellac–sunflower oil oleogel (2 wt% shellac) was prepared, and then an oleogel emulsion (20 wt%, w/o) was formed. This oleogel emulsion (SE) was used in spreads, chocolate paste, and cakes.[48] Classic 4/4 sponge cakes were prepared with 20% oleogel emulsion against a control, and a panel of 20 people evaluated the cakes for volume, cell size uniformity, crumbliness, sponginess, moistness, and stickiness properties. Sensory data showed that SE cake was comparable with the control sample, except for cell size uniformity and crumbliness. It was indicated that the uneven distribution of air in the batter due to the lack of solid lipid crystals caused the outcome.

Hazelnut oil–beeswax and sunflower wax oleogels (5 wt%) were prepared and used in cookie formulation at 24.07 wt% usage levels against a control sample prepared with commercial bakery shortening.[49] A trained panel of 12 people defined and used 3 cookie surface properties (glossiness, color, smoothness), 5 cross-sectional texture properties (compactness, pore distribution, crust thickness, interior color, color difference), 5 mouth properties (bite hardness, fracturability, sandiness, dispersibility, meltdown) sensed during mastication, and 2 flavor properties (flavor, sweetness) to profile the textural and flavor properties of the cookie samples. Furthermore, 200 volunteer consumers tested the samples for appearance, texture, flavor, smell, and acceptability with a 5-point hedonic scale. Also, instrumental texture measurements for the cookies were completed. This integrated study proved that wax oleogel cookies were almost identical to the standard cookies. The panel results indicated that there are no significant differences among the samples for the attributes evaluated. Interestingly, consumers liked the oleogel cookies more than the control sample, possibly due to enhanced compactness and fracturability. This study is a good example of understanding consumer behavior in sensory terms and the related instrumental data.

In another study, hydroxypropyl methylcellulose (HPMC) was utilized to structure sunflower oil into solid-like oleogels through foam-templates, and the prepared oleogels were used as a shortening replacer in muffins at 25, 50, 75, and 100% weight replacement levels.[50] Instrumental hardness, springiness, cohesiveness, and chewiness were measured. This study lacked sensory evaluation analyses. The results indicated that at 50% oleogel replacement level, the muffins had almost equal soft and chewy texture values, but the consumer data were missing. Although the study was very detailed on batter and muffin instrumental analyses, the absence of sensory analyses and consumer tests was evident when making an overall conclusion.

A very interesting study utilized camellia oil-based oleogels prepared with tea polyphenol–palmitate (Tp–palmitate) particles (2.5%, m/v) and varying citrus pectin concentrations of 1.5, 2.5, 3.5, and 4.5% (m/v) using an emulsion-templated method.[51] The prepared oleogels were used in classic 4/4 sponge cakes in a formula of 100 g wheat flour, 100 g liquid whole egg, 100 g powdered sugar, 1.3 g baking powder, and 100 g butter or oleogels. Seventy untrained assessors used a 7-point hedonic scale to evaluate the crust color, crumb color, odor, flavor, and texture of the cakes prepared. A picture of the prepared cakes and their sensory results in spider-web graphical format can be observed in Figure 12.1. Compared with the control sample prepared with butter, all oleogel cakes had lower sensory scores, and especially the sensory score decrease was linearly correlated with the pectin content in the oleogel formulation. While a decrease in texture score was notable, odor scores were not very different. A significant negative correlation between hardness and cake quality was also evident. The overall quality score of the control cakes was 32.03, while that of the cakes prepared with oleogels had an overall quality score between 21 and 27. Overall, oleogels prepared with food-grade polymers were found good enough in cake formulations to replace butter partially.

The same research group utilized very similar oleogels prepared with 3.0% tea polyphenol ester (Tp-ester) particles with four fatty acid chain length (Tp-laurate [C12], Tp-myristate [C14], Tp-palmitate [C16], and Tp-stearate [C18]), and 2.5% citrus pectin.[52] The prepared oleogels were used in cookie production as a fat replacer at 25, 50, 75, and 100% replacement levels for

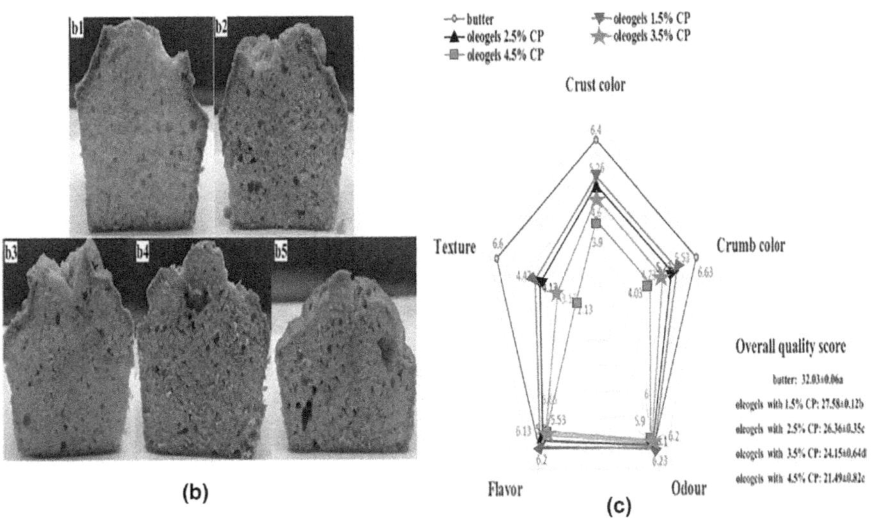

Figure 12.1 The photographs (b) and comparative sensorial parameters (c) of the cake products [(b) photographs of cake products prepared using butter (b1), oleogels with 1.5% (b2), 2.5%, (b3), 3.5% (b4), and 4.5% (b5) (m/v) CP, respectively. (c) Comparative sensorial parameters of cakes prepared using butter, oleogels with 1.5%, 2.5%, 3.5%, and 4.5% (m/v) CP, respectively].
Reproduced from ref. 51 with permission from Elsevier, Copyright 2019.

butter. In addition to many other analyses, the cookies were assessed by 60 untrained and semi-trained panelists on a 9-point hedonic scale for the acceptance degree of appearance, aroma, crispiness, taste, and overall acceptability. The sensory scores of each Tp-ester oleogel cookies decreased with the increasing replacement level. Compared with the butter cookies, a decrease in the evaluation scores of the oleogel cookies was observed, but the decrease of the cookies made with Tp-palmitate oleogels and Tp-stearate oleogels at the replacement level of 25% and 50% was not significant.

Another study aimed to prepare zero-*trans* cake shortening (ZTC) from a chemically interesterified blend of 66.416% palm stearin, 33.584% canola oil, and 2% glycerol monopalmitate.[53] Although they did not call the ZTC an oleogel, it was a two-component oleogel. They utilized a standard cake formulation with two commercial shortenings and the ZTC. Sixty untrained panelists and a 9-point hedonic scale were used to evaluate the overall acceptance, taste, color, fat congelation, odor and texture. The ZTC-made cake had the second softest sample and the highest overall sensory acceptability. Furthermore, it was shown that level of softness is not the only factor affecting sensory acceptability, and softness was not directly related to batter rheology.

Pehlivanoglu *et al.* prepared vegetable oil–carnauba wax oleogels (5 wt%), and also oleogel emulsions including 18% water.[54] Both oleogels and watery-oleogels were used in cake formulations and cakes prepared according to the AACC 10–90 method. Fifteen untrained panelists used the 5-point hedonic scale to test cakes for crust color, crumb structure, taste, odor, chewiness, and overall quality. The highest acceptability score was given to the cake prepared with the oleogel made with the sunflower oil–cottonseed oil mixture. Furthermore, it was evident that with watery oleogels, calorie reduction was possible with acceptable sensory values.[54] More studies with different types of oleogels and emulgels for full or partial replacement of bakery shortenings are needed.

12.5.3 Sensory Properties of Comminuted Meat Products Containing Oleogels

Comminuted meat products are one of the most frequently applied categories for edible oleogels. Among many other studies, only those with sensory evaluations are considered below. Generally, various oleogels are used in these meat products to reduce total saturated and *trans* fatty acids, to provide omega and other specific fatty acids, and also to reduce total fat contents. In one study, an olive oil–water–soy protein (10 : 8 : 1) oleogel emulsion was replaced by 15, 25, 35, 45, and 55% of pork meat in venison salchichon sausages.[55] Among many analyses, consumer acceptance and preference tests were completed with 44 consumers. All samples had 5.0 mean acceptability (like) scores, and replacement up to 25% by an oleogel was found similar to the control sample. At this rate of replacement, saturated fatty acids reduced significantly and oleic acid increased two-fold.[55]

Canola oil–ethyl cellulose (EC)–sorbitan monostearate (SMS) oleogels were used to replace beef fat in frankfurters.[56] Sensory analysis was done with 12 semi-trained panelists with a 10 cm line scale for hardness, juiciness, and fattiness (*i.e.*, 1 = very soft to 10 = very hard). Substitution with canola oil doubled the hardness of the frankfurters relative to the control. The organogel prepared with 8% EC and 1.5 or 3.0% SMS resulted in a hardness value similar to the control sample by both sensory and texture profile analysis. Without SMS addition, sensory data indicated lower hardness values than regular canola oil but still higher than the control. Gels prepared using higher EC concentrations (12 and 14%) yielded meat products with a higher sensory hardness than the control.[56] The same research group extended their analysis with the same oleogel and found that sensory juiciness was reduced with liquid canola oil but increased with increasing ratio of the oleogel. The oiliness sensation was also enhanced with oleogel substitution. Overall, they concluded that the oleogel could replace vegetable oil in sausages successfully and could yield the nutritionally-modified fatty acid composition.[57]

Sunflower oil was structured with a 15 : 5 (wt%) monoglyceride:phytosterol mixture and used to replace pork backfat at 50% in frankfurter sausages.[58] Sensory analysis of the samples was accomplished with 38 untrained members and an unmarked continuous linear hedonic scale. The results showed that the control sample had higher values for hardness, brittleness, gumminess, and chewiness, but the cohesiveness and elasticity values were not different. Saturated fat reduction was possible with acceptable physicochemical, textural, and sensory characteristics.

Wolfer *et al.*[59] utilized soybean oil–rice bran wax (2.5 and 10 wt%) oleogels in frankfurter production at a 17.67% addition level against pork backfat and soybean oil controls. Ten trained panelists evaluated the samples with a 15-point line scale for cured frankfurter aroma, cured frankfurter flavor, chewiness, firmness, off-flavor, exterior darkness, interior darkness, and pinkness. The control sample was darker and redder; instrumental texture parameters were similar. Oleogel replacement did not influence the cured aroma, but the cured flavor was significantly reduced.[59]

In another study,[60] sunflower oil emulgels with pork skin and water (1 : 1.5 : 1.5) were prepared and used in replacement of pork back fat in Bologna-type sausages at 25, 50, 75, and 100% replacement levels. A hundred consumers made an acceptance test with a 9-point hedonic scale for color, aroma, flavor, texture, and overall acceptance. Furthermore, a check-all-that-apply questionnaire was completed to relate descriptors to appearance (pale color, ideal color, oily, and strange aspect), aroma (soft, rancid, fatty, pleasant), taste (pork, rancid, soft, strong, bitter, pleasant, unpleasant, and seasoning in the right amount) and texture (sandy, ideal, rubbery, hard, dry, difficult to chew, and juicy). Sensory data revealed that with up to 50% of fat replacement by an oleogel, no sensory defect was detected. By this rate of replacement, 29% total fat, 21% energy, and 10% cholesterol reduction were possible.[60]

In another study,[61] linseed oil was gelled with beeswax (8 wt%) and replaced for pork back fat in sausage formulation at 25 and 50% rates. Sensory analysis was made with 32 panelists for overall acceptability, appearance, odor, hardness, juiciness, taste, and global perception with a 7-point hedonic scale. For all attributes, the control sample had significantly higher scores. More research to improve the sensory quality was suggested, especially for appearance and color.[61] Linseed oil–beeswax and linseed oil-γ-oryzanol + β-sitosterol (60 : 40) oleogels were prepared and replaced for pork back fat in sausage formulation at 20 and 40% by weight.[62] Sensory analyses were performed by 28 panelists for overall acceptability by a 7-point hedonic scale. The control sample was more acceptable, but samples with oleogels at 40% replacement level were more acceptable than those of the 20% replacement levels. Especially the textural parameters needed to be improved for enhanced consumer acceptability.

Pintado and Cofrades[63] utilized olive oil + chia oil–beeswax oleogels (10 wt%) and emulgels (45 wt% oils) in fat replacement for dry fermented meat sausage. The replacement level for oleogel and emulgel were 7.5% and 15%. The sensory attributes (general appearance, odor, flavor, texture, and overall acceptability) were evaluated on a 10-point scale, 0 being considered as "dislike strongly" and 10 as "like strongly" with 30 assessors. The control sample had higher scores for all attributes, and further studies to improve the sensory quality were suggested.[63]

Chicken sausages were prepared with a fat replacement for pork back fat by 41.9% level with a soybean oil–rice bran wax (2.5 and 10 wt%) oleogel.[64] A trained panel of 13 assessors used a 15 cm unstructured line scale to assess the sensory terms of Bologna aroma, other aromas, texture, moistness, Bologna flavor, other flavors, and color on stored samples at 21, 49, 70, and 98 days. The fat replacement had no significant effect on texture profile parameters and incisor peak force during storage, and there was no sensory difference for aroma, flavor, texture, and moistness attributes. The color of the full-fat sample was more intense and more liked, expectedly.

Beef pates/patties and similar products are the other meat product group in which animal fat was replaced by some oleogels. An olive oil–linseed oil–fish oil mixture (44.39-37.87-17.74%) and ethyl cellulose (11 g/100 g oleogel) and sorbitan monostearate (3.67 g/100 g oleogel) were used to prepare the oleogels.[65] Also, the same oil mixture and beeswax (11 g/100 g oleogel) were used to prepare the other oleogel. Both oleogels were used to substitute backfat totally or partially in liver pates. Hedonic sensory analysis with 40 panelists on a 10-point scale was used for odor, color, texture, and overall acceptability. The fat replacement did not affect the emulsion stability, color, and texture, but the sensory properties in the ethyl cellulose oleogel-substituted sample were lower. Beeswax oleogel substitution had no effect on the sensory properties, and all samples had 'neutral' sensory acceptability indicating a promising result for fatty acid modification in pates.

In another study,[66] canola oil–hydroxypropyl methylcellulose (4 wt%) foam-templated oleogels were used to replace (50 and 100% w/w) animal fat in meat patties. Thirty assessors evaluated the cooked meat patties for visual appearance, color, flavor, taste, tenderness, juiciness, and overall acceptability with a 9-point hedonic scale. Also, instrumental texture and other analyses were completed. The firmness and sensory properties of the meat patties are presented in Figure 12.2.

The instrumental firmness values of the patties with oleogels were lower than both the control and canola oil samples (Figure 12.2). Similarly, sensory properties of tenderness, juiciness, and overall acceptability were significantly different, and these were correlated with the instrumental firmness values. It was acknowledged that to get a soft texture, the entrapment of fat and moisture was necessary, and reduced cooking loss by the use of oleogels contributed to the tender texture. The highest overall sensory acceptability was observed in a sample with 50% fat replacement. However, oleogel replacement did not significantly affect the visual appearance, color, flavor, and taste, and a change of the ratio of saturated to unsaturated fatty acids was achieved from 0.73 to 0.18 to enhance the meat patties nutritionally.[66]

In a similar study,[67] a soybean oil oleogel prepared with ethyl cellulose : avicel RC-591 : α-cellulose (67.0 : 16.5 : 16.5) fully replaced pork back fat in beef patties. Forty untrained panelists rated the sample pairs for their preference of taste, color, texture, fat sensation, and overall acceptability on a 10 cm structured graphical hedonic scale. Generally, the sensory acceptability scores of oleogel-replaced samples were lower than the control, mainly due to a darker and less red color.[67] Another study[68] utilized a linseed

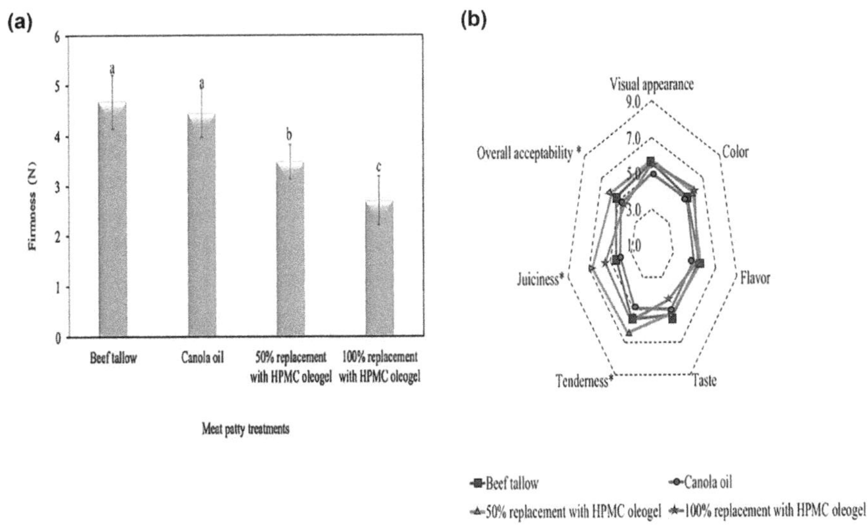

Figure 12.2 Effect of beef tallow replacement with hydroxypropyl methylcellulose (HPMC) oleogels on the firmness at room temperature (a) and sensory properties (b) of meat patties.
Reproduced from ref. 66 with permission from Elsevier, Copyright 2019.

oil–beeswax (8 wt%) oleogel in fat replacement up to 30 and 60% in beef pate samples. Sensory evaluation with 13 taster and 7-point hedonic scale was completed for the visual aspect, smell, texture, flavor, and global evaluation. Oleogel incorporation resulted in decreased hardness and adhesivity, and the control sample had higher sensory scores for all attributes. The sample with 30% fat replacement had a combined score for 'liked' indicating acceptability with significantly improved unsaturated fatty acid composition.[68]

Another group of meat products in which fat was replaced by oleogels are the burgers. Sesame oil–beeswax (10 wt%) oleogels replaced animal fat in a beef burger at 25 and 50% levels, and proximate composition, texture, color, fat absorption, cooking loss, shrinkage, microstructure, oxidative stability, and sensory acceptance were measured.[69] Approximately 11% cooking loss reduction and 1.6% fat absorption reduction were achieved as positive results of fat replacement, and overall sensory acceptability was the highest in the 50% fat replaced sample. Martins *et al.*[70] used a linseed oil–oryzanol + sitosterol (60 : 40) oleogel (8 wt%) to replace the fat at 25 and 75% replacements in hamburgers. Cooked samples were tested by 16 panelists with a 7-point hedonic scale for smell, tenderness, juiciness, taste, and global evaluation attributes. The sample with 25% fat replacement was found acceptable, but for all sensory attributes, the control sample had the highest scores.[70] A hydrogelled emulsion composed of 12.5% chia oil, 12.5% linseed oil, 70% water, and 1% polysorbate 80 was used to replace pork fat in burgers at 20, 40, 60, 80, and 100% replacement levels.[71] One hundred consumers tested the cooked burgers for overall acceptability with a 9-point scale. The technological properties were suitable after fat replacement, but hardness and chewiness scores increased due to an increased protein : lipid ratio. Sensory data proved that fat replacement up to 60% could be possible without any large reduction in the sensory attributes.[71]

Gomez-Estaca *et al.*[72] prepared olive : linseed : fish oil (44.39 : 37.87 : 17.74%) oleogels with ethyl cellulose and beeswax, in the same way, as explained in their previous publications.[65] The burgers included 4% meat, 8% water/ice mixture, 1% salt, 1% spice mixture, and 6% fat, for each of the oleogels. Odor, color, texture, and overall acceptability were evaluated on a 10-point scale by 40 people. Burgers, including ethyl cellulose oleogels, suffered lipid oxidation and softer texture. Higher sensory acceptability ratings were present for the burgers made with the beeswax oleogel, but burgers with ethyl cellulose oleogels had scores below the neutrality point.[72] In a continuation study, Gomez-Estaca *et al.*,[73] prepared the same ethyl cellulose and beeswax oleogels, but this time 0.2 g curcumin/100 g oleogel was incorporated into the oleogels. Then, the oleogels were used in replacement of pork fat at a 6% fat addition level in burgers. Odor, color, texture, and overall acceptability were evaluated on a 10-point scale with 40 consumers. The beeswax oleogel-replaced sample was found technologically adequate and sensorially acceptable, but the acceptance of color was reduced due to the yellow color provided by curcumin. Curcumin also has enhanced oxidative stability. Lately, soybean oil–ethyl cellulose (6 wt%) and soybean oil–ethyl cellulose (2 wt%) + adipic acid

(4 wt%) oleogels were prepared and used in 50% fat replacement in beef burgers.[74] Odor, taste, texture, color, and overall acceptability of the samples were assessed by 20 semi-trained panelists using a 10-point hedonic scale. The burger sample formulated with 2% ethyl cellulose and 4% adipic acid oleogel exhibited a satisfactory texture profile and acceptable color and organoleptic characteristics.[74]

Overall, various oleogels were used in comminuted meat products to reduce saturated and *trans* fatty acids very successfully. The sensory quality of the most products was almost acceptable. Usually, problems with a softer texture, color changes, and oxidation were encountered. To improve the overall quality, the type of product and the amount of animal fat that is being replaced must be selected after pre-trials. In addition, an appropriate oleogel or emulgel must be selected based on the textural, sensory, and technological properties, and the consumer expectations of the meat product produced.

12.5.4 Sensory Properties of Margarine/Spread Products Containing Oleogels

Margarine and spreads are commonly consumed fat products all over the world. Most of these products are emulsions containing not less than 80% fat. There are many types of margarine and spreads on the market, like regular, stick, whipped, soft tub, diet, liquid, *trans*-free, low calorie, whipped spreads, and butter blends. The emulsions are composed of oil and aqueous phases. The aqueous phase may include water, milk or milk products, salt, potassium chloride, and antimicrobials (sorbic and benzoic acids and their salts). The oil phase may include antioxidants (tocopherols, propyl, octyl, and dodecyl gallates, butylated hydroxytoluene, butylated hydroxyanisole, ascorbyl palmitate, and ascorbyl stearate) and metal scavengers (lecithin, isopropyl citrate, and calcium disodium ethylenediaminetetraacetic acid), flavoring substances (diacetyl, lactones, butyric acid and its esters, ketones, aldehydes), vitamins (vitamin A and D), and colorants (β-carotene, natural extracts). The type and amounts of each ingredient depend on the product type and regulatory issues. The type, level, composition, and crystal habit of the oil/fat used in the margarine formulation determine the functional properties such as consistency, texture, spreadability, oil separation, melting behavior, appearance, color, mouthfeel, taste, product shelf-life, and stability. Generally, margarine requires a solid fat index (SFI) of around 10–20 at serving temperature. Furthermore, polymorphism is important, and oils with β' polymorph crystals are preferred. The margarine processing parameters like formulating the two phases, emulsification, cooling and crystallization, working and tempering, packaging, storage, and distribution were also effective in product quality.[1,2,6]

There is a limited number of studies in the literature reporting about margarine/spreads prepared with oleogels. Hwang *et al.*[75] used some oleogels in actual laboratory-scale margarine production. Unfortunately, this study did not include any sensory analysis, but the instrumental texture and other measurements were completed. They used sunflower (SW), candelilla (CW), and rice bran wax (RX) oleogels as the fatty phase (80%) and produced regular margarine. CW margarine showed phase separation, while the other two were stable and firm enough. Especially SW margarine was similar to commercial products in terms of texture and melting behavior, but sensory studies were missing. The same research team extended their margarine study with 12 different vegetable oils and SW at 3, 5, and 7 wt% concentration.[76] The oleogel : water ratio was 80 : 20, and regular margarine formulations were used. Fatty acid composition of the fatty phase was ineffective in firmness, while total polar compounds in the oil inversely affected margarine firmness. Unfortunately, there was no sensory evaluation in this study as well.[76]

In our laboratory, olive oil–beeswax emulgels with Xanthan gum, Tween20/80, salt, potassium sorbate, and citric acid ingredients as margarine-like products were developed. Although there was no sensory study, the emulgels were shown to be quite similar to commercial margarine samples in their textural and thermal properties. The emulgels were stable over 90 days of storage for texture, color, and oxidation.[77]

High-oleic sunflower oil–candelilla wax, monoglyceride, and other ingredients as the lipid phase and regular water phase were used to prepare margarine with low saturated fatty acid contents.[78] All main chemical, physical, and structural analyses were completed in addition to storage stability tests. Furthermore, a sensory acceptance test with 120 panelists for appearance, aroma, taste, texture, overall impression, and spreadability with a 9 cm non-structured hedonic scale was accomplished. Also, an 'intent to purchase' test was evaluated using the 5-point scale, which ranged from 1 = certainly would not buy, to 5 = would certainly buy the product. The sensory scores for commercial margarine (CM) and oleogel margarine (OM) were compared over storage time. Scores of all CM were above 6 points, indicating consumer like. The spreadability of the OM samples was higher than CM. Other attributes were found to be a little lower, but still acceptable. The scores for overall taste, texture, and impression were lower (>4.0). The flavor scores were the lowest in the OMs, and this may have greatly influenced the score of the overall impression. Further studies to improve sensory quality were suggested.[78]

Overall, studies reporting the sensory properties and consumer preferences for margarine and spread-type products containing oleogels are quite limited, and therefore, more research is encouraged in this area.

12.5.5 Sensory Properties of Chocolate/Confectionery Products Containing Oleogels

Confectionary broadly defines sweet foods and is divided into two, baker's and sugar confectionery. This section is dedicated to chocolate and chocolate confectionery products. Chocolate is a product of the beans from the *Theobroma cacao* tree. The beans of the tree yield three main materials for many uses; chocolate liquor, cocoa butter, and cocoa powder. Chocolate is a multi-typed manufactured good, but its main ingredients are chocolate liquor (15–42%), cocoa butter (8–20%), sugar (40–50%), skim milk or other solids (1–15%), and lecithin (0.4%). Chocolates are exceptionally defined for their texture, flavor, and melting behavior. The distinguishing property of chocolate is being solid at room temperature and yet rapidly melting in the mouth. Cocoa butter is the substance responsible for this property due to its specific fatty acid composition and distribution within triglycerides. Cocoa butter contains 98% triglycerides, 1% free fatty acids, 0.3–0.5% diglycerides, 0.1% monoglycerides, 0.2% sterols, 150–250 ppm tocopherols, and 0.05-0.13% phospholipids. Stearic acid (36%), oleic acid (34%), and palmitic acid (25%) are the dominant fatty acids, and most oleic acid is esterified at the sn-2 position. Generally, palmitic–oleic–stearic (36–42%), stearic–oleic–stearic (23–29%), and palmitic–oleic–palmitic (13–19%) configurations are dominant. The characteristic brittle non-greasy texture, rapid melting in the mouth, and excellent quality maintenance at room temperature is the sum result of the cocoa butter fatty acid composition and TAGS configurations.[1,79,80]

Some fat sources have been used to replace cocoa butter in confectionery products. Generally, the term 'cocoa butter alternatives' was suggested for these fats, and they were classified as 'cocoa butter equivalents – CBE', 'cocoa butter replacers – CBR', and 'cocoa butter substitutes – CBS'. There is no need here to go further into these sources, but a very limited number of studies utilized oleogels as cocoa butter replacers and the studies with sensory analyses are discussed below.[79,80]

Shellac wax oleogel (1.5 wt%) in rapeseed oil was prepared and then emulsified with water to obtain a shellac wax oleogel emulsion. A commercial oil binder (Palsgaard Oil Binder 01) was used in the standard chocolate paste as the control, and the emulgel has replaced the binder at 100% of oil binder and around 27% of palm oil with repressed oil.[48] Chocolate pastes are prepared from cocoa powder and sugar dispersions in continuous oil medium. Their main usage purposes were to prevent sedimentation of dispersed particles and oil separation. Storage at 30 °C for 4 weeks showed no oiling out with the emulgel. Since the SFI was quite low in oleogel replaced paste, the paste was suggested not to have a greasy taste. Although the rheological data suggested feasibility, no sensory study was completed in this study.[48]

In another study,[81] hydroxypropylmethyl cellulose and xanthan gum were used to prepare olive and sunflower oil oleogels and the oleogels were used in the formulations of chocolate spreads. There was a control spread made with coconut fat (C); spread made with 50% coconut fat and 50% olive oil oleogel (CO); spread made with 50% coconut fat and 50% sunflower oil oleogel (CS); spread made with olive oil oleogel (O); spread made with sunflower oil oleogel (*S*). The CO and CS samples included 15 g of oleogels and coconut butter, while the O and S samples included only 30 g of oleogels. The rest of the ingredients (sugar, water, skimmed milk powder, and cocoa powder) were the same for all samples. In this very detailed study, among many analyses, the sensory Free Choice Profile analysis was conducted with 20 untrained consumers. In the first session, each consumer generated his/her own describing terms for the spreads, and the differences among the three samples were generated by a Repertory Grid Method (RGM). The appearance, taste, aroma, and texture were the main attributes evaluated. Then, each consumer used their own list of terms by rating the intensity of each term on a 10 cm unstructured line scale. The Free Choice Profile data were analyzed by Generalized Procrustes Analysis (GPA). This sensory technique is advantageous to collect information on the spontaneous sensations that occur when the product is consumed. The 20-consumer panel generated about 191 different sensory descriptors related to the appearance, taste, aroma, and texture of the samples. After the Generalized Procluster Analysis (GPA), a total variance of about 78.51% was obtained. This indicated the variability of the sensory attributes shown by the first two dimensions. Researchers provided a GPA mapping with all 191 descriptors shown in different colors for the different samples, but to obtain information was rather hard from this very crowded map, and a distribution map was obtained based on the frequency of mention of the most representative attributes of each dimension. The two-dimensional GPA map is shown in Figure 12.3.[81]

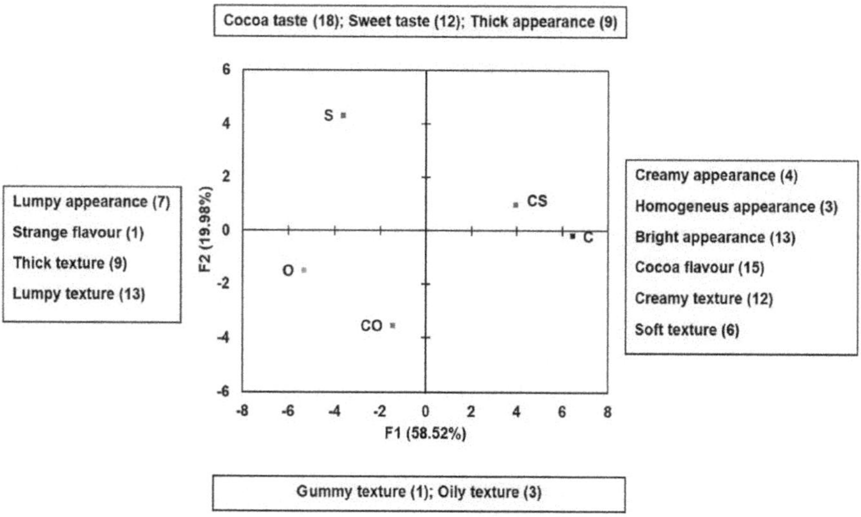

Figure 12.3 Representation of the samples in the two-dimensional GPA diagram where the main descriptors correlated with the first two dimensions of the average space are reflected [control spread made with coconut fat (C); spread made with 50% coconut fat and 50% olive oleogel (CO); spread made with 50% coconut fat and 50% sunflower oleogel (CS); spread made with olive oleogel (O); spread made with sunflower oleogel (*S*)]. Reproduced from ref. 81 with permission from Elsevier, Copyright 2021.

On the positive side of the X-axis, dimension 1 represents 58.52% of the variance related to the appearance and texture terms. The terms "creamy appearance" and "bright appearance", "creamy texture," and "cocoa flavor" were related to the spreads C and CS. On the contrary, the O and S samples were characterized with attributes such as "lumpy appearance", "lumpy texture", and "thick texture". It was thought that the undissolved particles in the O and S spreads caused the "lumpy appearance" and "lumpy texture". Around 19.98% of the variance was represented with Dimension 2 on the Y-axis. The "gummy texture" and "oily texture" were related to spreads made with olive oil oleogels. However, on the positive side of the Y-axis, desirable characteristics for spreads related to flavor as "cocoa taste" and "sweet taste" were described and related to spreads made with sunflower oil oleogels. Overall, the sunflower oil oleogel and coconut fat mixture (50% each) had sensory attributes similar to the control sample with a healthier fatty acid composition.[81]

It is quite clear that oleogels and emulgels can be used in formulations of various chocolate confectionery products, but consumer expectations and buying decisions in particular must be researched before reaching a meaningful verdict on the possibility of commercialization.

12.5.6 Sensory Properties of Dairy Products Containing Oleogels

Some of the important functional and sensory properties of dairy products arise from the presence of milk fat. Milk fat is mainly composed of triglycerides (98%) and minor components, including free fatty acids, phospholipids, cerebrosides, sterol esters, tocopherols, pigments, and flavor compounds. Also, health-promoting compounds like medium-chain fatty acids, conjugated linoleic acid, sphingomyelin, butyric acid, ether lipids, and vitamins are provided with milk fat or its products like cream and butter. The saturated fatty acids of palmitic, myristic, stearic, lauric, and butyric compromise around 70% of all fatty acids, and the rest includes oleic and linoleic acids. Butterfat has around 210–250 saponification number, 30–33 iodine number, 0.4% unsaponifiable matters, and a melting range of −30 °C to 37 °C. The solid : liquid ratio at a given temperature determines its textural and rheological properties. Generally, butter shows a fine and close texture, with a firm and waxy body, plastic rheology and good spreadability. It's very special flavor is due to the presence of diacetyl, dimethylsulfide, and several lactones, esters, ketones, aldehydes, and free fatty acids.[82,83]

Different oleogels were used in dairy products to replace the expensive butterfat. The most common dairy products where oleogels replaced butter are cheeses, cream cheese, and ice cream. Except for one study, there was no sensory evaluation of the products formulated. Most of the time, nutritional composition data, instrumental texture, and microstructural data were provided. Obviously, there is research needed for sensory evaluations and consumer tests. In one available study,[84] soybean oil–rice bran wax (RBW) oleogels at 8 and 10 wt%, and ethylcellulose (EC) oleogels at 10 wt% were prepared and used against control samples, including full-fat and fat-free commercial cream cheeses. Also, control samples with butterfat and soybean oil were prepared in the same way. The prepared cream cheese samples were then analyzed for composition, microscopy, texture, rheology, fatty acid composition, and sensory analysis. The structural networks of the control and oleogel cream cheeses were found quite to be similar using confocal laser scanning microscopy. Although the oleogel cream cheese samples had comparable hardness values with the control, the liquid oil added sample had the highest hardness values. While the spreadability and stickiness values were similar, the adhesiveness values were lower in the oleogel cream cheese sample. Generally, samples prepared with RBW were mimicking the control sample more successfully for spreadability, hardness, stickiness, and adhesiveness results. The thermal stability of the oleogel cream cheese samples was found to be higher. The RBW oleogel cream cheese was selected with full-fat commercial control cream cheese to make a sensory analysis. Twenty participants evaluated the hardness, spreadability, flavor, mouthfeel, sweetness, and bitterness with a 5-point 'just about right' test to determine if the attribute was too much (score of "5"), too little (score of "1"), or "just about right" (score of "3"). There was no significant difference between the RBW and the control cream cheese for hardness, spreadability, and mouthfeel. In contrast, the flavor and bitterness for the RBW sample were found to be too strong. The fatty acid profile of the oleogel cream cheeses was improved.[84] More research to replace fully or partly the milk fat in dairy products such as different cheeses, yoghurt, quark, and others with detailed sensory evaluation and consumer tests are needed.

12.5.7 Sensory Properties of Other Food Products Containing Oleogels

To control oil migration in the tahini halva, *in situ* oleogelation approaches were used during tahini halva production to limit the mobility of sesame oil.[85] Tahini halva is a sweet food prepared by mixing sugar, water, citric or tartaric acid, emulsifier, and soapwort (*Saponaria officinalis*) extract with tahini (sesame paste) to get a solid, fibrous, and crunchy mass. It is common as sugar confectionary in the Mediterranean, the Middle East, and Balkan countries. The halva contains around 55–65% sesame oil in its formulation, and oil migration during storage is encountered as a problem. The loss of natural fibrous texture, mechanical stress, and temperature fluctuations might be the reasons for oil syneresis. To resolve the problem, some stabilizers, hydrocolloids, and hydrogenated fats have been added as additives. To immobilize the migrating sesame oil, sunflower, shellac, and beeswax were added at 1, 3, and 5 wt% during halva manufacturing.[85] Analytical, textural, and sensory analyses were completed. Sunflower and shellac wax additions at 3 and 5 wt% almost totally resolved the oil migration problem. The Flavor/Texture Profile (F/TP) analysis done with ten panelists indicated that the negative attribute 'released oil' was significantly lower in wax-added samples. Although there were small differences among samples for other taste and aroma attributes, they all were in a positive dimension. In addition, the consumer preferences of the wax-added halva samples were in the order of sunflower wax (3%) > shellac wax (3%) > beeswax (3%). In a continuation study, the same halva products were stored for 6 months at 25 and 35 °C. During the storage, significantly enhanced oil leakage was evident in the control sample, while halvas prepared with added waxes were quite well stable. Furthermore, consumer tests indicated that among the samples stored for 6 months, halvas prepared with added sunflower wax and hydrogenated palm stearin had higher hedonic scores.[86]

For instantly fried noodles, soybean oil–carnauba wax oleogels (5 and 10 wt%) were used as a frying medium.[87] Although several instrumental and chemical analyses were completed on the noodles, the sensory evaluations were missing in this study. Hence, it was not possible to totally estimate whether oleogels would be a suitable medium for food frying. Then, some preliminary research data with 5 wt% sunflower wax–rapeseed oil oleogels used to make French fries was published.[88] The French fries with the oleogel were a little more firm than the control. Also, frying with the oleogel resulted in 1% lower fat absorption. A sensory paired comparison test showed that samples fried in the oleogel were defined as less oily on the product surface tested with fingers and less oily in mouthfeel. Furthermore, there was no significant difference in the taste attributes between the samples. More studies are needed for similar applications.

12.6 Aromatics Profile of Oleogels and Food Products Containing Oleogels

It is well known that if a volatile compound can be vaporized at a temperature close to room temperature, and stimulates olfactory nasal nerves, it could probably produce a sensation called smell. Usually, smells with the 'liked' profile are called aroma and smells with the unliked profile are called odor in food science. Furthermore, an aroma or odor is not directly related to the concentration of a particular volatile compound in the food sample; rather it is related to its odor threshold value. The odor threshold is defined as the lowest stimulus (concentration) capable of producing a sensation. Hence, concentration alone is meaningless, but the threshold value of a specific compound within a known media (matrix) is needed to determine how important that specific compound is in a sample's perceived aroma. There are numerous examples where a volatile compound with a high concentration contributes slightly to the perceived aroma due to its very high odor threshold value, or *vice versa*.[27,30]

Logically, the aroma of an oleogel is the result of the volatiles present in the oil used to prepare it, and the volatiles present in the organogelators used. Although refined vegetable oils offer limited aromatic volatiles, virgin or cold-pressed oils are very rich in aromatic compounds. There is enough literature about the volatile constituents of various vegetable oils.[89] Contrarily, the volatile compositions of various organogelators are very limited in the literature. The main manufacturers of plant waxes (Kahlwax, Koster Keunen *etc.*) provide product descriptions on their web pages. For sunflower wax, beeswax, and carnauba wax, they only provide 'odor is faint' or 'odor is soft/characteristics' in the product statement sheets that come with the purchased product. Likewise, there is almost no study in the literature reporting the aromatic volatile compositions of these common plant waxes used in oleogel preparations. In an exciting study,[90] the volatile compositions and sensory evaluations of synthetic polyethylene (PE) or Fischer–Tropsch (FT) waxes were provided. The five waxes were sensorially defined with 'fatty, soapy/citrus-like, vinegar-like, sweaty/cheesy, and tallowy' descriptors. Moreover, 27 aromatic volatiles in PE waxes and 37 aromatic volatiles in FT waxes were identified. The most common volatiles were heptanal, nonanal, octanal, acetic acid, butanoic acid, 1-nonanol, 1-decanol, phenol, lactones, sotolone, and carvacrol. The aroma definitions of volatiles were very well correlated with the sensory descriptors developed by the panel.[90] Although there is no similar study for the plant waxes used as organogelators, beeswax ethanolic extract was analyzed for its volatile composition.[91] Around 82 compounds were identified and quantified belonging to different chemical classes as aldehydes, hydrocarbons, alcohols, esters, ketones, organic acids, terpenes, and derivatives. Quantitatively, aldehydes were the main representative class of the volatiles. Particularly, linear aldehydes as heptanal, octanal, nonanal, decanal, and dodecanal were common.

Ethyl cellulose (EC) oleogels are tasteless, transparent, and the strength of the gel is easily manipulated by selecting the composition and manufacturing processes.[92] However, during EC oleogel preparation the oil–EC mixture must be heated up to 140 °C to dissolve the EC, and at that temperature some oil oxidation would occur. Excessive heating times during EC oleogel preparation were shown to result in both oil oxidation and breakdown of EC and other additives.[93] Beyond oxidation caused volatiles, there is no data about the naturally present aromatic volatiles in EC organogelators. It was also stated that the EC oleogel used in a beef frankfurter resulted in 'chemical off-flavors', but the compound behind the sensation was not determined.[57] For the other common food-grade organogelators (saturated mono- and di-glycerides, ceramides, fatty alcohols, γ-oryzanol + sitosterols, lecithin, shellac, waxes and wax esters, sorbitan derivatives, and others), the literature is lacking in aromatic volatile composition data.

In a study,[94] sunflower oil–monoglyceride gels and emulgels with added water and co-surfactant (lecithin) were prepared. In the same formula, lemon oil (0.82 wt%) was also added, and the headspace CG–MS volatile analysis was completed. It was shown that the lamellar structure of monoglyceride affects the diffusion rate of limonene from oil into the water phase, and the aroma retention power was enhanced as the monoglyceride concentration was enhanced.[94] In another study,[43] hazelnut oil–beeswax and sunflower wax (5 wt%) oleogels were prepared with added commercial flavors (0.5 wt%, strawberry, banana, and butter). During 3 months of storage, volatile compositions were analyzed. It was shown that the main volatile compounds identified in the strawberry flavor added sample were ethyl acetate, ethyl butanoate, ethyl-2-methyl butanoate, D-limonene, and ethyl caproate; the banana flavor added samples mainly included isoamyl acetate, isoamyl valerianate, and ethyl acetate; and the butter-flavor added sample included 2,3-butanedione and 3-hydroxy-2-butanone. Furthermore, these added volatiles were also determined at the end of 3 months of storage, indicating the good volatile aromatic retention potential of the wax oleogels.[43] In a similar study,[44] virgin olive oil and the same waxes, and enriched flavor oleogels were produced and stored. Similarly, the good preservation ability of the wax network for the volatiles was quantified. Furthermore, the samples were well protected against oil oxidation. These two studies pointed out that wax oleogels with any added aromatic profile could be produced successfully, because the volatiles were very well retained in the gel matrix. Hence, aromatic oleogels and/or food products could be prepared.[44] In another study,[95] β-sitosterol : monoglyceride mixtures (10 : 0, 6 : 4, 4 : 6 and 0 : 10, wt%) were used as the organogelator to prepare sunflower oil oleogels (10 wt%). Then, a model volatile mixture (2,3-butanedione, ethyl butyrate, *cis*-3-hexenol, myrcene) was added into the melted oleogels and then gelled again. Finally, the headspace volatile analysis was completed. It was shown that oleogels showed a significantly controlled release of the volatiles, and were tunable *via* the microstructure of phytosterol-based oleogels under both dynamic and static conditions. This finding was attributed to the larger crystals with porosity in the oleogels, which facilitate the mass transfer of volatiles into the headspace. Overall, the oleogels could be a good matrix to control the release rate of aromatic volatiles in food products.[95]

Canola oil–carnauba wax (0–15 wt%), and grapeseed oil–beeswax (0–15 wt%) oleogels were prepared[96] and put into sealed vials before exposing them to heat incubation at 180 °C for 0, 30, 60, or 90 min and at 60 °C for 0, 2, 4, or 6 days. After that, the headspace volatile analysis was completed to follow the course of oil oxidation. The volatiles 2-butenal, 1-penten-3-ol, 2-pentenal, hexanal, heptanal, and 2,4-heptadienal were analyzed, and the amount of the volatiles was increased with increasing oxidation duration and temperature. In both oleogels, the rate of lipid oxidation was lower compared with the control, but the proportion of wax was not significantly effective against oxidation.[96] In a similar study,[97] sunflower oil–rice bran wax oleogels (10 wt%) were prepared and used in cream cheese production. Hexanal and nonanal as secondary lipid oxidation products were analyzed by a dynamic headspace GC/MS system for two weeks of storage. The volatiles in the oleogels were higher than the sunflower oil, but the difference between oleogel cream cheese and the control cheese was not significant. Hence, minimal thermal degradation during oleogel preparation and enough oxidative stability in cream cheeses during storage suggested the possibility of fatty acid modification in the cream cheeses by oleogel additions.[97]

Overall, the literature lacks the aromatic volatile compositions of most common organogelators used in edible oleogels, and also studies reporting the volatile composition of oleogels and foods containing oleogels are very limited. Furthermore, studies indicating the interrelationship of volatile compounds and sensory analyses (sensory descriptive analyses and consumer tests) for oleogels and foods containing oleogels are very rare, and an immediate research challenge is determined.

12.7 Integration of Analytical and Sensory Data for Oleogel Optimization

Oleogels/emulgels have been prepared to replace the solid fats fully or partially in various food products. The main aim was to design equally functional fat with less saturated and no *trans* fatty acids. Secondarily, improving the fatty acid profile of the products with cheaper oil sources has been an aim. Depending on the target food product, different vegetable oil sources, organogelators, additives, and preparation techniques have been implemented to reasonably get the most suitable oleogel. Since in different food products, natural solid fats are expected to exert different functionalities, the designed oleogel is expected to provide similar functionalities. To reach this main goal, many research efforts have been completed.[3,4,9–11,23,24] We propose an integrated approach as summarized in Figure 12.4 to take sensory and consumer data into account during the optimization process for an oleogel for a specific purpose in food products.

In an oleogel or food containing oleogel, several physical, chemical, and instrumental analyses were usually completed to determine and compare product specifications. Also, competence with codex standards or legal requirements could be controlled with these collected data. Even if the products met all legal and codex requirements and provided technical properties within acceptable ranges, the ultimate success of a new product will still be decided by the ultimate line, namely the consumers. Since consumer tests can only provide global descriptions or requirements of consumer demands or expectations, the basic consumer language needs to be translated into sensory terms. Hence, after the common analyses for oleogels or foods containing oleogels, it would be suggested to make a detailed sensory descriptive analysis. From the final point to the oleogel preparation step, the interrelationship could be determined and managed for successful formulations. Once a candidate product is prepared, and all tests are completed, the consumer buying decision and preference/acceptance test results would be a good starting point to evaluate backwards for optimization. Consumer hedonic scores with moderate or improvable attributes must first be correlated with the sensory descriptive terms. These terms could indicate what

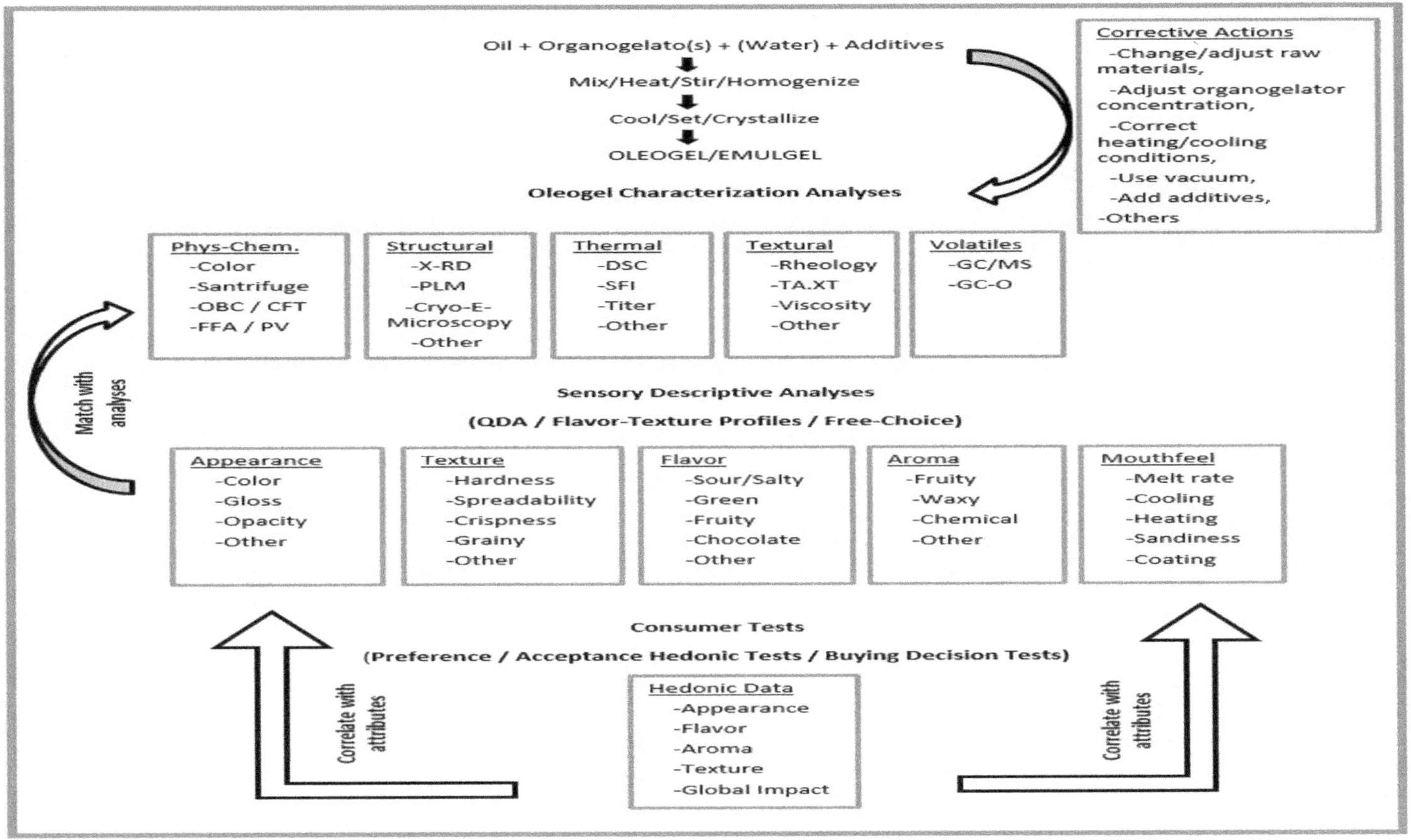

Figure 12.4 Proposed approach to interrelate sensory/consumer tests and instrumental analyses to adjust the process parameters for oleogels.

should be improved in the product as measurable attributes. Furthermore, these attributes could also be correlated with other data produced from analytical and instrumental measurements. Finally, one can reach several verdicts about what actually must be improved within product formulations or process parameters to reach the required levels that the consumer would accept. As an example, theoretically, one can propose that with selected vegetable oil and ethyl cellulose (EC), a successful oleogel was prepared as a shortening alternative for cookie production. After preparation and completion of all analyses against the control sample (say with regular bakery shortening), the consumer tests indicated that the consumers had accepted all attributes except product smell. Then, going one step back, one could determine the descriptive aroma properties to observe the low scores related to product smell. If the 'chemical-like' and 'rancid' scores were very high, the consumer refusal could easily be understood. Then, the peroxide value, volatiles compositions, and other analytical data could be screened to find out the possible reasons for the high scores for those descriptors. Finally, one can decide what corrective actions should be taken to overcome the problem. In this case, one can use a new EC (*i.e.*, deodorized EC) with a less intense 'chemical-like' aroma or completely change the organogelator to another one. Also, during oleogel preparation, a vacuum or a neutral gas atmosphere could be implemented to prevent oil oxidation to lower the 'rancid' score. Similar integrated approaches could be implemented for diverse kinds of oleogels and foods containing oleogels to satisfy the consumer expectations. There are, of course, many other duties to be completed for commercialization, but this chapter deals with sensory properties and their importance for oleogels and oleogel-used foods, as discussed.

12.8 Conclusions and Outlook

Since a considerable number of studies regarding the oleogel preparation, characterization, and food product applications are accumulated in the literature, the next step would be successful commercialization of these products. Of course, more research efforts for the clinical and health effects of oleogel consumption to provide safety data, resolving the regulatory issues, and economic feasibility are needed. Besides, any developed oleogel or the food products containing the oleogel must be accepted by the ultimate line, the consumers, before success in the marketplace. This chapter provides the current status of studies regarding sensory evaluations and aromatic compositions of different oleogels and foods containing oleogels. A great deal of integrated research is needed for successful product development and commercialization.

References

1. R. D. O'Brien, *Fats and Oils: Formulating and Processing for Applications*, CRC Press, New York, 3rd edn, 2000.

2. S. Nas and H. Y. Gökalp, *Bitkisel Yağ Teknolojisi (Vegetable Oil Technology)*, Sidas Medya Pub, İzmir, Turkey, 2017.
3. L. S. K. Dassanayake, D. R. Kodali and S. Ueno, *Curr. Opin. Colloid Interface Sci.*, 2011, **16**, 432.
4. S. S. Sagiri, M. Samateh and G. John, in *Edible Oil Structuring: Concept, Methods and Applications*, ed. A. R. Patel, Royal Soc. Chem., Chambridge, UK, 2018, ch. 2, pp. 25–52.
5. W. W. Nawar, *in Food Chemistry*, ed. O. R. Fennema, Marcel Dekker, Inc. New york, 2nd. edn, 1984, pp. 139–244.
6. M. M. Chrysam, in *Bailey's Industrial Oil & Fat Products*, ed. F. Shahidi, Wiley-Intersience Pub, New Jersey, US, 6th edn, 2005, pp. 33–82.
7. AOCS, *Method Cj 2-95, Official Methods and Recommended Practices of the AOCS*, 6th. edn, 2012.
8. E. D. Co and A. G. Marangoni, *J. Am. Oil Chem. Soc.*, 2012, **89**, 749.
9. A. G. Marangoni, J. P. M. van Duynhoven, N. C. Acevedo, R. A. Nicholson and A. R. Patel, *Soft Matter*, 2020, **16**, 289.
10. M. Pernetti, K. F. van Malssen, E. Flöter and A. Bot, *Curr. Opin. Colloid Interface Sci.*, 2007, **12**, 221.
11. M. A. Rogers, *Food Res. Int.*, 2009, **42**, 747.
12. D. Tang and A. G. Marangoni, *Adv. Colloid Interface Sci.*, 2006, **128–130**, 257.
13. T. G. Mezger, *Applied Rheology*, Anton Paar GmbH, Austria, 2014.
14. A. G. Marangoni and N. Garti, in *Edible Oleogels: Structure and Health Implication*, ed. A. G. Marangoni and N. Garti, AOCS Press, Urbana, 2011, ch. 1, pp. 1–17.
15. A. R. Patel, in *Edible Oil Structuring: Concept, Methods and Applications*, ed. A. R. Patel, Royal Soc. Chem. Press, 2018, ch. 1, pp. 3–24.
16. J. F. Toro-Vazquez, J. Morales-Rueda, E. Dibildox-Alvarado, M. Charo-Alonso, M. Alonzo-Macias and M. M. Gonzalez-Chavez, *J. Amr. Oil Chem. Soc.*, 2007, **84**, 989.
17. A. J. Martins, A. A. Vicente, R. L. Cunha and M. A. Cerqueira, *Food Funct.*, 2018, **9**, 758.
18. A. Singh, F. I. Auzanneau and M. A. Rogers, *Food Res. Int.*, 2017, **97**, 307.
19. E. Yılmaz and S. Ok, in *Edible Oil Structuring: Concept, Methods and Applications*, ed. A. R. Patel, Royal Soc. Chem. Press, 2018, ch. 11, pp. 217–249.
20. D. Libster, A. Aserin and N. Garti, in *Edible Oleogels: Structure and Health Implications*, ed. A. G. Marangoni and N. Garti, AOCS Press, Champaign, US, 2011, ch. 11, pp. 235–270.
21. A. R. Patel, *Alternative Routes to Oil Structuring*, Springer Inter. Pub., New York, 2015.
22. K. D. Mattice and A. G. Marangoni, in *Edible Oil Structuring: Concept, Methods and Applications*, ed. A. R. Patel, Royal Soc. Chem. Press, 2018, ch. 4, pp. 71–94.
23. H. Pehlivanoğlu, M. Demirci, O. S. Toker, N. Konar, S. Karasu and O. Sagdic, *Crit. Rev. Food Sci. Nutr.*, 2018, **58**, 1330.

24. A. Puşcaş, V. Mureşan, C. Socaciu and S. Muste, *Foods*, 2020, **9**, 70.
25. A. J. Martins, A. A. Vicente, L. M. Pastrana and M. A. Cerqueira, *Food Sci. Human Welln.*, 2020, **9**, 31.
26. H.-S. Hwang, *Biocatal. Agric. Biotechnol.*, 2020, **26**, 101657.
27. M. Meilgaard, G. V. Civille and B. T. Carr, *Sensory Evaluation Techniques*, CRC Press, Boca Raton, US, 1991.
28. A. Szydłowska-Czerniak, A. Tułodziecka, M. Momot and B. Stawicka, *J Am. Oil Chem. Soc.*, 2019, **96**, 405.
29. P. J. White, in *Introduction to Fats and Oils Technology*, ed. R. D. O'Brien, W. E. Farr and P. J. Wan, AOCS Press, Champaign, IL, 2nd edn, 2000, ch. 18, pp. 341–370.
30. H. Stone, R. N. Bleibaum and H. A. Thomas, *Sensory Evaluation Practices*, Elsevier, Amsterdam, 4th edn, 2012.
31. J. Yang and J. Lee, *Foods*, 2019, **8**, 54.
32. K. Warmer and T. Nelson, *J Am. Oil Chem. Soc.*, 1996, **73**, 157.
33. G. V. Civille and B. G. Lyon, *Aroma and Flavor Lexicon for Sensory Evaluation: Terms, Definitions, References, and Examples*, ASTM Data Series Publication DS 66, ASTM, PA, US, 1996.
34. AOCS, *Method Cg 2-83*, AOCS Press, Champaign, IL, US, 1997.
35. N. J. Dubost, R. L. Shewfelt and R. R. Eitenmiller, *J. Food Qual.*, 2002, **26**, 27.
36. B. B. Ronn, G. Hyldig, L. Wienberg, K. B. Qvist and A. M. Laustsenb, *Food Qual. Pref.*, 1998, **9**, 187.
37. M. Michicich, Z. Vickers, M. C. Martini and C. B. Labat, *Food Qual. Pref.*, 1999, **10**, 147.
38. C. C. Akoh, B. H. Kim, R. L. Shewfelt and H. Lee, *J. Food Sci.*, 2005, **70**, s406.
39. A. C. Mosca, J. L. Rocha, G. Sala, F. Velde and M. A. Stieger, *Food Hydrocolloids*, 2012, **27**, 448.
40. F. Angerosa, in *Handbook of Olive Oil: Analysis and Properties*, ed. J. Harwood and R. Aparicio, Aspen Pub., New York, 2013, ch. 14, pp. 523–556.
41. H. R. Moskowitz, in *Food Product Development: From Concept to Marketplace*, ed. E. Graf and I. S. Saguy, Aspen Pub., Inc., MA, US, 1990, ch. 8, pp. 157–188.
42. E. Yılmaz and M. Öğütcü, *RSC Adv.*, 2015, **5**, 50259.
43. E. Yılmaz, M. Öğütcü and Y. Karagül Yüceer, *J. Food Sci.*, 2015, **80**, S2035.
44. E. Yılmaz, M. Öğütcü and O. Güneser, *J. Am. Oil Chem. Soc.*, 2015, **92**, 1429.
45. E. Yılmaz, M. Öğütcü and N. Arifoğlu, *J. Oleo Sci.*, 2015, **64**, 1049.
46. E. Yılmaz, E. Keskin Uslu and B. Toksöz, *J. Oleo Sci.*, 2020, **69**, 1317.
47. M. C. Lee, X. Jiang, J. T. Brenna and A. Abbaspourrad, *Food Funct.*, 2018, **9**, 5598.
48. A. R. Patel, P. S. Rajarethinem, A. Grędowska, O. Turhan, A. Lesaffer, W. H. De Vos and D. Van de Walle, *Food Funct.*, 2014, **5**, 645.
49. E. Yilmaz and M. Öğütcü, *Food Funct.*, 2015, **6**, 1194.

50. I. K. Oh and S. Lee, *Food Hydrocolloids*, 2018, **77**, 796.
51. S.-Z. Luo, X.-F. Hu, Y.-J. Jia, L.-H. Pan, Z. Zheng, Y.-Y. Zhao, D.-D. Mu, X.-Y. Zhong and S.-T. Jiang, *Food Hydrocolloids*, 2019, **95**, 76.
52. L.-H. Pan, X.-L. Wu, S.-Z. Luo, H.-Y. He and J.-P. Luo, *J. Food Sci.*, 2020, **85**, 2461.
53. Z. Saghafi, M. H. Naeli, M. Bahmaei, M. Tabibiazar and A. Zargaraan, *J. Food Meas. Charact.*, 2019, **13**, 3040.
54. H. Pehlivanoglu, G. Ozulku, R. M. Yildirim, M. Demirci, O. S. Toker and O. Sagdic, *J Food Process Preserv.*, 2018, e13621.
55. M. C. Utrilla, A. G. Ruiz and A. Soriano, *Meat Sci.*, 2014, **97**, 575.
56. S. Barbut, J. Wood and A. G. Marangoni, *Meat Sci.*, 2016, **122**, 155.
57. S. Barbut, J. Wood and A. G. Marangoni, *J. Food Sci.*, 2016, **81**, C183.
58. D. Kouzounis, A. Lazaridou and E. Katsanidis, *Meat Sci.*, 2017, **130**, 38.
59. T. L. Wolfer, N. C. Acevedo, K. J. Prusa and J. G. Sebranek, *Meat Sci.*, 2018, **145**, 352.
60. S. L. da Silva, J. T. Amaral, M. Ribeiro, E. E. Sebastiano, C. Vargas, F. L. Franzen, G. Schneider, J. M. Lorenzo, L. L. M. Fries, A. J. Cichoski and P. C. B. Campagnol, *Meat Sci.*, 2019, **149**, 141.
61. D. Franco, A. J. Martins, M. López-Pedrouso, L. Purriños, M. A. Cerqueira, A. A. Vicente, L. M. Pastrana, C. Zapata and J. M. Lorenzo, *Foods*, 2019, **8**, 366.
62. D. Franco, A. J. Martins, M. López-Pedrouso, M. A. Cerqueira, L. Purriños, L. M. Pastrana, A. A. Vicente, C. Zapata and J. M. Lorenzo, *J. Sci. Food Agric.*, 2020, **100**, 218.
63. T. Pintado and S. Cofrades, *Foods*, 2020, **9**, 830.
64. R. Tarté, J. S. Paulus, N. C. Acevedo, K. J. Prusa and S.-L. Lee, *LWT - Food Sci. Technol.*, 2020, **131**, 109659.
65. J. Gómez-Estaca, A. M. Herrero, B. Herranz, M. D. Álvarez, F. Jiménez-Colmenero and S. Cofrades, *Food Hydrocolloids*, 2019, **87**, 960.
66. I. Oh, J. H. Lee, H. G. Lee and S. Lee, *Food Res. Int.*, 2019, **122**, 566.
67. B. M. Ferrer-González, I. García-Martínez and A. Totosaus, *Grasas y Aceites*, 2019, **70**, e320.
68. A. J. Martins, J. M. Lorenzo, D. Franco, M. Pateiro, R. Domínguez, P. E. S. Munekata, L. M. Pastrana, A. A. Vicente, R. L. Cunha and M. A. Cerqueira, *Gels*, 2020, **6**, 17.
69. M. Moghtadaei, N. Soltanizadeh and S. A. H. Goli, *Food Res. Int.*, 2018, **108**, 368.
70. A. J. Martins, J. M. Lorenzo, D. Franco, A. A. Vicente, R. L. Cunha, L. M. Pastrana, J. Quiñones and M. A. Cerqueira, *Eur. J. Lipid Sci. Technol.*, 2019, **121**, 1900111.
71. R. T. Heck, E. Saldaña, J. M. Lorenzo, L. P. Correa, M. B. Fagundes, A. J. Cichoski, C. R. de Menezes, R. Wagner and P. C. B. Campagnol, *Meat Sci.*, 2019, **156**, 174.
72. J. Gómez-Estaca, T. Pintado, F. Jiménez-Colmenero and S. Cofrades, *Food Bioproc. Technol.*, 2019, **12**, 1068.

73. J. Gómez-Estaca, T. Pintado, F. Jiménez-Colmenero and S. Cofrades, *LWT – Food Sci. Technol.*, 2020, **119**, 108909.
74. L. Adili, L. Roufegarinejad, M. Tabibiazar, H. Hamishehkar and A. Alizadeh, *LWT – Food Sci. Technol.*, 2020, **126**, 109277.
75. H. S. Hwang, M. Singh and E. L. Bakota, *J. Am. Oil Chem. Soc.*, 2013, **90**, 1705.
76. H. S. Hwang, M. Singh, J. K. Winkler-Moser, E. L. Bakota and S. X. Liu, *J. Food Sci.*, 2014, **79**, C1926.
77. M. Öğütcü, N. Arifoğlu and E. Yılmaz, *J. Am. Oil Chem. Soc.*, 2015, **92**, 459.
78. T. L. T. da Silva, K. F. Chaves, G. D. Fernandes, J. B. Rodrigues, H. M. A. Bolini and D. B. Arellano, *J. Am. Oil Chem. Soc.*, 2018, **95**, 673.
79. R. E. Wainwright, in *Bailey's Industrial Oil & Fat Products vol. 3*, ed. Y. H. Hui, Wiley-Intersience Pub., New York, 1996, ch. 9, pp. 353–408.
80. G. Talbot, in *Industrial Chocolate Manufacture and Use*, ed. S. T. Beckett, Blackwell Publishing Ltd., West Sussex, UK, 4th edn, 2009, ch. 19, pp. 415–433.
81. S. Bascuas, M. Espert, E. Llorca, A. Quiles, A. Salvador and I. Hernando, *LWT - Food Sci. Technol.*, 2021, **135**, 110228.
82. D. Hettinga, in *Bailey's Industrial Oil & Fat Products vol. 3*, ed. Y. H. Hui, Wiley-Intersience Pub., New York, 1996, ch. 1, pp. 1–64.
83. R. M. Kerr, X. Tombokan, S. Ghosh and S. Martini, *J. Agric. Food Chem.*, 2011, **59**, 2689.
84. H. L. Bemer, M. Limbaugh, E. D. Cramer, W. J. Harper and F. Maleky, *Food Res. Int.*, 2016, **85**, 67.
85. M. Öğütcü, N. Arifoğlu and E. Yılmaz, *Eur. J. Lipid Sci. Technol.*, 2017, **119**, 1600189.
86. M. Öğütcü, N. Arifoğlu, R. Temizkan and E. Yılmaz, *La Riv. Ital. Del. Sos. Gras.*, 2018, **95**, 95.
87. J. Lim, S. Jeong, I. M. Oh and S. Lee, *LWT - Food Sci. Technol.*, 2017, **84**, 788.
88. B. Matthäus, M. Schubert, N. Erlenbusch, I. Smit, L. Weber and S. Nikolay, *INFORM*, 2020, **31**, 22–26.
89. M. F. Ramadan, *Cold Pressed Oils: Green Technology, Bioactive Compounds, Functionality, and Applications*, Academic Press, 1st edn, 2020, p. 774.
90. P. Denk, E. Ortner and A. Buettner, *Int. J. Adhes. Adhes.*, 2019, **95**, 102406.
91. A. Felicioli, G. Cilia, S. Mancini, B. Turchi, G. Galaverna, M. Cirlini, D. Cerri and F. Fratini, *Food Biosci.*, 2019, **29**, 102.
92. K. D. Mattice and A. G. Marangoni, in *Edible Oil Structuring: Concept, Methods and Applications*, ed. R. Ashok, Patel, Royal Soc. Chem., Chambridge, UK, 2018, ch. 12, pp. 250–276.
93. A. J. Gravelle, S. Barbut and A. G. Marangoni, *Food Res. Int.*, 2012, **48**, 578.
94. S. Calligaris, S. Da Pieve, G. Arrighetti and L. Barba, *Food Res. Int.*, 2010, **43**, 671.
95. D.-X. Yang, X.-W. Chen and X.-Q. Yang, *J. Sci. Food Agric.*, 2018, **98**, 582.
96. B. Yi, M. J. Kim, S. Y. Lee and J. Lee, *Food Sci. Biotechnol.*, 2017, **26**, 79.
97. C. Park, H. L. Bemer and F. Maleky, *J. Am. Oil Chem. Soc.*, 2018, **95**, 1267.

Subject Index

AA. *See* adipic acid (AA)
acceptable daily intake (ADI), 160
ADI. *See* acceptable daily intake (ADI)
adipic acid (AA), 151
AFM. *See* atomic force microscopy (AFM)
American Oil Chemists' Society (AOCS), 320
American Society for Testing and Materials (ASTM), 320
AMF. *See* anhydrous milk fat (AMF)
anhydrous milk fat (AMF), 58–59
AOCS. *See* American Oil Chemists' Society (AOCS)
ASTM. *See* American Society for Testing and Materials (ASTM)
atomic force microscopy (AFM), 169, 171, 181, 184

beeswax (BW), 237, 241, 248, 255, 286, 288
berry wax (BEW), 237, 239
BEW. *See* berry wax (BEW)
biopolymer-based oleogels, 113
    food applications of, 100–104
    oil binding capacity and oil loss, 97–98
    rheological properties of, 98–99
    sensory evaluation of, 101, 105
    textural analysis of, 99–100
β-lactoglobulin (BLG), 12, 13, 15
BLG. *See* β-lactoglobulin (BLG)
bovine serum albumin (BSA), 12, 13, 15
BSA. *See* bovine serum albumin (BSA)
BW. *See* beeswax (BW)

CAGR. *See* compound annual growth rate (CAGR)
candelilla wax (CW), 98, 101, 237, 239–240, 245–248, 286, 290, 296, 297
Candida antarctica lipase, 262, 265, 266
canola protein isolates (CPIs), 106, 113
    cake batter and cake properties, 109–111
    LVR, 108
    oil biding capacity of, 107–108
    oil droplets, 108
    viscoelastic properties, 108–109
cardiovascular diseases (CD), 164
carnauba wax (CRW), 237, 240
carnauba wax (CW), 105, 150
CAS. *See* Chemical Abstracts Number (CAS)
CBS. *See* commercial bakery shortening (CBS)
CD. *See* cardiovascular diseases (CD)
CGC. *See* critical gelling concentration (CGC)
CHDs. *See* coronary heart diseases (CHDs)
Chemical Abstracts Number (CAS), 164
Cinnamomum camphora, 130

CNPs. *See* crystalline nanoplatelets (CNPs)
commercial bakery shortening (CBS), 151
compound annual growth rate (CAGR), 105
controlled shear deformation (CSD), 176
controlled shear stress (CSS), 176
coronary heart diseases (CHDs), 157
CPIs. *See* canola protein isolates (CPIs)
CPP. *See* critical packing parameter (CPP)
critical gelling concentration (CGC), 172
critical packing parameter (CPP), 216
crude rice bran oil (RBO), 161
CRW. *See* carnauba wax (CRW)
crystalline nanoplatelets (CNPs), 157
CSD. *See* controlled shear deformation (CSD)
CSS. *See* controlled shear stress (CSS)
CW. *See* candelilla wax (CW); carnauba wax (CW)

Dacryodes edulis, 130
dairy products
  food applications
    ice crystals, 190–191
    milk fat, 191
    multiphase system, 190
    yoghurts, 191–192
  oleogel products, sensory properties of, 338–339
DGs. *See* diglycerides (DGs)
differential scanning calorimetry (DSC), 43, 69, 124, 173, 263
diglycerides (DGs), 262
DMTA. *See* dynamic mechanic thermal analysis (DMTA)
double emulsions
  applications, 26–27
  inner and outer phases, 22
  O/W/O
    evaluating, methods for, 25–26
    microstructure of, 22–23
    preparation of, 22–23
    stabilization of, 24–25
  types of, 21
DSC. *See* differential scanning calorimetry (DSC)
dynamic mechanic thermal analysis (DMTA), 173

EC. *See* ethylcellulose (EC)
edible oleogels
  analytical and sensory data, integration of, 343–345
  aromatics profile
    canola oil–carnauba wax, 342
    EC oleogels, 341
    FT waxes, 341
    grapeseed oil–beeswax, 342
    hazelnut oil–beeswax and sunflower wax, 342
    odor threshold, 340
    PE waxes, 341
    β-sitosterol : monoglyceride mixtures, 342
    sunflower oil–monoglyceride gels and emulgels, 341–342
    virgin olive oil, 342
    volatile compositions, 341
  edible fats/oils, sensory assessment of
    affective testing methods, 322–323
    AOCS method, 320
    categories, 322
    descriptive analysis methods, 322–323

edible oleogels (*continued*)
discriminating tests, 322
QDA, 323–324
sensory evaluation, 319–320
texture properties of, 322
natural plastic fats, 316
oil structuring, 318–319
oleogel products, sensory properties of
beef pates/patties, 331
beeswax oleogel, 331
canola oil, 330
chocolate/confectionery products, 335–338
corn oil–beeswax oleogels, 326
dairy products, 338–339
fat replacement, 331, 333–334
fish oil–beeswax oleogels, 326
French fries, 340
hazelnut oil–beeswax, 327
HO oleogels, 324
HPMC, 327, 332
linseed oil, 331
margarine/spread products, 334–335
meat products, 329, 333
QDA test, 324–325
sensory attributes, 326
shellac–sunflower oil oleogel, 327
soybean oil oleogel, 332
sunflower oil emulgels, 330
sunflower wax oleogels, 327
tahini halva, 339
Tp-palmitate, 329
Tp-stearate, 329
VOO oleogels, 324
ZTC, 329
solid and semi-solid fats, physical properties of
fat, melting behavior of, 317
functional properties of, 317
long-chain fatty acids, 317
polymorphic forms, 317
TAGS, 316
solid edible fats, 316
trans-fatty acids, 315–316
EFSA. *See* European Food Safety Authority (EFSA)
emulsion-templated indirect oleogelation, 96–97
ethylcellulose (EC), 175, 222
ethylcellulose oleogels
added surfactants and mechanical reversibility, 301
compositional effects
mechanical strength of, 299
MGC, 299
oleic acid and oleyl alcohol, 300
organic solvents, 299
surfactant molecules, 300–301
thermoreversible plastic materials, 298
in foods, 303
processing conditions, effects of, 301–303
Euphorbia antisyphilitica, 239
European Food Safety Authority (EFSA), 160, 240

faba bean protein isolates (FBPIs), 106, 113
cake batter and cake properties, 109–111
LVR, 108
oil biding capacity of, 107–108
oil droplets, 108
viscoelastic properties, 108–109

fatty acid composition
cocoa butter, 123
formulate *trans*-fat-free products, 122
hydrogenated oils, 121
mango kernel fat, 123
palm kernel and coconut oils, 123
fatty alcohol oleogels
aliphatic hydrocarbons, 141
application of
CBS, 151
cookies, sensory properties of, 151
oil-based material, 150
policosanol, 151–152
semi-finished materials, 150
soybean oil-oleogels, 151
microstructural characteristics of, 148–149
rheological characteristics and modeling of
crystal network, 143
modified fractal model, 143
policosanol as gelator, 143–148
use of, 142
FBPIs. *See* faba bean protein isolates (FBPIs)
Federal Food, Drug, and Cosmetic Act (FFDCA), 163
FFA. *See* free fatty acids (FFA)
FFAl. *See* free fatty alcohols (FFAl)
FFDCA. *See* Federal Food, Drug, and Cosmetic Act (FFDCA)
FHSBO. *See* fully hydrogenated soybean oil (FHSBO)
Fischer–Tropsch (FT) waxes, 341
foam-templated indirect oleogelation, 96
food grade systems
forming oil-in-water (O/W) emulsions, 2
organogel-like systems, 2
food systems
fatty alcohols, application of, 150–153
food grade systems
forming oil-in-water (O/W) emulsions, 2
organogel-like systems, 2
oleogels, 229–230
vegetable wax oleogels, 150–153
free fatty acids (FFA), 282, 307
free fatty alcohols (FFAl), 307
fruit wax (FW), 237, 240
FT waxes. *See* Fischer–Tropsch (FT) waxes
fully crystallized mixtures (FCMs), 45
fully hydrogenated soybean oil (FHSBO), 246, 291
FW. *See* fruit wax (FW)

Generalized Procrustes Analysis (GPA), 337
generally recognized as safe (GRAS), 54, 150, 206
glycerol monooleate (GMO), 301
glycerolysis-structured lipid systems, 278–279
1,3-DGs and 1,2-DGs, 262
edible oils, 262
food applications
back extrusion testing, 275–277
DSC melting profile, 276
margarine, characteristics of, 276, 278
trans-free lipid systems, 275
MGs, 261–262
oil composition, effects of
cottonseed systems, 266
crystallization regions, 266
diolein crystals, 270
fat crystal network, 268
fatty acid composition, 268

glycerolysis-structured lipid systems (*continued*)
HOAO system, 266
linoleic acid content, 274–275
monopalmitin, 270
native oil, 271
oleic acid contents, 273–274
plant-based oil, 274
SFC–temperature profile, 273
tigernut oil system, 268
reaction conditions, effects of
glycerol : TG molar ratio, 263–264
monoolein and monopalmitin production, 263–264
oil binding capabilities, 264
R. miehei lipase, 265
glyceryl stearate (GS), 147
GMO. *See* glycerol monooleate (GMO)
GPA. *See* Generalized Procrustes Analysis (GPA)
GRAS. *See* generally recognized as safe (GRAS)
GS. *See* glyceryl stearate (GS)

hazelnut oil (HO), 324
HC. *See* hydrocarbons (HC)
HDL cholesterol. *See* high-density lipoprotein (HDL) cholesterol
HE. *See* heated emulsion (HE)
heated emulsion (HE), 108
high-density lipoprotein (HDL) cholesterol, 2, 92
high-intensity ultrasound (HIU)
acoustic waves, 53–54
application
baked products, 84
chocolate, 84, 86
emulsions, 86
crystallization, 54
fat composition, 82, 84
fat crystallization, 55
high-saturated fats, 54
all-purpose shortening, 59–60
AMF, 58–59
cocoa butter and pure triacylglycerols, 55, 58
lauric-based fats, 61–62
palm oil, 60–61
sonication effect on, 55–57
internal factors, 54
lipid materials, 86–87
lipid oxidation, 78–80
low-saturated fats, 54
composition, effect of, 68
palm-based fat, 67–68
soft stearin, 67
sonication effect on, 62–64
soybean-based fat, 62, 65–67
oleogels, physicochemical properties of
binary and ternary oleogels, 69–72
crystallization process, 69
mono-component oleogels, 72–78
structuring edible oils, 69
physicochemical properties, 53
shaved agitation, 82
sonication conditions, 82
SSHE, 80–82
stages, 55
supercooling, 82
high-oleic sunflower oil (HOSO), 69
high pressure liquid chromatography (HPLC), 4
HIU. *See* high-intensity ultrasound (HIU)
HO. *See* hazelnut oil (HO)
HOSO. *See* high-oleic sunflower oil (HOSO)
HPLC. *See* high pressure liquid chromatography (HPLC)

HPMC. *See* hydroxypropyl methylcellulose (HPMC)
12-HSA. *See* (*R*)-12-hydroxystearic acid (12-HSA)
HSFO. *See* hydrogenated sunflower oil (HSFO)
hydrocarbons (HC), 282
hydrogel-templated indirect oleogelation, 97
hydrogenated sunflower oil (HSFO), 126
hydroxypropyl methylcellulose (HPMC), 96, 327

Irvingia gabonensis, 130
3iTT. *See* three-interval thixotropy tests (3iTT)

LC. *See* lecithin (LC)
LDL cholesterol. *See* low-density lipoprotein (LDL) cholesterol
lecithin (LC), 204–205
linear viscoelastic region (LVR), 99, 284
LMWGs. *See* low molecular weight gelators (LMWGs)
low-density lipoprotein (LDL) cholesterol, 2, 92
low molecular weight gelators (LMWGs), 140, 318
LVR. *See* linear viscoelastic region (LVR)

MAG. *See* monoacylglycerols (MAG)
MD. *See* molecular dynamics (MD)
MDP. *See* Mettler Dropping Point (MDP)
Mettler Dropping Point (MDP), 124
MGC. *See* minimal gelling concentration (MGC)
MGs. *See* monoglycerides (MGs)
microencapsulated oil powder
    advantages of, 31
    emulsification, 32
    food industry, applications in, 38
    freeze-drying, 37–38
    multidimensional functional properties, 27
    spray-drying, 33–37
    structure of
        categories, 32
        internal phase, 31
        size and shape of, 31
        types of, 31
        wall materials, 31
minimal gelling concentration (MGC), 287, 299
molecular dynamics (MD), 169
monoacylglycerol oleogels
    compositional effects, 303–305
    in foods, 306–307
    processing conditions, effects of, 305–306
monoacylglycerols (MAG), 247
mono-component oleogels
    crystal–crystal interactions, 78
    elastic properties, 77
    harder structure, 75
    low cooling rate, 77–78
    MGs concentration, effects of, 73
    non-sonicated samples, 77
    physical properties of, 75–76
    polarized light microscopy images and visual appearance of, 74–75
    propolis wax, 72–73
monoglycerides (MGs), 140, 204–205
Moringa oleifera, 128, 129
Myrica cerifera, 240

natural waxes
    gelators, 236
    TAG–wax hybrid fat systems
        bakery applications, hybrid systems in, 254–255
        categories, 245
        hybrid emulsions, 252–254
        hybrid shortenings, 245–252

natural waxes (*continued*)
triacylglycerol hardstock, 243–244
trans fats, 235–236
vegetal and insect sources
BEW, 239
chemical components, 236–237
chemical composition and physical properties of, 237–238
CRW, 240
CW, use of, 239–240
FW or bayberry wax, 240
legislation, 241–243
RBW, 240
SHW, 241
SW, 240
YBW and WBW, 240–241
natural wax oleogels
compositional effect
components of, 286
gelator concentration, effect of, 290
MGC, 287–288
oil composition, effect of, 292
in foods, 296–298
processing conditions
3iTT, 294
cooling rate, 292–293
irreversible junction zones, 296
reversible junction zones, 296
shearing effects, 293
structural loss, 296
NOALs. *See* no-observed-adverse-effect levels (NOALs)
no-observed-adverse-effect levels (NOALs), 164

OBC. *See* oil binding capacity (OBC)
oil binding capacity (OBC), 71, 172
centrifugation-based method, 180
compression forces, 177
fat-continuous products, 177
filter paper procedure, 180
modified filter paper method, 179–180
oil loss, 179
palm oil shortening, 178
standard filter paper method, 179–180
oil capsules, 46
oil foams
aqueous and non-aqueous foams, 38–39
benefits and application, 45–46
crystal properties, influence of
crystal network generation, 45
molecular structure, 41
polymorphism and phase behavior, 42–45
size and shape, 41–42
edible-gas-in-oil systems, 39
liquid foams, 38
preparation of, 39–40
stabilization of, 40–41
structure of, 39
oil-in-water-in-oil (O/W/O) double emulsions, 21, 46
applications, 26–27
evaluating, methods for, 25–26
microstructure of, 22–23
overview of, 27–30
preparation of
hydrophilic emulsifier, 22–23
phase inversion, 24
popular methods for, 22–23
stabilization of, 24–25
oil structuring
biopolymer-based oleogels, characterization of
food applications of, 100–104
oil binding capacity and oil loss, 97–98

rheology of, 98–99
sensory evaluation, 101, 105
textural analysis of, 99–100
drawbacks, 91
emulsion-templated oleogels, development of
cakes, texture analysis of, 111–112
faba bean and canola protein isolate-based oleogels, 107–111
FBPI and CPI oleogels, 106–107
oilseed proteins, 106
plant proteins, 105–106
indirect oleogelation
emulsion-templated, 96–97
foam-templated, 96
hydrogel-templated, 97
oleogelation
direct and indirect approaches, 93
small molecules and biopolymers, 94–96
potential health implication of
LDL cholesterol, 92
oleogels, 92–93
saturated and *trans* fat, 92
oleogelation, 40, 92–93, 112
biopolymer-based oleogels, characterization of, 113
food applications of, 100–104
oil binding capacity and oil loss, 97–98
rheological properties of, 98–99
sensory evaluation of, 101, 105
textural analysis of, 99–100
direct approach, 93
EC and polar molecules, 282
emulsion-templated oleogels, development of
cakes, texture analysis of, 111–112
faba bean and canola protein isolate-based oleogels, 107–111
FBPI and CPI oleogels, 106–107
oilseed proteins, 106
plant proteins, 105–106
food systems, 229–230
indirect approach, 93
emulsion-templated, 96–97
foam-templated, 96
hydrogel-templated, 97
MG crystals, 282
plastic fats, 281, 283
ethylcellulose oleogels, 298–303
fatty acids and fatty alcohols, 307–309
monoacylglycerol oleogels, 303–307
natural wax oleogels, 286–298
rheological methods
3iTT, 285
compression-based methods, 284
empirical methods, 283–284
extrusion-based methods, 284
force–time curve, 284
fundamental rheological methods, 284
SAOS methods, 284
viscoelastic materials, 285
saturated and *trans* fats, 282
small molecules and biopolymers, 94–96
wax components, 282

organogels
fatty alcohol
aliphatic hydrocarbons, 141
application of, 150–153
microstructural characteristics of, 148–149
rheological characteristics and modeling of, 142–148
use of, 142
LMW, 140–141
semi-solid fats, 153
vegetable wax oleogels, 150–153
γ-oryzanol
crude RBO, 161
hypothetical oryzanol production, 161–162
ternary mixture, 169–170
oryzanol–sitosterol structuring system, 193
binary mixture, physical chemistry of
binary β-sitosterol and γ-oryzanol system, 166–168
binary sub-systems, 166
two single-phase, 167
characterization, methods of
CGC, 172–173
firmness and deformation, 174–177
gel's behavior, 177
microstructure, 180–181
OBC, 178–180
sol–gel transition, 173–174
digestion
lipase activity, 165
phytosterol esters, 165–166
plant sterols (PS), chemical structure of, 165
risk factors, 164
TAGs, release and bioavailability of, 165
food applications
confectionery products, 187–188
dairy products, 190–192
direct consumption, margarine and spreads for, 189–190
meat products, 188–189
sitosterol/oryzanol oleogels, 186–187
hydrogenated plant oils, 158
legislation
GRAS, 163–164
novel food applications, 163
overview of, 164
phytosterols and esters, 164
MAGs and DAGs, 160
oil structuring system, 160
oleogelator classes, classification of, 158–159
γ-oryzanol
crude RBO, 161
hypothetical oryzanol production, 161–162
palm oil, 158
β-sitosterol and phytosterols, 162
solvent composition, influence of
emulsion firmness, 182
ferulic acid moiety, 183
gel–sol transition, 184
intra- to inter-tube interactions, 183–184
network interactions, 184–185
oleic acid and tocopherol acetate, 185–186
polar components, type of, 185
sterol/sterol-ester oleogel properties, 181–182
ternary mixture
β-sitosterol/γ-oryzanol system, 169–170

DSC dissolution temperatures, 170
gel–sol transition, 171–172
tubules, nucleation and growth of, 171
O/W emulsion, 32
O/W/O double emulsions. *See* oil-in-water-in-oil (O/W/O) double emulsions
vegetal oils. *See* vegetal oils

palm oil (PO)
advantageous features, 1
formation of *trans* double-bonds, 2
high-saturated fats, 60–61
RSPO, 1–2
partially crystallized mixtures (PCMs), 45
partially hydrogenated oil (PHO), 92
PC. *See* phosphatidylcholine (PC)
PG. *See* polymeric gelators (PG)
PHO. *See* partially hydrogenated oil (PHO)
phosphatidylcholine (PC), 216
phytosterols, 162
plant sterols (PS), 165
plastic fats, 281, 283
ethylcellulose oleogels
added surfactants and mechanical reversibility, 301
compositional effects, 298–301
in foods, 303
processing conditions, effects of, 301–303
fatty acids and fatty alcohols, 307–309
monoacylglycerol oleogels
compositional effects, 303–305
in foods, 306–307
processing conditions, effects of, 305–306
natural wax oleogels
compositional effect, 286–292
in foods, 296–298
processing conditions, 292–296
PLM. *See* polarized light microscopy (PLM)
PO. *See* palm oil (PO)
polarized light microscopy (PLM), 70
policosanol, 142, 151–152
fiber-like crystals, 148–149
rheological properties, 143–148
polyethylene (PE) waxes, 341
polyglycerol polyricinoleate (PGPR), 24
polymeric gelators (PG), 318
propolis wax (PW), 72–73, 246
PS. *See* plant sterols (PS)

QDA. *See* Quantitative Descriptive Analysis (QDA)
Quantitative Descriptive Analysis (QDA), 323
quantum satis, 241

(*R*)-12-hydroxystearic acid (12-HSA), 204
rapeseed oil, 3
RBO. *See* crude rice bran oil (RBO)
RBW. *See* rice bran wax (RBW)
Repertory Grid Method (RGM), 337
RGM. *See* Repertory Grid Method (RGM)
Rhizomucor miehei, 265
Rhus verniciflua, 239
rice bran wax (RBW), 150, 152, 237, 240, 286–287, 298
RMSD. *See* root mean square deviation (RMSD)
root mean square deviation (RMSD), 308
Roundtable on Sustainable Palm Oil (RSPO), 1–2
RSPO. *See* Roundtable on Sustainable Palm Oil (RSPO)

SANS. *See* small-angle neutron scattering (SANS)
SAOS. *See* small amplitude oscillatory shear (SAOS)
saturated fatty acids (SFAs), 2, 54, 235
SAXS. *See* small-angle X-ray scattering (SAXS)
scanning electron microscopy (SEM), 132, 169
SCFE. *See* supercritical fluid extraction (SCFE)
scraped surface heat exchanger (SSHE), 80, 189
SDS. *See* sodium dodecyl sulfate (SDS)
SEM. *See* scanning electron microscopy (SEM)
SFAs. *See* saturated fatty acids (SFAs)
SFC. *See* solid fat content (SFC)
SFI. *See* solid fat index (SFI)
shellac wax (SHW), 237, 241
SHW. *See* shellac wax (SHW)
β-sitosterol, 162, 169–170, 181, 187, 189
small amplitude oscillatory shear (SAOS), 284
small-angle neutron scattering (SANS), 169
small-angle X-ray scattering (SAXS), 169
SMS. *See* sorbitan monostearate (SMS)
sodium dodecyl sulfate (SDS), 3
solid fat content (SFC), 61, 119, 214, 254
    chemical interesterification, 129
    HSFO, 126
    hydrogenated shortenings, profiles of, 127
    lauric fat, 128
    margarine formulation, 126, 129
    vs. temperature profiles for HSFO, 126–127
    trans-isomers, 128
solid fat index (SFI), 317
sorbitan monostearate (SMS), 330
SO/SA. *See* stearyl alcohol/stearic acid (SO/SA)
SSHE. *See* scraped surface heat exchanger (SSHE)
stearyl alcohol/stearic acid (SO/SA), 307
sterols, 162
StStSt. *See* tristearin (StStSt)
sunflower protein isolate
    citric acid, 6
    globulins and albumins, 4
    HPLC chromatogram of, 5
    hydrodynamic radius, 6
    optical detection, 4
    pH, function of, 5
    protein content, impact of, 8–9
    SDS-PAGE technique, 4–5
    solid–liquid separation, 4
sunflower wax (SW), 175, 204, 236, 237, 240
supercritical fluid extraction (SCFE), 161
SW. *See* sunflower wax (SW)

TAGs. *See* triacylglycerols (TAGs)
TAGS. *See* triacylglycerides (TAGS)
TAG–wax hybrid fat systems
    bakery applications, hybrid systems in, 254–255
    categories, 245
    hybrid emulsions
        water droplets, 254
        wax-triglyceride hard-stock fat, application of, 253
        W/O emulsions, 252–253
    hybrid shortenings
        dynamic cooling, 248–252
        static cooling, 245–248
    triacylglycerol hardstock, 243–244
TEA. *See* total energy intake (TEA)
TFACs. *See* *trans*-fatty acid chains (TFACs)
Theobroma cacao, 336
three-interval thixotropy tests (3iTT), 285

total energy intake (TEA), 2
trans-fatty acid chains (TFACs), 2
triacylglycerides (TAGS), 201
triacylglycerols (TAGs), 55, 91
 cocoa butter, 125–126
 formulate *trans*-fat-free products, 125
 palm kernel and coconut oils, 125
 sunflower oil, hydrogenation processes in, 123–124
tristearin (StStSt), 68
tropical fats
 chemical composition
  fatty acid composition, 121–123
  triacylglycerol composition, 123–126
 crystallization behavior, 129–130
 functional and nutritional properties of, 120
 groups, 120
 microstructure, 130–131
 palm oil and coconut oil, 120
 polymorphic behavior
  β crystals, 134
  forms, 133–134
  melting temperature, 134–135
  non-hydrogenated shortening, 135
  palm stearin blend, 136
  plastic fat, 133
  SAXS and WAXS regions, 134–135
  short and long spacings, 134–135
 rheological properties, 132–133
 semi-solid edible vegetable fats, 119
 SFC
  chemical interesterification, 129
  HSFO, 126
  hydrogenated shortenings, profiles of, 127
  lauric fat, 128
  margarine formulation, 126, 129
  vs. temperature profiles for HSFO, 126–127
  trans-isomers, 128
 trans-fat, 120, 136

UE. *See* unheated emulsion (UE)
ultra small angle X-ray scattering (USAXS), 132
unheated emulsion (UE), 107–108
USAXS. *See* ultra small angle X-ray scattering (USAXS)

vegetable oils
 crystallized system, 202
 CVD, risk of, 202
 food systems, application of oleogels in, 229–230
 lecithin, molecular self-assembly of
  β polymorph, 221
  hydrophobic or alkyl tails, 216
  MG–LC interactions, 216–217, 219–220
  MG per mole of LC, 217, 221
  phosphatidylcholine, 216
  primary hydroxyl group, 216
  secondary hydroxyl group, 216
 MG crystallization, ethylcellulose effect on
  canola oil–EC–MG systems, 225–226
  EC–MG interaction, 229
  polymers, 223
  polymorphic transition, 226
  SKB and PKB, 223–226
  substitution degree (SD), 223

vegetable oils (*continued*)
molecular self-assembly
MGs and LC, 204–205
rice bran wax, 203
three-dimensional microstructure, 203
monoglycerides as gelator molecules
emulsions and foams, formation of, 206
inverse lamellar structure, 207
Lα and sub-α phases, 208
micelles, 207
molecular self-assembly, 206
–OH group, 205
reversible phase transitions, 208
sub-α1 ⇄ sub-α2 transition, 210
wide-angle and small-angle, 207–208
oleogels, monoglycerides in
cooling and heating thermograms, 210–211
Lα mesophase, 210
β phase formation, 214
β polymorph, 215
sub-α or sub-α1 phase, 210–212
sub-α phase crystallization, 212
viscoelastic properties, 214
TAGS, 201–202
trans fatty acids, 202–203
vegetal oils
dried emulsions, redispersion of, 8
droplet size distributions, 8–9
emulsion, 17–18
adsorbed molecules, 13
adsorbed protein layers, 13
biliquid foams, as a function of droplet fraction, 13–15
BSA and BLG, 13
characterization, 6–7
drying, 8
dynamic light scattering, 13
freeze-drying, 15–17
osmotic resistance, 11
preparation, 6
interfacial coverage, 9–10
interfacial tension, 8
osmotic pressure measurements, 7–8
protein quantification, 7
sunflower protein isolate
citric acid, 6
HPLC chromatogram of, 5
hydrodynamic radius, 6
optical detection, 4
pH, function of, 5
SDS-PAGE technique, 4–5
virgin olive oil (VOO), 324
VOO. *See* virgin olive oil (VOO)

water-in-oil-in-water (W/O/W), 21
wax esters (WE), 282
WBW. *See* white beeswax (WBW)
WE. *See* wax esters (WE)
white beeswax (WBW), 237, 240–241
WHO. *See* World Health Organization (WHO)
World Health Organization (WHO), 2
W/O/W. *See* water-in-oil-in-water (W/O/W)

YBW. *See* yellow beeswax (YBW)
yellow beeswax (YBW), 237, 240–241

zero-*trans* cake shortening (ZTC), 329
ZTC. *See* zero-*trans* cake shortening (ZTC)